AF800833

Sociocultural Explorations of Science Education

Volume 27

Series Editors
Catherine Milne, Steinhardt School of Education
New York University, New York, NY, USA
Christina Siry, University of Luxembourg
Ossining, NY, USA

The series is unique in focusing on the publication of scholarly works that employ social and cultural perspectives as foundations for research and other scholarly activities in the three fields implied in its title: science education, education, and social studies of science.

The aim of the series is to promote transdisciplinary approaches to scholarship in science education that address important topics in the science education including the teaching and learning of science, social studies of science, public understanding of science, science/technology and human values, science and literacy, ecojustice and science, indigenous studies and science and the role of materiality in science and science education. Cultural Studies of Science Education, the book series explicitly aims at establishing such bridges and at building new communities at the interface of currently distinct discourses. In this way, the current almost exclusive focus on science education on school learning would be expanded becoming instead a focus on science education as a cultural, cross-age, cross-class, and cross-disciplinary phenomenon.

The book series is conceived as a parallel to the journal Cultural Studies of Science Education, opening up avenues for publishing works that do not fit into the limited amount of space and topics that can be covered within the same text. Book proposals for this series may be submitted to the Publishing Editor: Claudia Acuna E-mail: Claudia.Acuna@springer.com

More information about this series at https://link.springer.com/bookseries/16729

Anders Jakobsson • Pia Nygård Larsson
Annika Karlsson
Editors

Translanguaging in Science Education

Springer

Editors
Anders Jakobsson
Science, Mathematics & Society
Malmö University
Malmö, Sweden

Pia Nygård Larsson
Culture, Language & Media
Malmö University
Malmö, Sweden

Annika Karlsson
Science, Mathematics and Society
Malmö University
Malmö, Sweden

ISSN 2731-0248 ISSN 2731-0256 (electronic)
Sociocultural Explorations of Science Education
ISBN 978-3-030-82972-8 ISBN 978-3-030-82973-5 (eBook)
https://doi.org/10.1007/978-3-030-82973-5

© The Editor(s) (if applicable) and The Author(s), under exclusive license to Springer Nature Switzerland AG 2021
This work is subject to copyright. All rights are solely and exclusively licensed by the Publisher, whether the whole or part of the material is concerned, specifically the rights of translation, reprinting, reuse of illustrations, recitation, broadcasting, reproduction on microfilms or in any other physical way, and transmission or information storage and retrieval, electronic adaptation, computer software, or by similar or dissimilar methodology now known or hereafter developed.
The use of general descriptive names, registered names, trademarks, service marks, etc. in this publication does not imply, even in the absence of a specific statement, that such names are exempt from the relevant protective laws and regulations and therefore free for general use.
The publisher, the authors and the editors are safe to assume that the advice and information in this book are believed to be true and accurate at the date of publication. Neither the publisher nor the authors or the editors give a warranty, expressed or implied, with respect to the material contained herein or for any errors or omissions that may have been made. The publisher remains neutral with regard to jurisdictional claims in published maps and institutional affiliations.

This Springer imprint is published by the registered company Springer Nature Switzerland AG
The registered company address is: Gewerbestrasse 11, 6330 Cham, Switzerland

Contents

Chapter 1
Translanguaging as a Pedagogical Strategy for Enhancing Multilingual Science Students' Learning in Different Educational Contexts

Anders Jakobsson, Pia Nygård Larsson, and Annika Karlsson

Abstract The purpose of this introductory chapter is to present and discuss translanguaging as a pedagogical strategy to facilitate and expand science learning for multilingual students. In the chapter, challenges and opportunities in multilingual science classrooms are discussed, with a special focus on specific language usage in science education and the challenging literacy requirements placed on students so that they can succeed in science studies. This also includes students' multilingual and multimodal meaning-making and in what ways their dialogic communication can constitute resources in teaching and learning. Further, the processes of inclusion and exclusion within science education are discussed. The latter perspective is addressed by using translanguaging theories that are strongly bound to issues of language ideologies and social justice. Another important aim of this first chapter is to introduce the 11 research studies reported on in the following chapters.

The 35 authors of this book, consisting of an international group of researchers from the areas of science education and applied linguistics, explore translanguaging practices in multilingual science classrooms in Hong Kong, Lebanon, Luxembourg, South Africa, Sweden, and the United States. These practices provide rich examples of translanguaging strategies in various educational contexts, including primary, middle, secondary levels, higher education, and after-school programs.

Keywords Multilingual science students · Multilingual and multimodal meaning-making · Pedagogical strategies · Translanguaging

---

A. Jakobsson (✉) · P. Nygård Larsson · A. Karlsson
Malmö University, Malmö, Sweden
e-mail: anders.jakobsson@mau.se; pia.nygard-larsson@mau.se; annika.karlsson@mau.se

© The Author(s), under exclusive license to Springer Nature Switzerland AG 2021
A. Jakobsson et al. (eds.), *Translanguaging in Science Education*, Sociocultural Explorations of Science Education 27,
https://doi.org/10.1007/978-3-030-82973-5_1

1.1 Introduction

In this book, researchers from around the world explore how *translanguaging* as a pedagogical strategy or method can facilitate and expand science learning for multilingual students in different educational contexts. Most of the chapters in the book indicate that a majority of multilingual students worldwide, in different school forms and at different levels, have a common experience of learning science in a language of instruction that is not always the most appropriate for the individual. This often means that these students are forced to develop understanding and knowledge of abstract and complex scientific concepts and theories with the help of their second language, which strongly implies that they become disadvantaged in their learning processes. However, several studies have also demonstrated that teachers strive to act accordingly by creating new learning opportunities that include other language resources, such as the native home language of the students. In some educational contexts, this means that multilingual students are allowed and even encouraged to use and freely draw on their entire language repertoire in a dynamic and pragmatic way (García & Wei, 2014). In such translanguaging classrooms, students are supported to use all available language and other multimodal resources in order to enhance their learning opportunities, to develop an understanding of the subject content and the specific language used in science teaching. In recent years, a growing number of studies in science education and in other areas indicate that such learning situations may have a positive effect on students' understanding and learning (e.g., Harman et al., 2020; Karlsson et al., 2019; Licona & Kelly, 2020; Poza, 2018).

Traditionally, multilingual students have frequently been described from a deficit perspective or as problems to be addressed, often based on the argument of a presumed lack of skills in the language of instruction (e.g., Cummins, 2000). This is especially true for minoritized multilingual students in contexts dominated by monolingual educational policies (e.g., Probyn, 2019). However, translanguaging as a pedagogical practice has increasingly contributed to the legitimization of minoritized multilingual students' use of all their language resources as meaning-making tools in the classrooms, which also reinforces their engagement and identity formation as learners (e.g., García & Wei, 2014). In the present book, 11 individual research studies or projects provide rich examples of analyses of such translanguaging situations in different educational contexts across national and cultural boundaries, from preschool to upper secondary school and tertiary education. The aim has been to focus on and analyze the opportunities and potential obstacles that arise as a consequence of using translanguaging strategies in science education. Many of these studies also explore how translanguaging perspectives may be combined with other means of supporting or scaffolding students when approaching and appropriating ways of doing and learning science.

1.2 Science Education from Multilingual and Cultural Perspectives

Ever since the 1970s, an increasing number of studies in the area of science education have focused on the pedagogical obstacles and limitations that risk leading to a large proportion of students all over the world becoming uninterested or feeling excluded from science. Researchers have emphasized the risk that the special character of the subject area leads to students experiencing science as difficult, abstract, authoritarian, nonrelevant, alienating, and an area that is not for them (e.g., Banks & Banks, 2019; Lemke, 2001; Osborne et al., 2003; Riegle-Crumb et al., 2011; Van Horne & Bell, 2017). Within this context, it can be noted that it is about 20 years since Lemke (2001) argued that it is no longer sustainable to imagine science learning as an equal process for all students, only demanding logical thinking and application, but rather that cultural, language, and class differences constitute significant aspects of students' interest and learning in science. In our increasingly globalized societies, this argument may be even more valid today than it was 20 years ago.

Recently, however, many studies within science education have explored the specific language usage and the challenging literacy requirements placed on students in order to succeed in science studies at different levels of the education system (e.g., Buxton & Lee, 2014; Serder & Jakobsson, 2016). Most of these studies indicate that the teachers' language use in the classroom is crucial for how students generally perceive and understand the subject. For example, if the teaching consciously focuses on science from a language and literacy perspective, where new concepts, expressions and different discursive languages are made explicit, it increases the opportunity for students to actively take part in the teaching (e.g., Karlsson et al., 2020; Nygård Larsson, 2018). Others have pointed out that students' usage of various language repertoires during explorative group conversations benefit their understanding and participation in science instruction (e.g., Jakobsson et al., 2009; Martin-Beltrán et al., 2017; Msimanga & Lelliott, 2014). However, studies also identify the great challenges in multilingual classroom activities and the need to address the complexity of science teaching and learning in order to avoid simplifying or misunderstanding content and language (Karlsson et al., 2020). Nonetheless, researchers also stress the importance of avoiding an overly narrow literacy perspective that may lead to students being prevented from engaging in the exploratory and argumentative practices of doing science and making sense of scientific phenomena (e.g., Charamba, 2020; Harman et al., 2020; Langman, 2014; Licona & Kelly, 2020).

An important conclusion in most of all these studies is that science teaching and learning can be regarded as a social literacy practice and a disciplinary discourse with specific ways of doing, thinking, talking, reading, and writing that clearly differ from daily perspectives on the world (Nygård Larsson & Jakobsson, 2020). In these science literacy practices, different modes become intertwined, such as talk, images, models, graphs, symbols, and gestures. That is, in multilingual science classrooms there exist a complex interplay between different student languages;

daily, academic, and disciplinary-specific discourses; and different modalities. Thus, the multimodal nature of translanguaging creates the need for an exploration across modes in educational practice and in research (García & Wei, 2014). Consequently, a number of studies about translanguaging have explored this complexity and the affordances of science students' multilingual and multimodal meaning-making and dialogic communication (e.g., Siry & Gorges, 2020; Ünsal et al., 2018; Wu & Lin, 2019). The various chapters in this book constitute examples of this type of study.

1.3 Processes of Inclusion and Exclusion Within Science Education

From a sociocultural perspective, language is considered a crucial tool for learning, alongside other multimodal means of expression. However, this perspective does not understand language as a neutral tool; instead, it is strongly connected to issues of identities, discourses, culture, power relations, social class, gender, the relative status of different languages, and attitudes toward multilingualism (e.g., Cummins, 2014; Jakobsson & Davidsson, 2012). This implies that all of these factors also have an impact on the various processes of inclusion and exclusion in the teaching of students. Thus, for a number of reasons – that go beyond the above-discussed literacy aspects – students may benefit from being allowed and encouraged to use all their available language resources in science classrooms. This broader perspective is problematized and addressed by translanguaging theories, which are strongly bound to issues of language ideologies, social justice, and the problematization of monolingual ideologies and monoglossic norms in education and society (e.g., García & Wei, 2014). Consequently, a number of translanguaging studies within science education point to the fact that if multilingual students' diverse cultural and linguistic backgrounds and previous experiences are taken into account and included in science teaching and learning, their engagement, identity work, and empowerment in science become positively affected (e.g., Lemmi et al., 2019; and the studies in this volume).

However, in a variety of contexts, qualitative science education has usually been reserved for the dominant social classes, which helps to preserve their privileged linguistic and cultural practices (e.g., Salloum & BouJaoude, 2020). This tension between educational policy and practice is especially visible in post-colonial contexts (McKinney, 2017). In such contexts, language ideologies, monolingual education policies, and school practices have served to limit educational and societal opportunities in a continued coloniality, which leads to processes of exclusion and considerably lower performance of the multilingual students (e.g., Probyn, 2019; and Chaps. 10–12 in this volume).

1.4 Eleven Research Studies on Translanguaging in Science Education

An increasingly important issue in culturally and linguistically diverse contexts is whether the teaching is organized in ways that offer minoritized multilingual students equal conditions for learning. It is in relation to these conditions that the 35 authors of the 11 studies in this book aim to explore and examine the opportunities and obstacles that a translanguaging strategy in science education may offer. Several of the authors present situations where the use of translanguaging as a pedagogy obviously supports multilingual students' language and knowledge development, and thus contribute to increased learning opportunities in science. Others point out situations in which the approach seems to strengthen and affirm students' cultural and multilingual identities and their sense of belonging, thus contributing to an inclusive science education. However, several chapters also assert that translanguaging strategies are considerably more sophisticated and complex than just encouraging students to use different languages.

A common feature of all the chapters in the book is the exploration of translanguaging as a pedagogical tool that may contribute to the framing of science learning in linguistically and culturally diverse classrooms and thereby to the empowerment of multilingual students. The theoretically and empirically rich studies in this book can be presented and organized in a number of different ways. In structuring the book, the editors have been guided by the dominating ideas and themes of the chapters. However, the chapters contain varied and complex contextual circumstances and conditions, research results, perspectives, and orientations that go beyond these themes.

Chapters 2, 3, and 4 have a special focus on informal and spontaneous translanguaging practices, and on how dialogic and open-ended interactions can inform and transform science teaching or assessment. Chapters 5, 6, and 7 explore the various functions of translanguaging in relation to social roles and academic purposes, especially students' development of understanding of discourse and scientific knowledge. This is also highlighted in Chaps. 8 and 9, which report on interventional studies of translanguaging as a means of supporting the fundamental epistemic and semantic practices of science learning. These themes are also present in Chaps. 10, 11, and 12, which also take a strong stance in relation to issues of language ideology and social justice, problematizing monolingual language ideologies and colonial remnants that create unequal conditions for the students and contribute to their exclusion from access to quality education.

In the second chapter, *Translanguaging within an integrated framework for multilingual science meaning-making*, Cory Buxton, Ruth Harman, Lourdes Cardozo-Gaibisso and Max Vazquez Dominguez explore translanguaging interactions in two contexts in the United States: (1) conversations about science and language during family interviews that occurred during a series of Steps to College through Science bilingual family workshops; and (2) students' written responses on a bilingual (English–Spanish) and multimodal science assessment, designed and administered by members of the research team in the middle and high school classrooms of project teachers. Through a systemic functional linguistic (SFL) discourse analysis, the authors explore how different evidence for translanguaging in these two contexts can deepen our understanding of the role of translanguaging in science meaning-making. Their findings show that translanguaging occurs in more restrictive ways within formal assessment contexts than it does in the more dialogic contexts of the family workshops, where family members build relationships and negotiate meanings in collaborative ways. Thus, as the discussion moves from an informal space to a formal one, the authors invite readers to think about how joint construction and dialogic meaning making could inform more traditional classroom and assessment practices.

In the third chapter, *Translanguaging for STEM learning: Exploring tertiary learning contexts*, Juliet Langman, Jorge Solís, Lina Martin-Corredor, Nguyen Dao, and Karla Garza Garza contribute to a growing body of translanguaging research in higher education. The authors explore how unplanned, spontaneous translanguaging occurs in the context of large lecture classes in the United States, where multilingual students informally and formally translanguage to appropriate STEM concepts and processes and engage in identity work. The authors also examine faculty researchers' and students' perspectives and language ideologies on the potential value of unplanned and planned translanguaging spaces in university contexts with large numbers of multilingual speakers with varied past educational experiences. The data reveal that when some students are given the educational space to do so, translanguaging (English–Spanish) practices bubble up, even in settings where all oral and written educational materials are in English. Interview data reveal that the students have mixed feelings about their own language proficiency when it comes to academics, and that they make choices about participation in class on the basis of those self-perceptions. The study also shows that there is little awareness of how faculty (and students) may think about leveraging their shared multilingual linguistic repertoires in higher education and that the context remains primarily viewed through a lens of monolingual English education.

In Chap. 4, *Young children's transmodal participation in science investigations: Drawing on a diversity of resources for meaning-making*, Christina Siry, Sara Wilmes, Kerstin te Heesen, Doriana Sportelli, and Sandy Heinericy share their examination of how 4- to 6-year-old children in a highly culturally and linguistically diverse kindergarten class in the multilingual country of Luxembourg work to employ and animate meaning-making and communicative resources. The context of the study is when the children conduct investigations with worms, using a variety of approaches including digital microscopes. The authors use video-based critical ethnography to examine and interpret the children's multilingual and multimodal (transmodal) interactions and the various resources students draw on. The analysis reveals that dialogic classroom structures supported children's transmodaling, which the authors conceptualize as the fluid use of diverse semiotic resources, including language resources. The analysis also reveals the ways in which dialogic open-ended classroom structures mediated the creation of spaces for transmodaling and, in turn, these transmodal spaces mediated students' agentic actions. The findings illustrate how the students were allowed to draw on a range of resources in interaction around science, and how it is therefore imperative to work toward structures that create transmodal spaces for culturally and linguistically diverse students.

In Chap. 5, *Translanguaging in the science classroom: Drawing on students' full linguistic repertoires in bi/multilingual settings*, Catherine Lemmi, Greses Pérez, and Bryan A. Brown explore how translanguaging is enacted in a fourth-grade bilingual science classroom in the United States, and how translanguaging acts serve purposes for the teacher and her students within this classroom. A vignette is analyzed from an elementary science class in which a teacher and four students engage in translanguaging, when talking about an experiment to determine the properties of soils. The conversation analysis identifies and describes the purposes accomplished by each utterance and provides insights of how the speakers interact with one another. The result shows that the students engaged in translanguaging (English–Spanish) for a variety of social and academic purposes. In the exchanges studied, the authors found three main purposes that were accomplished: (1) making comparisons between science and lived-experiences, (2) asking and answering procedural questions, and (3) using a scientific nomenclature. The episodes give a glimpse of hybrid translingual practices in a science classroom and the authors discuss the significance of these acts in terms of their value for the development of scientific notions and linguistic goals. In addition, the authors provide recommendations for science teachers on how to facilitate opportunities for students to draw on their various language repertoires when verbalizing their scientific thinking.

In Chap. 6, *The affordances of leveraging multilingual repertoires in scientific reasoning among resettled refugee teens: Functions of translanguaging in scientific reasoning*, Shannon Mary Daniel, Minjung Ryu, Mavreen Rose Santa-Ana Tuvilla, and Casey Elizabeth Wright illuminate the affordances of refugee teenagers leveraging their full linguistic repertoires as they reason about scientific phenomena in a multilingual after-school program focused on the impacts of climate change on human life, while developing the students' critical literacy skills. The young people in the program are from the linguistically diverse Chin state in Myanmar and have lived in the United States for various lengths of time. The study provides specific examples of the functions of the students' translanguaging practices in relation to their science learning, including attention to peer-scaffolding of scientific meaning-making, as well as support to enhance students' engagement in doing science. Based on discourse analysis of four selected events, the authors show how the participants translanguage across several languages to explain science ideas to one another, discuss cross-cutting concepts such as causal relationships, provide task-related support, and make sure each group member understands. The chapter concludes with a discussion of how educators can support young people in drawing upon their various languages, even when educators do not share the students' fluency in these languages.

In Chap. 7, *Students' multilingual negotiations of science in third space*, Annika Karlsson, Pia Nygård Larsson, and Anders Jakobsson explore the ways in which students' multilingual negotiations in a science classroom in a middle school in Sweden (grades 4–6) may contribute to the development of the students' subject knowledge and language use. This includes focusing on whether and how the movements between different languages and discourses have an impact on this development. The negotiations are illustrated as movements in *multilingual discursive loops*, which includes dynamic movements between "national" languages (Swedish and Arabic) and discursive languages (every-day and scientific discourses), with support from multimodal resources. The authors describe these movements in a model and conclude that the students' language and conceptual development are facilitated by providing opportunities for joint meaning-making and opening up the hybrid negotiation space (third space) that constitutes a translanguaging science classroom. The authors argue that the model, as an analytical tool, offers opportunities for science teachers at different levels, as well as teacher students, to deepen their understanding of the complex situation that multilingual students experience when creating meaning and understanding of the science content, thus contributing to an awareness that science education is a communicative language activity.

In Chap. 8, *Translanguaging, trans-semiotizing, and trans-registering in a culturally and linguistically diverse science classroom*, Peichang (Emily) He and Angel M. Y. Lin investigate how a science teacher engages students in a diverse multilingual secondary school (grade 7) in Hong Kong, and how the teacher guides the students to improve their academic literacy and facilitate their use of multilingual and multimodal resources. For this, the teacher uses and adapts a thematic pattern-based concept + language mapping (CLM) pedagogy, designed by the researchers. The authors address two research questions: (1) How do translanguaging, trans-semiotizing, and trans-registering as spontaneous scaffolding strategies facilitate meaning-making in the multilingual science classroom? (2) How is academic literacy development facilitated by trans-registering in the integrated science lessons? The result indicates that knowledge construction was developed based on the simultaneous meaning-making of different modes – verbal mode, visual mode, and actionable mode – and the students' different languages (Cantonese and Putonghua) constituted just one of the semiotic resources used. After five science lessons, the students reported that the activities and materials were interesting and made learning easier and, according to the teacher, the CLM pedagogy helped to increase student participation and supported science teaching and learning. Furthermore, the test results demonstrated that students had made progress in the development of both content knowledge and academic language.

In Chap. 9, *Translanguaging in middle school science: Written arguments about issues of biodiversity*, Peter Licona and Gregory J. Kelly investigate how translanguaging as a pedagogy can support students' engagement in the epistemic and discourse-intensive science practices. These practices include learning ways of formulating evidence in specific genres leading to the development of scientific argumentation. In order to investigate how science education is enacted in a dual-language (English–Spanish) education setting in middle school (grade 7) in the United States, a curriculum was co-constructed by one of the authors and the teacher to promote students' spoken and written arguments about societal issues of biodiversity. To support the development of the relevant epistemic practices, the teacher helped students to construct and evaluate arguments using the scientific argumentation framework *claims-evidence-reasoning*, as well as translanguaging as a strategy. The analysis of the students' written arguments demonstrated ways in which the communicative function of the academic task was accomplished, and often enhanced, by the use of translanguaging among the student writers. Furthermore, the authors noted that translanguaging as a pedagogy builds on students' assets and opens up the potential for developing emergent bilinguals' identities and interest in learning science.

In Chap. 10, *Contradictions confronting hybrid spaces for translanguaging in the Lebanese context: A CHAT perspective*, Sara Salloum describes the Lebanese educational system, with science taught and assessed in a foreign language (English or French). Deploying a foreign language in science instruction has raised concerns regarding Lebanon's low performance on international comparative assessments. Even though learning a foreign language is considered a main objective of the curriculum to reflect Lebanon's multicultural and multilingual image, it is a key factor leading to school dropout amongst disadvantaged groups. To better understand such tensions, Salloum utilizes cultural historical activity theory (CHAT) to conceptually and empirically identify tensions as 'contradictions' within the Lebanese context. An important implication is that reform needs to pay close attention to social justice and equity issues, and that opening up hybrid spaces for translanguaging needs to address tensions among students' everyday language and experiences and the specialized content and discourse of canonical science. Salloum concludes that a content literacy model and translanguaging can hold promise in Lebanese multilingual science classrooms as these open up hybrid spaces in which students' communicative resources are deployed for constructing deeper meanings in science. However, translanguaging, especially in terms of deploying the home language, can go against strongly held monoglossic notions in science education, which would create tensions against its enactment.

In Chap. 11, *Translanguaging in science education in South African classrooms: Challenging constraining ideologies for science teacher education*, Annemarie Hattingh, Carolyn McKinney, Audrey Msimanga, Margie Probyn, and Robyn Tyler challenge the anglonormative and monoglossic language ideologies that, in reality, mean that 9 out of South Africa's 11 official languages are not recognized as languages of teaching and learning from grade 4 in South African schools. The authors explore the translanguaging practices in multilingual science classrooms and draw from their different research projects to offer complementary perspectives on teaching and learning science. The results from the studies illustrate the argument being made for an acceptance of translanguaging practices to counter the educational disadvantages, which usually result in low performance on national and international assessments. The empirical examples are presented in four vignettes: (1) the grade 4 transition year from home language isiXhosa instruction to English and how this constrains learners' participation and meaning-making; (2) and (3) translanguaging practices in small group work and teacher-led whole-class talk and how these engage learners in meaning-making and scaffolds learning; and (4) how written translation activities enable learners to deepen their conceptual understanding. Collectively, these vignettes offer insights for science teacher education and into how to inform a conscious translanguaging pedagogy.

In Chap. 12, *Leveraging multilingualism to support science education through translanguaging pedagogy*, Erasmos Charamba describes how the students' meaning-making and language use is affected when three chemistry teachers introduce translanguaging as a pedagogy into tenth-grade chemistry classes in a secondary school in South Africa. The data were gathered through lesson observations, video recordings, and interviews. During the interviews, the researcher used video elicitation techniques by showing the participants selected video footage in which the students were actively translanguaging. Using empirical examples, the chapter illustrates how the students use their linguistic repertoire as they interact with each other during collaborative work. The results show the students' ways of speaking, writing, and participating in a multilingual science class and how translanguaging scaffolds science learning. In this way, translanguaging offered the teachers and students opportunities to access scientific content through the linguistic resources and communicative repertoire they bring to the science classrooms. The author concludes by encouraging teachers to move away from the current monolingual bias, which in most cases results in social injustice and academic underachievement by students taught in a language other than their mother tongue.

Funding This work was financially supported by the research program, Disciplinary Literacy and Inclusive Teaching (LIT), Malmö University.

References

Banks, J. A., & Banks, C. A. M. (Eds.). (2019). *Multicultural education: Issues and perspectives*. Wiley.

Buxton, C. A., & Lee, O. (2014). English learners in science education. In N. Lederman & S. Abell (Eds.), *Handbook of research on science education, volume II* (pp. 204–222). Routledge.

Charamba, E. (2020). Translanguaging in a multilingual class: A study of the relation between students' languages and epistemological access in science. *International Journal of Science Education, 42*(11), 1779–1798.

Cummins, J. (2000). *Language, power & pedagogy: Bilingual children in the crossfire*. Multilingual Matters.

Cummins, J. (2014). Beyond language: Academic communication and student success. *Linguistics and Education, 26*, 145–154.

García, O., & Wei, L. (2014). *Translanguaging: Language, bilingualism and education*. Palgrave Macmillan.

Harman, R., Buxton, C., Cardozo-Gaibisso, L., Jiang, L., & Bui, K. (2020). Culturally sustaining systemic functional linguistics praxis in science classrooms. *Language and Education*. https://doi.org/10.1080/09500782.2020.1782425

Jakobsson, A., & Davidsson, E. (2012). Using sociocultural frameworks to understand the significance of interactions at science and technology centers and museums. In E. Davidsson & A. Jakobsson (Eds.), *Understanding interactions at science centers and museums* (pp. 3–21). Brill Sense.

Jakobsson, A., Mäkitalo, Å., & Säljö, R. (2009). Conceptions of knowledge in research on students' understanding of the greenhouse effect: Methodological positions and their consequences for representations of knowing. *Science Education, 93*(6), 978–995.

Karlsson, A., Nygård Larsson, P., & Jakobsson, A. (2019). Multilingual students' use of translanguaging in science classrooms. *International Journal of Science Education, 41*(15), 2049–2069.
Karlsson, A., Nygård Larsson, P., & Jakobsson, A. (2020). The continuity of learning in a translanguaging science classroom. *Cultural Studies of Science Education, 15*(1), 1–25.
Langman, J. (2014). Translanguaging, identity, and learning: Science teachers as engaged language planners. *Language Policy, 13*(2), 183–200.
Lemke, J. L. (2001). Articulating communities: Sociocultural perspectives on science education. *Journal of Research in Science Teaching, 38*(3), 296–316.
Lemmi, C., Brown, B. A., Wild, A., Zummo, L., & Sedlacek, Q. (2019). Language ideologies in science education. *Science Education, 103*(4), 854–874.
Licona, P. R., & Kelly, G. J. (2020). Translanguaging in a middle school science classroom: Constructing scientific arguments in English and Spanish. *Cultural Studies of Science Education, 15*(2), 485–510.
Martin-Beltrán, M., Daniel, S., Peercy, M., & Silverman, R. (2017). Developing a zone of relevance: Emergent bilinguals' use of social, linguistic, and cognitive support in peer-led literacy discussions. *International Multilingual Research Journal, 11*(3), 152–166.
McKinney, C. (2017). *Language and power in post-colonial schooling: Ideologies in practice*. Routledge.
Msimanga, A., & Lelliott, A. (2014). Talking science in multilingual contexts in South Africa: Possibilities and challenges for engagement in learners home languages in high school classrooms. *International Journal of Science Education, 36*(7), 1159–1183.
Nygård Larsson, P. (2018). 'We're talking about mobility': Discourse strategies for promoting disciplinary knowledge and language in educational contexts. *Linguistics and Education, 48*, 61–75.
Nygård Larsson, P., & Jakobsson, A. (2020). Meaning-making in science from the perspective of students' hybrid language use. *International Journal of Science and Mathematics Education, 18*(5), 811–830.
Osborne, J., Simon, S., & Collins, S. (2003). Attitudes towards science: A review of the literature and its implications. *International Journal of Science Education, 25*(9), 1049–1079.
Poza, L. E. (2018). The language of ciencia: Translanguaging and learning in a bilingual science classroom. *International Journal of Bilingual Education and Bilingualism, 21*(1), 1–19.
Probyn, M. (2019). Pedagogical translanguaging and the construction of science knowledge in a multilingual South African classroom: Challenging monoglossic/post-colonial orthodoxies. *Classroom Discourse, 10*(3–4), 216–236.
Riegle-Crumb, C., Moore, C., & Ramos-Wada, A. (2011). Who wants to have a career in science or math? Exploring adolescents' future aspirations by gender and race/ethnicity. *Science Education, 95*(3), 458–476.
Salloum, S., & BouJaoude, S. (2020). Understanding interactions in multilingual science classrooms through Cultural-Historical Activity Theory (CHAT): What do contradictions tell us? *International Journal of Science and Mathematics Education*. https://doi.org/10.1007/s10763-020-10114-5
Serder, M., & Jakobsson, A. (2016). Language games and meaning as used in student encounters with scientific literacy test items. *Science Education, 100*(2), 321–343.
Siry, C., & Gorges, A. (2020). Young students' diverse resources for meaning making in science: Learning from multilingual contexts. *International Journal of Science Education, 42*(14), 2364–2386.
Ünsal, Z., Jakobson, B., Wickman, P., & Molander, B. (2018). Gesticulating science: Emergent bilingual students' use of gestures. *Journal of Research in Science Teaching, 55*(1), 121–144.
Van Horne, K., & Bell, P. (2017). Youth disciplinary identification during participation in contemporary project-based science investigations in school. *Journal of the Learning Sciences, 26*(3), 437–476.
Wu, Y., & Lin, A. M. (2019). Translanguaging and trans-semiotising in a CLIL biology class in Hong Kong: Whole-body sense-making in the flow of knowledge co-making. *Classroom Discourse, 10*(3–4), 252–273.

Chapter 2
Translanguaging Within an Integrated Framework for Multilingual Science Meaning Making

Cory Buxton, Ruth Harman, Lourdes Cardozo-Gaibisso, and Max Vazquez Dominguez

Abstract Translanguaging is emerging as a significant pedagogical tool to help science educators frame learning in linguistically and culturally diverse classrooms. To effectively support and challenge multilingual science learners, however, translanguaging must be conceptualized as more than an isolated philosophy or instructional strategy. Instead, we position translanguaging as one component of an integrated model of culturally and linguistically sustaining disciplinary meaning making. In this study, translanguaging interactions are explored in two contexts: (1) conversations about science and language during family interviews that occurred during a series of Steps to College through Science bilingual family workshops and (2) students' written responses on a multilingual and multimodal science assessment, designed and administered by members of the research team in the classrooms of project teachers. Through a systemic functional linguistic (SFL) discourse analysis, we explore how differential evidence of translanguaging in these two contexts can deepen our understanding of the role of translanguaging in support of science meaning making.

Keywords Translanguaging · Science education · Multilingual learners · Family engagement · Multilingual assessment · Systemic functional linguistics

C. Buxton (✉)
Oregon State University, Corvallis, OR, USA
e-mail: buxtonc@oregonstate.edu

R. Harman
University of Georgia, Athens, GA, USA
e-mail: rharman@uga.edu

L. Cardozo-Gaibisso
Mississippi State University, Starkville, MS, USA

M. V. Dominguez
University of North Georgia, Dahlonega, GA, USA
e-mail: Max.VazquezDominguez@ung.edu

© The Author(s), under exclusive license to Springer Nature Switzerland AG 2021

A. Jakobsson et al. (eds.), *Translanguaging in Science Education*, Sociocultural Explorations of Science Education 27,
https://doi.org/10.1007/978-3-030-82973-5_2

2.1 Introduction

A new wave of frameworks and standards to shape science education has emerged across the globe in recent years. From the European Commission's (2015) *Framework for Science Education for Responsible Citizenship* to the Canada 2067 (2017) *STEM Learning Roadmap*, to the *Framework for K-12 Science Education* (NRC, 2012) and the resulting *Next Generation Science Standards* (NGSS Lead States, 2013) in the United States, these documents emphasize students' participation in investigations that promote engagement with each other's thinking through multimodal representation and communication. However, as sociologist Karl Maton has argued (2014), our education systems are not well prepared to support inquiry-based processes and cumulative knowledge building for students, especially for those coming from culturally and linguistically minoritized backgrounds.

To address these challenges, translanguaging (García & Wei, 2014) has been proposed as an effective theoretical and pedagogical lens for understanding the ways in which bilingual learners draw from all their available semiotic resources to make meaning. This meaning making is accomplished through a combination of linguistic (e.g., written and spoken words) and non-linguistic (e.g., gestures) resources (Kusters et al., 2017; Blackledge & Creese, 2017). Otheguy et al. (2015) defined translanguaging as "the deployment of a speaker's full linguistic repertoire without regard for watchful adherence to the socially and politically defined boundaries of named (and usually national and state) languages" (p. 281).

In science classrooms, much of the translanguaging research has focused on students' reading and writing about science content (e.g., Poza, 2018; Espinaoza, 2016). One consequence of this focus has been an outsized attention to the role of multilingual vocabulary in science meaning making (Groves, 2016). While reading, writing, and vocabulary development are, of course, important to science learning, translanguaging research can do more to explicitly support the broader epistemic practices at the heart of new science standards, such as analyzing and interpreting data, developing and using models, or designing solutions (National Academies, 2018). Thus, while translanguaging is emerging as a significant pedagogical tool to help educators frame learning in diverse classrooms, we argue that to effectively support and challenge multilingual science learners, translanguaging must be conceptualized as more than an isolated philosophy or instructional strategy. Instead, we position translanguaging as one component of an integrated model of culturally and linguistically sustaining disciplinary meaning making. As Allard (2017) highlights in her ethnographic analyses of translanguaging practices in a high school context, broad ecological support of multilingual meaning making needs to be in place for translanguaging to be effective, especially in the context of new immigration receiving regions, such as many parts of the United States.

In the current study, we use data from a large US National Science Foundation-funded research and development project called the Language-rich Inquiry Science with English Language through Biotechnology (LISELL-B). This project, which concluded in 2018, explored how middle school and high school science teachers, working with students in grades 6–10, responded to changing student demographics and changing science standards as they participated in a multifaceted professional learning program (Buxton & Caswell, 2020). The primary aim of LISELL-B was to support teachers, multilingual students, and families in dynamically integrating science investigation practices with use of the language of science (Buxton et al., 2019). Fifty teachers from 10 schools located in the Southeast United States participated in our professional learning over a three-year period. These educators, predominantly white monolingual English speakers who were teaching an English-only science curriculum, engaged in professional learning in five distinct contexts, each of which focused attention on different aspects of how to meet the needs and build on the assets of multilingual learners in ways that were specific to supporting meaning making in science. These professional learning contexts were (1) an annual week-long Teacher Professional Learning Institute focused on negotiating common understandings of science and language; (2) an annual two-week-long Student Summer Enrichment Academy for multilingual learners focused on dynamically experimenting with the practices learned during the teacher institute; (3) a series of annual Steps to College through Science Bilingual Family Workshops focused on teachers, students, and families doing science together and learning about STEM careers; (4) a series of annual Teachers Exploring Student Writing Workshops focused on understanding students' emergent science meaning making through analysis of their writing; and (5) on-demand classroom support when requested by teachers, focused on co-planning and co-teaching by project staff to integrate the project practices into secondary science classrooms (for more details, see Buxton et al., 2015).

Our focus in this study is on translanguaging interactions in two of these contexts: (1) conversations about science and language during family interviews that occurred during the Steps to College through Science Bilingual Family Workshops and (2) students' written responses on a project-designed multilingual and multimodal science assessment, designed and administered by members of the research team in the classrooms of all project teachers. Through a Systemic Functional Linguistic (SFL) discourse analysis, we explore how differential evidence of translanguaging in these two contexts can deepen our understanding of the role of translanguaging in science meaning making.

2.2 Conceptual Framework: Language, Culture, and Knowledge-Building Through Science

Our Culturally Sustaining-SFL for Science (CS-SFL: Science) framework is informed by three theoretical constructs that give rise to a set of instructional practices that simultaneously support students' language development, cultural sustenance, and knowledge building through science (Harman et al., 2020). While our focus is on multilingual students' science learning, we believe that all students can benefit from a deeper integration of language development, cultural sustenance, and knowledge building in the context of science learning. The theoretical constructs that inform our framework are as follows:

Language Development for Building Semantic Structures Our view of language development is functional in nature and is informed by Systemic Functional Linguistics (Halliday & Matthiesen, 2004). We see language as a system that offers pliable choices to make meaning for specific purposes and audiences across disciplinary and social contexts. In our view, language is always implicated in knowledge building as it is through language that we build both semantic and epistemic structures. With current shifting expectations for how scientific reasoning and inquiry should unfold during instruction, we argue that a parallel shift is required in how science classroom discourse is conceptualized (Halliday, 2004). For example, teachers need to support students in understanding both how and why to make shifts in language use, such as when students need to switch from language used for peer negotiation in a lab group, to oral explanation of their findings used to respond to teacher questions, to written explanation used to write up a lab report. This view of language led us to integrate specific linguistic practices related to building students' semantic power into workshops and assessments. These practices included generating a multilingual meaning making environment; making genre choices based on topic, purpose, and audience; and providing flexible language scaffolding based on specific science learning goals.

Cultural Sustenance for Integrating Diverse Forms of Knowledge The CS-SFL: Science framework applies and extends understandings from Culturally Sustaining Pedagogies (CSP; Paris, 2012; Paris & Alim, 2017) to support students in remixing their own linguistic and cultural knowledge with that of NGSS. That is, they are encouraged to bring together physical, cultural, and linguistic knowledge from community and school. From traditional funds of knowledge perspective (González et al., 2006), cultural relevance requires seeking out the knowledge and skills that students are gaining in their homes and communities and then looking for connections that can be drawn from that home knowledge to support understanding of the school curriculum. Adopting the broader, culturally sustaining perspective requires educators to actively work in collaboration with their students and students' families to maintain and further develop students' and communities' cultural and linguistic resources. In the CS-SFL: Science framework, we push this idea still further to explore how students recursively take what they are learning in school and apply it to make new meaning of their out-of-school interests and passions.

Culturally sustaining science practices can serve to challenge the supposed divide between the academic discourse and practices often equated with intelligence and knowledge and the everyday linguistic, cultural, and epistemic practices associated with activities beyond the classroom. Thus, a pedagogy of translanguaging is a core culturally sustaining practice in that it normalizes multilingual and multicultural perspectives across activities by valuing and using the full range of linguistic, multi-semiotic, and cultural resources that students bring to the learning environment (García & Wei, 2014).

Knowledge-Building for Epistemic Power The new wave of science standards represents a fundamental shift in thinking about how best to apprentice all students, including multilingual learners, into scientific knowledge-building through observation, reasoning, and argumentation (Turkan et al., 2014). For example, all students should engage in a range of age-appropriate science practices that are linguistically demanding, as they gain an integrated view of science and language through shared experiences with phenomena as the basis for disciplinary meaning making (Lee, 2017). However, business-as-usual K-12 science instruction for multilingual learners in the US context has continued to focus on learning discrete concepts as facts to be memorized, often in separate classes for second language learners (Bunch, 2013). Further, typical language supports for multilingual learners in science (e.g., word walls, sentence starters, graphic organizers) tend to front-load the teaching of technical and disciplinary vocabulary as separate from participation in scientific exploration and classroom discourse that is central to disciplinary knowledge building (MacDonald et al., 2017). Still, updated science standards provide a renewed opportunity for science teachers to innovate around the role and timing of semiotic scaffolding so that instruction can center rich meaning making resources (e.g., multimodal graphs, verbal reports, various modes of representation) scaffolded for multilingual learners as needed (Jakobson & Axelsson, 2017).

In summary, our CS-SFL: Science framework responds to the dynamic range of science practices that students need to develop through science learning. To achieve this goal, translanguaging is positioned as one of the key practices for enacting the framework.

2.3 Methods and Data Sources

In this study we used two distinct data sets from two project contexts to focus on the semiotic choices of multilingual students and families engaged in science learning: family interviews conducted during our Steps to College through Science bilingual family workshops and student responses on our project-designed constructed response science assessments. We analyzed these data sets to answer the following research questions:

1. What translanguaging practices do multilingual learners and their families enact when connecting their family and community cultural backgrounds with their developing knowledge of science and science careers during family interviews?
2. What translanguaging practices do multilingual secondary school students enact when engaged in sensemaking on a project-designed bilingual constructed response assessment?

We explored whether and how the learning interactions in these contexts supported the three key tenets of the CS-SFL: Science framework: language development, cultural sustenance, and science knowledge generation. We do not attempt to compare or contrast the practices in the two sites, but, rather, to explore the diverse ways that translanguaging was enacted.

2.3.1 Data Set 1: Family Interviews

The first data set comes from family interviews gathered after the project's Steps to College through Science bilingual family workshops (Kirmaci et al., 2019). These workshops consisted of an ongoing series of half-day science experiences for science teachers and multilingual families and were hosted at different post-secondary education institutions. In each family workshop, participants rotated through three sessions: one focused on STEM research and careers led by professors and/or staff at the host institution; a second session focused on family conversations on how to navigate and get ready for college and careers; and a third session focused on where families and teachers engaged together as co-learners in science inquiry, highlighting science practices and the language of science through hands-on investigations. Workshops ended with a shared meal and informal conversation. All sessions were conducted bilingually in English and Spanish. All families who participated in the project workshops were asked to engage in a family interview that was conducted during the last workshop of each year. The interview took the form of a multilingual interactive card game wherein parents and children took turns drawing cards and then asking and answering the questions provided. For this study, interviews from all families who consented to participate during a 2-year period (2014–2016) were audio recorded and transcribed (Year 1: 12 family interviews and Year 2: 23 family interviews).

To explore how the translanguaging practices of our family workshop groups connected to our theoretical framework, we conducted a thematic coding of the data within the three categories of language use, culturally sustaining pedagogies, and knowledge building. Within these categories, we analyzed the types of translanguaging enacted by the family group in terms of how parent(s) and child(ren) co-constructed meaning and what meanings emerged from these joint constructions around science investigations and careers. We used an SFL ideational analysis, especially of the logical meta-function within the ideational system, to explore how the turns of talk of the parents and children interacted to make meaning. In general, as Eggins and Slade (2006) discuss, when responding to a previous turn of talk, the next speaker negotiates the proposition set up by the previous speaker by continuing what was introduced by the first speaker through elaboration (clarifying, exemplifying, or restating); extension (offering additional or contrasting information), or enhancement (qualifying information by providing details of time or place, etc.). We found this SFL analysis helpful in seeing how the parents and children negotiated meaning making through their turns of talk in the translanguaging context of their family conversations about science.

2.3.2 Data Set 2: Student Assessments

All students in the classes of the science teachers who participated in the project were asked to complete a project-developed bilingual constructed response assessment at the start and end of each school year. Each year of the project, approximately 1500 student assessments were administered, of which roughly one third were from students who spoke Spanish as their home language. The second data set for this study consisted of a sample of 38 assessments drawn randomly from the assessments collected during the 2016 post-assessment from multilingual students who self-identified as Hispanic, named Spanish as their primary home language, and included at least some writing in more than one language on their assessment. We further selected 10 of the response items on the assessment, analyzing each of these 10 items on the sample of 38 student assessments. This yielded a total of 276 individual responses out of a possible 380. We note that it was common for students to leave some items blank on our assessment, which was both lengthy and quite different from the assessments the students typically experienced. A sample assessment scenario is presented in Appendix 2.1, and brief descriptions of the other scenarios and of the 10 focal response items selected for this study are presented in Appendix 2.2.

Our bilingual (English-Spanish) multimodal constructed response assessments were designed to allow bilingual students to demonstrate their science meaning making (Buxton et al., 2019). In tune with Cowie (2013), our assessments focused on students' ability to construct "argumentation, model building, and explanation as central to learning and knowing science" (p. 474). The assessments served two specific purposes within the project. First, they served as a formative tool for our project team, allowing us to make changes in our pedagogical design and curriculum materials based on what we were seeing in the student responses. Second, they served as an evaluative tool to show the impact of our design and development efforts with teachers and families. With those two purposes in mind, it was clear from the outset that our assessment needed to normalize and center bilingualism and multilingual resources, rather than normalizing monolingualism and treating bilingual resources as an accommodation. Thus, each of the seven scenarios presented in the assessment was developed around a science investigation that had everyday connections and included a range of multimodal resources to enhance science meaning making such as illustrations, tables and diagrams, as well as providing all information bilingually. Students were encouraged to read and write in English, Spanish, or any combination they preferred and were asked to record that choice for each item.

Elsewhere, we have described a range of both quantitative and qualitative analyses we have performed on the student assessment data for the formative and evaluative purposes we have mentioned (Kim et al., 2017; Cardozo-Gaibisso et al., 2019). For this comparative analysis of selected responses, we began by creating organizational categories (Maxwell, 2013). The three components of our CS-SFL: Science framework provided guidance for considering how students demonstrated language development, cultural connections, and science knowledge building. We then deepened this analysis through substantive categorization that described the range of translanguaging choices and practices our multilingual secondary school students enacted. These categories assisted us in identifying four profiles that described students' translanguaging choices in this specific compositional translanguaging context as we attempted to "explicitly identify the content of the person's statement or action" (Maxwell, 2013, p. 197).

2.4 Findings

As alluded to above, our findings from the two contexts of the family interviews and student assessments are very different. Translanguaging occurs in more restrictive ways within formal assessment contexts as opposed to in the more dialogic contexts of the family workshops where family members build relationships and negotiate meanings in collaborative ways. Thus, as our discussion moves from an informal to a formal space, we invite readers to think about how joint construction and dialogic meaning making could inform more traditional classroom and assessment practices. To do so, we begin by discussing our findings from the family interviews as the entry point to understand the translanguaging practices enacted through our CS-SFL: Science framework and then shift our attention to the student assessments.

2.5 Co-construction of Culturally Sustaining Practices in the Steps to College Through Science Bilingual Family Workshops

Fundamental to our view of culturally sustaining science practices is the idea that the goal of broadening participation in science is more than just an equity issue but is also a question of improving science itself by adding new perspectives and experiences (Harding, 1991). Thus, science teachers learning to sustain and incorporate an array of multilingual and multicultural perspectives with their students and communities should be positioned as part of a broader project to improve science as well as science education. In our analysis of the family interviews, we found rich instances where parents and children disrupted the view of science as located solely in formal learning spaces such as classrooms and laboratories. Instead they viewed themselves as engaging with science through work, through relationships with family, and through interactions with nature. In general, the parents and children co-constructed this culturally sustaining view of science through turns of talk that usually began with the parent sharing their widened perspective and the child elaborating or extending it.

In the case of one mother and daughter team, for example, Patricia, the mother, exhibited an expansive view of science as related to lived experiences with family and nature. She talked about her childhood and her experience of sowing seeds and watching them grow into food. Her daughter picked up on this perspective and gave an example of the continued use of science in their household. Both spoke in Spanish except for the pre-designed questions that Carolina chose to ask in English. All translations into English were done by members of our research team.

Carolina: Have you learned about science outside school?

Patricia: Sí, desde bien Chiquita... desde los cinco años para mi era ciencia ir al campo a las labores a sembrar. Mi papá me daba semillas de maíz y de frijol y haciamos con el azadón haciamos un hoyito en la tierra hasta terminar una hectárea o tres hectáreas; toda la tierra. Dos semillas de maíz y dos de frijol. Y cada tiempo íbamos a ver si ya había crecido y que tanto había crecido, si había llovido, si no había llovido, si necesitaba limpiarse la tierra de la hierba y para mi eso me gustaba porque yo quería saber si la tierra era buena para que las plantas crecieran grandotas. (*Yes, since I was very little... since I was five years old I went to the countryside to grow things, that is science for me. My father gave me corn and bean seeds and we used a hoe and with it we made holes in a whole acreage. Two seeds of corn and two of beans, all the land. Then, we went every now and then to see how much plants grew, if it rained and if it didn't, if we need to clear the land of weeds. I liked that because I wanted to know if the soil was good for the plants to grow big)*

Carolina: Pues pusimos unas plantas afuera de la casa y veíamos si había crecido o no. Y quería saber si unas habían crecido más que otras. Por que unas no crecieron y por que otras si. *(So now we put some plants outside home and we watched if they grow or not and we wanted to know if they grow more than the others and why).*

A micro-linguistic analysis of the excerpt above shows that Carolina enacted the same perspective as her mother about community science, using a parallel construction to that used by her mother. They both began by describing explorations they engaged in with family members and then explained the inquiry they were pursuing while engaged in the practical work of growing food. In other words, the close relational bond of mother and daughter supported the joint construction of science as an important culturally sustaining discipline not restricted to school.

Similarly, in the father and son discourse excerpt below, the father (Melvin) in Spanish expanded on the understanding of what science is by explaining how it connected to his work as a mechanic. In response to a question about knowing someone in the sciences, his son (Anthony) in English then summarized his father's description of his work as "mechanical science," following the father's lead in intertextually connecting the science activities in the workshops to the father's educational background and work.

Anthony: In the science workshops we have done experiments and visited labs together, can you tell me something you learned or something you remember about science in those workshops?

Melvin: Si correcto, eso me hace recordar que yo tomé una clase técnica, soy electricista, entonces yo puedo rebobinar motores y me hace recordar la fuerza magnética que hace un motor cada vez que la corriente circula alrededor el campo magnético que esto hace cuando la corriente llega a un motor eléctrico. (*Yes, right, that makes me think that I took a technical class, I am an electrician so I can rewind motors and that also makes me think of the magnetic force that every motor has every time current flows around the magnetic field which is made after the current flows in an electric motor).*

Melvin: Do you know someone who does work in science, if yes, how is science part of the job?

Anthony: Yes, because my dad is an electrician so he does a lot of mechanical science, things like that.

In extending Melvin's turn of talk, Anthony summarized in English what his father said by stating "he does a lot of mechanical science, things like that." Overall, these turns of talk function to convey and co-construct a familial understanding of science that extends beyond traditional views of who engages in science, in what contexts, and in what ways.

2.5.1 Co-construction of Scientific Meaning Making

Connected to the culturally sustaining approach that many families enacted in speaking about their experiences with science, family members also extended and elaborated their science knowledge through their engagement in the project workshops and subsequent dialogues. Thus, collective turns of talk around science knowledge building were enacted through discourse strategies that signaled cooperation in shared meaning making. Although there were times when parent or child signaled reluctance to engage in such discussion of science concepts because they felt uncertain about their knowledge, the more common pattern was for families to jointly construct an understanding of science by intertextually connecting their participation in investigations during the family workshops with their past experiences with science. Sometimes the families chose to speak only in Spanish and sometimes the children spoke in English and the parents in Spanish. Some chose to read the pre-designed questions in Spanish and some in English.

Analysis also showed that the science investigation practices in the family workshops played a crucial role for participants to activate their prior knowledge and to co construct new knowledge about science. In the following interaction, a mother, Monica, responded to her son's question by using highly concrete descriptions of one of the workshop's science investigations. Alberto then responded in a similar concrete way but in English. In his elaboration of his mother's talk, he both repeated and extended her explanation of the workshop investigation.

Alberto: Tell me about an experience you did in the workshops and then tell me how you solved it like a scientist.

Monica: Cuando estábamos con el vaso y la cuerda. Cuando estábamos con la cuerda y los diferentes tipos de movimiento y ondas de sonido y de calor. Fue muy interesante el saber que tienen longitud y tamaño y que eso las hace diferentes y que en eso depende si hacen un sonido alto o bajo. Eso no lo sabía y lo aprendí. (*When we worked with the cup and the piece of string. When we used the piece of string and we did the different wave movements and wave sounds and heat. It was very interesting to learn that waves have wavelengths, sizes and those make them different, which also results as a high or low sound. I did not know that and I learned it here.*)

Monica: Tell me about an experience you did in the workshops and then tell me how you solved it like a scientist.

Alberto: When we had the string and the cup, I had my hand too far away and the string was long and it made low sounds so it sounded like we went in a cave. Then, I pulled the cord and it was short and we ended up with a higher pitch and the sound was higher.

As evidenced above, Monica responded to the question by beginning with a concrete description of the sound waves investigation they conducted and then she moved into evaluative reasoning about what she learned through the investigation. Alberto, mirroring his mother's response, but in English, extended Monica's description by giving more details, such as how when his hand was farther away, the more the string vibrated, resulting in a lower pitch. He then expanded her explanation of their new collective understanding by describing specifically how his shortening of the string led to a rise in pitch. Overall, through discourse strategies of expansion and extension, Monica and Alberto's co-construction of their understanding of sound waves drew intertextually upon their family engagement in the workshop investigation and subsequent conversation.

In the second example of co-constructed science knowledge building, the daughter, Nancy, used her turn of talk to extend her mother's descriptions by restating in English key elements of what her mother had stated.

Nancy: At the science workshops we have done science experiments and visited science labs together, can you tell me something you did or something you learned in these workshops?

Antonia: En el que estamos ahora me llamo mucho la atención que elementos tan simples como la sal, el detergente, y el azúcar, la combinación entre ellos. Esos elementos químicos y utilizando una fruta tan sencilla como la fresa se puede obtener el ADN con elementos. (*The activity we just did was very interesting because of the simple elements like salt, detergent, and sugar. The combination of those elements. Those chemical elements mixed with a simple strawberry give as a result the DNA extraction.*).

Antonia: In the family workshop we did science experiments together, can you tell me something you did or something you learned?

Nancy: Well, I learned that different combinations of simple things that you can find in your house can let you see a person's DNA or just a simple fruit's DNA, like a strawberry. We experienced that you can find the strawberry DNA using dish detergent, salt and I cannot remember the other one.

In Antonia's turn of talk, she described a DNA extraction experiment and then explained how and why it interested her. She detailed the elements involved in the process. In her turn of talk, Nancy picked up on Antonia's description of the processes by providing a description of the elements such as dish detergent involved in the inquiry. Overall, Antonia's talk provided a more abstract reasoning about the processes, and Nancy complemented this by moving into more concrete details.

As can be seen in these examples, families co-constructed their science meaning making, as contributions from parents and children came together to generate new knowledge derived from shared experiences through discourse strategies of extension, elaboration, and repetition. Family members drew intertextually from their collective experiences in the family workshops, showing how they valued and learned from the collective science experience. Indeed, as we reported in another recent study (Harman et al., 2020), a multilingual and embodied approach to learning science seemed to disrupt the supposed gap between academic language of schooling and hybrid practices of minoritized multilingual communities (Flores & Schissel, 2014). Instead, we saw substantive science knowledge generation occurring in these community collaborative workshops.

2.5.2 Co-construction of Language Sustenance Through Translanguaging

A central goal of the family workshops was to represent science knowledge and practices not as an elitist domain but as a way of problem solving and understanding the natural world that is accessible to all. The family interview provided a safe space where family members could try out their articulation of different science concepts and practices without feeling assessed or judged.

In discussing the role of language in science, one father (Luis) and son (Sergio) explored the integration of what they learned in the investigation and the language of science they had learned.

Sergio: In the science workshops we have talked about academic language and how it is used in school. Tell me something you have learned about academic language and how it is used in these workshops.

Luis: Bueno, en los talleres nos han enseñado muchas palabras científicas como las partes de animales, los químicos, las recetas, los instrumentos, y cosas diferente como las protecciones contra los químicos y las pre-

> cauciones que uno tiene que tomar también. [...] aprendimos cómo trabaja la ciencia en las vidas normales o de los científicos y que es lo que tienen que hacer en situaciones donde extraen el DNA de objetos como huesos o la sangre o incluso fresas y en otros hemos trabajado con instrumentos de la temperatura del clima y también hemos trabajado con veterinarios. (*Well, in these workshops they have taught us many scientific words like animal parts, chemicals, prescriptions, instruments, and other different things like protections against chemicals and the precautions one has to have too. [...] We learned about how science works in our normal lives or about what scientists have to do in different situations where they extract DNA from objects like bones, blood, or even a strawberry, we have also worked with instruments that measure the temperature of the weather and we have worked with veterinarians*).

In their co-construction, Sergio and Luis pointed to the inextricable connection between scientific investigation and the language used to describe it. Together they co constructed the language of science as necessarily integrated into scientific inquiry.

In the following conversation among a mother (Antonia), daughter (Nancy), and son (Ignacio), there is similar evidence that the family was engaging in science discourse based on their efforts to describe the investigations they participated in. That is, their embodied experiences doing science supported both their access to and desire to make use of a more specialized language of science. However, they also highlighted how different and complex that language can seem. Ignacio even said it was a totally separate language, enthusiastically giving examples to support this claim.

Antonia: In the family workshops we talked about academic vocabulary and how language is used in the school. Tell me something you learned about academic language and what you learned in the workshop.

Nancy: Well you can use academic languages in the school in science and that would be good to use in science because that would give you more vocabulary, more experience using all that vocabulary.

Ignacio: The periodic table, the way they call some atoms. It is really interesting how they call other stuff.

Antonia: Scientists use language in a very unique way, what can you tell me about how scientists use language?

Ignacio: Scientists use very hard language, I believe like for a scientist to learn the language is like to learn a whole new other language because it is really different than ours, it's really long words and stuff that you have to add to it.

Antonia: Me as a parent is more easy to remember DNA, dexo... dexo something. it is just an example. El lenguaje de los científicos es difícil. (*Scientific language is difficult*).

Ignacio: In chemistry we are learning how some compounds are mixed together and perhaps if you see two nonmetals that are together, they would be called like if one is oxygen and the other is nitrogen, and if there are two nitrogens then it is called di-nitrogen or tetra, it depends on the number of … that's an example of how scientists can use language.

In this case, Ignacio stressed the separate nature of scientific language and everyday language. Antonia extended this proposition by giving examples of words such as DNA that she learned through the investigation and then also restated what Ignacio said about the language of science being difficult.

In this discussion of the findings from the family workshop interviews, we have pointed to the value of opening up community dialogue about scientific language, culture, and knowledge building as much as possible. In this way, we hoped to support families in expanding their understanding of science by availing themselves of whatever linguistic and other semiotic choices they found most useful to convey meaning. Across the examples we have presented, we see families making organic connections between their own lived experiences and scientific concepts. The interviews also directly reflected the family workshop practices, which were designed to incorporate the linguistic and cultural repertoires of multilingual students and family members while making the richness of these resources visible to the students' teachers.

What is highlighted also through these examples is the importance of supporting multilingual learners in feeling comfortable and confident in moving recursively between concrete examples and more abstract reasoning in whatever language and register most useful to them in a given context. Our analysis showed that families sometimes shared emergent understandings of those concepts and practices, supporting each other in moving to deeper understanding. This points to the importance of embodied multi-generational learning as a key to supporting scientific understanding and also to the importance of dialogic environments such as the family conversations, where learners are encouraged to reflect on and describe what they have learned outside of the context of formal education or a formal assessment. However, because formal assessments remain central to how schools judge both teacher performance and student learning, projects that aim to broaden science learning must also consider the influence of assessments as well.

2.5.3 Translanguaging Practices in Our Bilingual Constructed Response Assessments

Although when we developed our project assessment in 2013, there was little written about how translanguaging pedagogies could be applied to the context of science learning and assessments, our project was based on the premise that languages are used for the purpose of meaning making and, thus, the time and place for appropriate use of language and register should largely be a student/learner decision, rather than a teacher or researcher one. Thus, we designed our assessments to provide students with the opportunity to use translanguaging practices to demonstrate science meaning making, leaving it to each student to decide how and when to use different languages.

Unlike the findings from the family interview analysis described above, the assessment analysis provided a more constrained context to identify examples of students leveraging aspects of our CS-SFL: Science framework, in terms of making explicit cultural connections or supporting each other in using fuller repertoires of their linguistic backgrounds. Thus what we were able to see in our analysis of the assessments was different, more limited, yet also complementary to what was seen in the family interviews. To describe the students' translanguaging practices that we identified in the assessments, we categorized the student answers into four broad translanguaging profiles. Although we identified four distinct profiles, these profiles did not occur with the same frequency, with the first two profiles occurring more frequently than the latter two in the set of student answers we analyzed.

As opposed to the dialogic nature of the family interview processes, these profiles highlight how the meaning making in the assessments tended to be intertextually connected to the assessment questions and texts we had designed. In other words, the assessment itself served as a linguistic and scientific support for multilingual learners (see Buxton et al., 2019). Within each profile, analysis showed that the focus in our project on inquiry and bilingual meaning making supported students in successfully constructing emergent scientific meaning on the assessments using concrete language in either Spanish or English. However, because of the restrictive nature of school-based assessments, students typically only brought school-based learning about phenomena to their responses and did not make connections to their cultural knowledge.

2.5.4 Profile 1: English Prevalent Translanguaging

Profile 1 included student answers which were predominantly written in English only. To be included in the sample, at least some use of Spanish was required, but in the case of profile 1 this use was limited and, may, for example, only have included an occasional Spanish language word or phrase. In this profile the use of Spanish was typically separate from, rather than integrated into, responses that were otherwise written in English.

The following data excerpts from the assessment are representative of a student who we categorized as adhering to Profile 1, where English was the prevalent language used to convey meaning. The first two examples typify responses from this student that were written completely in English and which expressed the level of science understanding roughly in line with the expectations of the middle school science curriculum that the student was being taught. The learner enacted scientific meaning making through use of concrete description that mirrors that of the language in the prompt. That is, the meaning making was supported through intertextual resourcing.

Q5 When Maria connects the rubber band the light bulb will not light up and there will be no complete circuit but when Maria connects the nail the light bulb will light up because it is a complete circuit.

Q20 The lip gloss doesn't have bee wax like the lipstick and the lip-gloss has less palm wax that the lipstick.

The next examples from this same student typify many of the responses in Profile 1 in that they use (English) language in ways that demonstrate an emergent understanding of the scientific concept but also indicate misconceptions and or do not clearly articulate the reasoning for their ideas. Drawing from our experience and interactions with students who participated in the project and, thus, completed the assessment, we can hypothesize that although students may have known the content, they felt that the expectation was for that content to be written in English, restricting their ability to fully elaborate their thinking.

Q8 The effect on putting yeast on bread dough is that the dough expands. The cause for that is that the yeast makes the bread dough turn brown.

Q10 The effect is that when Jorge pulls the rubber band back kinetic energy is shown because Jorge is using energy.

Q24 My observation supports my hypothesis because I put that people wear light clothes because the temperature is lower and my observation says the same.

The final examples we share from this student are typical of the occasional use of Spanish that is representative of students in Profile 1. The use of Spanish is in a short stand-alone phrase representing a relevant but incomplete thought that is not connected to other writing in English.

Q3 La bombilla se enciende. (*The light bulb turns on.)*

Q4 El clavo va a hacer un circuito completo. *(The nail will make a complete circuit.)*

From these examples, we can see that this student, representative of students we placed in Profile 1, wrote both in English and Spanish, but predominantly in English to convey meaning. We see a tendency for the learners in this profile to keep to descriptions of the phenomenon and refrain from moving to more abstract reasoning about it. There was not clear evidence for how these students made the choice of when to switch briefly to Spanish in the midst of their writing which was otherwise in English.

2.5.5 Profile 2: Spanish Prevalent Translanguaging

Profile 2 consisted of student answers written predominantly in Spanish only. As with Profile 1, the responses in Profile 2 were mostly monolingual Spanish with occasional English words or phrases included. As with Profile 1, the use of the second language was typically separate from, rather than integrated into, responses otherwise written in the dominant language.

Unlike Profile 1, however, many students who fell into Profile 2 began the assessment writing in English but then shifted into Spanish after the first few questions. The following data excerpts are representative of a student who adhered to Profile 2, where Spanish was the dominant language used to convey meaning.

The first two examples typify responses from the start of the assessment that were written completely in English. These responses are incomplete and fall short of expressing the level of science understanding that is roughly in line with the expectations of the middle school science curriculum the student was being taught.

> Q3 The light bulb will turn on.

> Q8 The dough got more bigger because the dough absorbed the water.

The next examples from this student typify many of the responses in Profile 2 in that they shift to using Spanish in ways that help them to demonstrate an emergent understanding of the scientific concept while still indicating misconceptions or failing to clearly articulate their reasoning for their ideas. That is, whether in Spanish or English the student's science understanding is still developing and does not seem to depend on the language used. We can hypothesize that if students in this profile wrote their answers predominately in Spanish, while the context in which those students learn science is a monolingual English one, much of the science content and inquiry practices would have been inaccessible to the student due to a schooling context which did not provide the linguistic support or multilingual environment needed.

> Q10 Cuando Jorge jala la resortera hace fuerza y cuando la suelta hace potential energy. (*When Jorge pulls the slingshot there is force, and when he releases it there is potential energy.*)

> Q24 Por que refleja la luz y el rojo oscuro la absorbe. (*Because it reflects light and dark red absorbs it.*)

From the examples above, we can see that this student, representative of our Profile 2, wrote both in English and Spanish but shifted from English to Spanish after the first few questions. Like in Profile 1, we do not have clear evidence for why the student made the choice of when to switch to Spanish. We suggest that students in this profile came to realize that they could convey meaning on these answers more effectively in Spanish even though they were accustomed to writing in English only on assignments and assessments in their school science class.

2.5.6 *Profile 3: Integrated Translanguaging*

Profile 3 consisted of students whose answers more fully combined Spanish and English in the responses to given questions. This was distinct from profiles 1 and 2; English and Spanish were largely kept separate. Responses in this profile seemed to switch between languages more intentionally for purposes related to describing their science meaning making. Some students in Profile 3 began writing in English and others began writing in Spanish but then gravitated to hybrid forms to more fully express their ideas.

In the case of our example student, they first responded in Spanish and then responded to the following question in English:

> Q3 Si metes un clavo en a el circuito se electrisara el clavo y no va a funcionar el circuito. (*If you include a nail in the circuit it will electrified the nail and the circuit will not work*).

> Q4 The nail transform to black or brown the clips don't want to work anymore.

This student then shifted to responses that combined English and Spanish as the assessment progressed. Ideationally, the student drew on Spanish to describe two action verbs integral to what was occurring in the investigation. Without access to translanguaging, the student might have become stuck in describing the complete process.

> Q5 The circuit then connect the rubber band se va a deshacer the rubber band I think the nail se va a pegar al circuit. *(The rubber band will come apart I think the nail will stick to the circuit.)*

> Q21 The lipstick for tener brillo I think has aceite. (*The lipstick to have gloss I think has oil.)*

Students in Profile 3 continued to shift their language choices from question to question, such as in this example where the student shifted back to all Spanish for one of the final responses.

> Q21 El hilo se va a romper si lo metes en el agua. (*The piece of thread will break if you put it in water.*)

This example of Profile 3 shows how a student sought to demonstrate understanding and elaborate on scientific concepts without feeling linguistically constrained by the monolingual practices typical of their school science instruction. In other words, a multilingual environment created by our use of bilingual materials and prompts supported this student in availing their full linguistic repertoires to make scientific meaning.

2.5.7 Profile 4: Translational Translanguaging

Profile 4 consisted of student answers that were written in both English and Spanish but representing direct translation of their ideas from one language into the other. That is, in Profile 4 the student is typically expressing the same idea twice, stating it once in English and once in Spanish. We have come to describe this as combining Spanish and English for translation purposes.

In the case of our example student for Profile 4, they typically first responded in Spanish and then restated the response in English.

> Q3 Que el foco se prendera/the light bulb will turn on.

> Q4 Le dara mas energia/it will give it more energy.

Q10 La piedra sale volando por la energia cinetica/the rock flies away because of the kinetic energy.

However, students in this profile did not frame every response in this alternating way. In the following examples from our profile 4 example, some responses of the student were just in Spanish and some were just in English.

Q8 Lo tiene que agregar porque es un ingrediente esencial que hace que la masa se expanda. (*You have to add it because it is an essential ingredient for the dough to expand.*)

Q20 Lipgloss does not need beeswax, and it doesn't need as many palm wax.

While we lack explicit evidence for why students in Profile 4 made the choice to engage in direct translation of their answers, we hypothesize that this is because students in our project rarely had an opportunity to engage with bilingual science resources in their science classes. We should not be surprised that lacking such experiences, students are not well prepared to understand why they are now being given such resources nor how they are expected to make use of them. Given that the assessment itself provided all information in both languages, some students might naturally conclude that this was the expectation for how they should respond as well. Other students, perhaps wishing to write in Spanish, but typically writing for teachers who speak English only, may have been thinking about their potential audience when engaging in this self-translation.

Overall, when considering the responses of multilingual students on our bilingual constructed response assessments, we found that students clustered into four profiles of language use. Profiles 1 and 2, representing English prevalent or Spanish prevalent responses were the predominant profiles, with profiles 3 and 4, representing more bilingual language use within and across responses being less common. We take from these findings the need to think about students enacting translanguaging practices across the assessment as a whole, rather than looking to understand students' translanguaging practices within their answers to individual questions. When students chose to answer the same question combining English and Spanish (as seen in Profiles 3 and 4), some seemed to do so for meaning making purposes, while others seemed to do so in line with a vision of providing translations to those who would read and score their assessments.

2.6 Discussion and Conclusions

In this study, we set out to investigate what and how translanguaging practices are enacted across two different spaces that were key components of a multiyear teacher professional development project. This project aimed to support multilingual learners more effectively in integrating their linguistic and cultural repertoires and backgrounds with their science learning experiences in middle and high school. Following Poza (2017), we believe that translanguaging provided "a lens through which to view the language practices of bilingual Latinx students as valuable, generative, and powerful, rather than in need of remediation" (p. 102). The two contexts were family conversations at the end of a series of bilingual family science workshops held at institutions of higher education, and students' responses on a bilingual constructed response science assessment given in their school science classes. Given that these two contexts were quite different from each other, we expected that the ways participants would take up and use the available translanguaging supports would look different and would perhaps raise complimentary questions and insights about how to support a translanguaging science classroom.

To answer our first research question about how multilingual families enact translanguaging practices in the context of interviews about family and community experience with science, we conducted a thematic and ideational SFL analysis of the families' multilingual conversations. We explored both how these families made collective meaning of the science topics and science careers they had learned about in the workshops and how they used translanguaging to position themselves and their communities in relation to science, education, and post-secondary careers. For example, through an ideational exploration of their multilingual texts, we began to see how families' multilingual meaning making functioned to support their construal of scientific concepts and the role of science in their lives, as well as how these patterns of translanguaging function to interweave cultural and school knowledge of science.

By taking an ecological perspective on translanguaging practices, our exploration of the dynamic multilingual meaning making of families points to the importance of informal community engagement around science inquiry as a key component for shaping translanguaging science learning environments. In other words, our findings point to the importance of translanguaging practices occurring beyond normative classroom learning, to include families and other community members both in and beyond school spaces. We believe that the dynamic multilingual and multigenerational science meaning making we witnessed in the family interviews could be extended into classroom practice by allowing for learner-initiated group inquiry projects that position students and guest family members as agentive science investigators and not just as learners of standard curriculum. In addition, our exploration of the family interview data shows that the patterns of translanguaging are connected intertextually to the family workshops, showing that the embodied community approach to learning encouraged families to use all their available communicative resources to engage together in meaning making about science.

Our second research question considered the translanguaging practices that multilingual students in grades 6–10 enacted when engaged in sensemaking on our bilingual constructed response assessments. We explored how multilingual students used a range of linguistic repertoires in their responses and the degree to which these fuller repertoires did or did not support students' demonstration of scientific understanding. However, given that students in our project were largely taught science by monolingual English-speaking teachers using English language instructional resources, we were not surprised that many bilingual students avoided or made only limited use of translanguaging resources available on the assessment.

Unlike the family interview context, in which students and their families were able to support each other in enacting oral translanguaging practices, in the case of our assessments, individual compositional (written) translanguaging presented many challenges. Likewise, in an often-cited meta-analysis of testing accommodations, Kieffer et al. (2009) examined the effectiveness and validity of specific testing accommodations to improve the performance of bilingual learners on large-scale assessments in mathematics and science. The researchers found that the only accommodation that consistently improved bilingual learners' performance was providing students with an English language dictionary. Others, who have raised questions about findings of this kind and why various first language supports and other linguistic accommodations do not seem to yield improved test performance, conclude that the explanation includes both academic language components (Duff, 2011) and socialization components (Hoff, 2006). In short, multilingual students in our context were rarely given opportunities to practice written translanguaging as part of their science learning, so it should not be surprising that most students struggled to productively use such resources in the context of a pre- and post-assessment.

If assessments could be conducted in ways that look more like our family workshops, centering multilingual group investigation, the meaning making might look and sound quite different from what we saw on our assessments. Because the design of our assessments was traditional in the sense of demanding individual knowledge and learning, and was administered in the regular classroom context, the students often fell into habituated school practices, with responses that gave little indication of the linguistic and cultural resources that they brought with them to the science classroom. We know that bilingual students learn best when leveraging their conceptual, linguistic, and cultural resources in making sense of the natural world (Rosebery et al., 2010). Further, culturally sustaining approaches to disciplinary instruction show the importance of supporting youth and their communities in maintaining and expanding their cultural and linguistic repertoires (Paris & Alim, 2017).

We also know that teachers need substantive support to create such opportunities for learners in their classrooms (Garcia & Wei, 2014). There is limited research exploring how robust translanguaging practices can take place in more formal instances, such as on written assessments, raising the question of how to jump from simply acknowledging the value of classroom translanguaging practices to adopting these practices in consequential contexts such as classroom assessments. In our project, we began exploring this need in our Teachers Exploring Student Writing Workshops, where project teachers partnered with members of the research team to explore their students' writing on the project assessments and other work samples that teachers brought to the workshops. In these workshops, the teachers often expressed the desire to do more to support their students in leveraging their full linguistic and cultural repertoires in the service of science learning, but they also shared that daily instruction rather that summative assessments seemed a more feasible context for these efforts. Garcia and Wei (2014) have made a similar point that "accepting translanguaging in assessment would require a change in epistemology that is beyond the limits of what most schools (and teachers) permit and value today" (p. 135). However, based on our findings from this study, we are using the next iteration of our project to design more dynamic assessments wherein pairs of students make meaning together through concept maps using whatever semiotic resources seem best to them.

Research on translanguaging has further shown that hegemonic valuing of the dominant language (English in our context) impacts how comfortable or ready bilingual learners will be in using their community linguistic resources (Major, 2018). In a study of a dual language classroom (English-Spanish) in the United States, Hamann et al. (2015) posited that the strong hierarchy between English and Spanish led to students' hesitation to privilege Spanish in their discourse. Such findings point to the need for educators to gain a more ecological perspective on multilingual classrooms. That is, we need to move from a view of translanguaging as a classroom instructional practice to one that sees translanguaging as part of an overarching framework that necessitates inclusion of community knowledge, dialogic interactions around science, and institutional policies that acknowledge and incorporate the lived experiences of multilingual learners. We have argued that translanguaging can be seen as one component of such an integrated model of culturally and linguistically sustaining disciplinary meaning making. In this way, we have come to think about translanguaging as the rationale for how, why, and when a teacher would make certain pedagogical choices around helping students build linguistic and cultural connections to scientific knowledge and practices.

Translanguaging takes place constantly in multilingual contexts during learning, whether educators recognize and support it or not. Still, translanguaging is only likely to be a productive resource for disciplinary learning when it is explicitly encouraged and supported as an academic (rather than just a sociocultural) resource. Findings from our study of family conversations and of student assessments, as well as from other studies in this volume, provide the field with needed guidance for what teachers should understand and be able to do in order to actually create robust translanguaging opportunities that support science learning. A particular challenge to this work is the need to disrupt the current rigid structures guiding summative assessments that limit science teachers' sense of agency to take instructional risks and to embrace ecological practices of multilingual meaning making. We believe that one productive direction is to promote increased opportunities for teacher-student-parent interactions that can empower teachers to extend themselves beyond their linguistic comfort zones as they learn to apply contextual and complex multi semiotic resources in their classroom learning spaces.

Appendices

Appendix 2.1: Sample Assessment Scenario

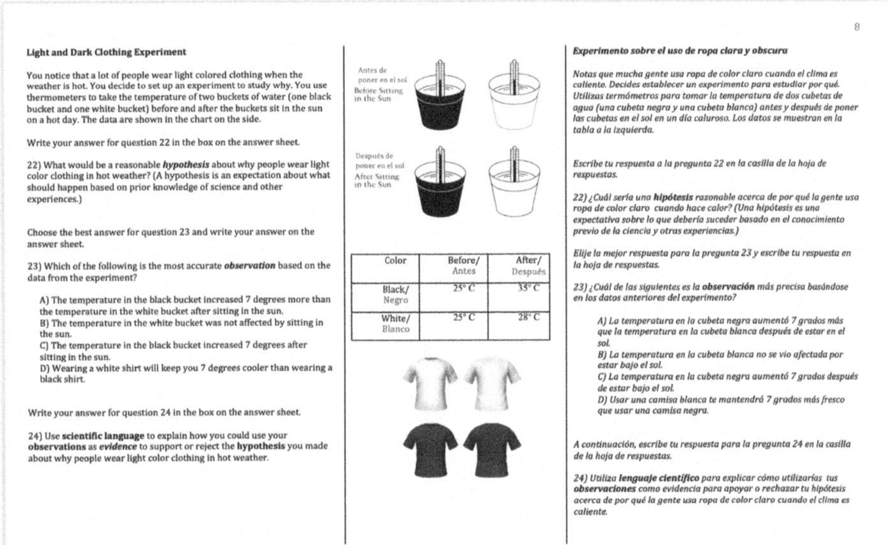

8

Light and Dark Clothing Experiment

You notice that a lot of people wear light colored clothing when the weather is hot. You decide to set up an experiment to study why. You use thermometers to take the temperature of two buckets of water (one black bucket and one white bucket) before and after the buckets sit in the sun on a hot day. The data are shown in the chart on the side.

Write your answer for question 22 in the box on the answer sheet.

22) What would be a reasonable ***hypothesis*** about why people wear light color clothing in hot weather? (A hypothesis is an expectation about what should happen based on prior knowledge of science and other experiences.)

Choose the best answer for question 23 and write your answer on the answer sheet.

23) Which of the following is the most accurate ***observation*** based on the data from the experiment?

A) The temperature in the black bucket increased 7 degrees more than the temperature in the white bucket after sitting in the sun.
B) The temperature in the white bucket was not affected by sitting in the sun.
C) The temperature in the black bucket increased 7 degrees after sitting in the sun.
D) Wearing a white shirt will keep you 7 degrees cooler than wearing a black shirt.

Write your answer for question 24 in the box on the answer sheet.

24) Use **scientific language** to explain how you could use your **observations** as ***evidence*** to support or reject the **hypothesis** you made about why people wear light color clothing in hot weather.

Antes de poner en el sol
Before Sitting in the Sun

Después de poner en el sol
After Sitting in the Sun

Color	Before/ Antes	After/ Después
Black/ Negro	25° C	35° C
White/ Blanco	25° C	28° C

Experimento sobre el uso de ropa clara y obscura

Notas que mucha gente usa ropa de color claro cuando el clima es caliente. Decides establecer un experimento para estudiar por qué. Utilizas termómetros para tomar la temperatura de dos cubetas de agua (una cubeta negra y una cubeta blanca) antes y después de poner las cubetas en el sol en un día caluroso. Los datos se muestran en la tabla a la izquierda.

Escribe tu respuesta a la pregunta 22 en la casilla de la hoja de respuestas.

22) ¿Cuál sería una **hipótesis** *razonable acerca de por qué la gente usa ropa de color claro cuando hace calor? (Una hipótesis es una expectativa sobre lo que debería suceder basado en el conocimiento previo de la ciencia y otras experiencias.)*

Elije la mejor respuesta para la pregunta 23 y escribe tu respuesta en la hoja de respuestas.

23) ¿Cuál de las siguientes es la **observación** *más precisa basándose en los datos anteriores del experimento?*

A) La temperatura en la cubeta negra aumentó 7 grados más que la temperatura en la cubeta blanca después de estar en el sol.
B) La temperatura en la cubeta blanca no se vio afectada por estar bajo el sol.
C) La temperatura en la cubeta negra aumentó 7 grados después de estar bajo el sol.
D) Usar una camisa blanca te mantendrá 7 grados más fresco que usar una camisa negra.

A continuación, escribe tu respuesta para la pregunta 24 en la casilla de la hoja de respuestas.

24) Utiliza ***lenguaje científico*** *para explicar cómo utilizarías tus* **observaciones** *como evidencia para apoyar o rechazar tu hipótesis acerca de por qué la gente usa ropa de color claro cuando el clima es caliente.*

Appendix 2.2: Description of Included Assessment Questions and Scenarios

Answers for assessment questions 3, 4, and 5 belong to the scenario called "Electric circuit experiment." Students are asked to describe cause and effect relationships as different items are added to an electrical circuit resulting in complete and incomplete circuits.

Answers for assessment question 8 come from the "Bread dough experiment" in which students are asked to describe the cause and effect relationships in the chemical reactions that occur when bread dough is rising.

Answers for assessment question 10 come from the scenario called the "Slingshot experiment" in which students are asked to explain the energy transformations between potential and kinetic energy when a ball is placed in and then launched from a slingshot.

Answers for assessment question 18 are taken from the scenario called "Plants need light experiment" in which students are asked to identify and describe the use of independent variables, dependent variables, and control conditions, in a student's science fair experiment about changing light conditions and plant growth.

Answers for assessment questions 20 and 21, come from the scenario called "Hardness of lipstick and lipgloss" in which students are asked to describe the relationships between a hypothesis, observations, and evidence in the context of a girl following recipes to make homemade lipstick and lipgloss.

Answers for assessment questions 22 and 24, come from the scenario called "Wearing light and dark colors experiment" in which students make and justify a hypothesis based on data provided as to why people may wish to wear light color clothing on a hot sunny day.

References

Allard, E. C. (2017). Re-examining teacher translanguaging: An ecological perspective. *Bilingual Research Journal, 40*(2), 116–130.

Blackledge, A., & Creese, A. (2017). Translanguaging and the body. *International Journal of Multilingualism, 14*(3), 1–19.

Bunch, G. C. (2013). Pedagogical language knowledge: Preparing mainstream teachers for English learners in the new standards era. *Review of Research in Education, 37*(1), 298–341.

Buxton, C., & Caswell, L. (2020). Next generation sheltered instruction to support English learners in secondary science classrooms. *Science Education, 104*(3), 555–580.

Buxton, C., Allexsaht-Snider, M., Kayumova, S., Aghasaleh, R., Choi, Y., & Cohen, A. (2015). Teacher agency and professional learning: Rethinking fidelity of implementation as multiplicities of enactment. *Journal of Research in Science Teaching, 52*(4), 489–502.

Buxton, C., Harman, R., Cardozo-Gaibisso, L., Jiang, L., Bui, K., & Allexsaht-Snider, M. (2019). Understanding science and language connections: New approaches to assessment with bilingual learners. *Research in Science Education, 49*(4), 977–988.

Canada 2067. (2017). *Canada 2067 STEM learning roadmap*. http://bit.ly/2FFtFdO. Accessed 28 Mar 2019.

Cardozo-Gaibisso, L., Kim, S., Buxton, C., & Cohen, A. (2019). Thinking beyond the score: Multidimensional analysis of student performance to inform the next generation of science assessments. *Journal of Research in Science Teaching, 57*(6), 856–878.

Cowie, B., Moreland, J., & Otrel-Cass, K. (2013). *Expanding notions of assessment for learning*. Sense.

Duff, P. (2011). Language socialization into academic discourse communities. *Annual Review of Applied Linguistics, 30*, 169–192.

Eggins, S., & Slade, D. (2006). *Analysing casual conversation*. Equinox.

Espinosa, C. M. (2016). Reclaiming bilingualism: Translanguaging in a science class. In O. García & T. Kleyn (Eds.), *Translanguaging with multilingual students: Learning from classroom moments* (pp. 174–192). Routledge.

European Commission. (2015). *Science education for responsible citizenship*. Authors.

Flores, N., & Schissel, J. L. (2014). Dynamic bilingualism as the norm: Envisioning a heteroglossic approach to standards-based reform. *TESOL Quarterly, 48*(3), 454–479.

García, O., & Wei, L. (2014). Language, bilingualism and education. In *Translanguaging: Language, bilingualism and education* (pp. 46–62). Palgrave Macmillan.

González, N., Moll, L. C., & Amanti, C. (2006). *Funds of knowledge: Theorizing practices in households, communities, and classrooms*. Routledge.

Groves, F. H. (2016). A longitudinal study of middle and secondary level science textbook vocabulary loads. *School Science and Mathematics, 116*(6), 320–325.

Halliday, M. A. K. (2004). *The language of science*. Continuum.

Halliday, M. A. K., & Matthiessen, C. M. I. M. (2004). *An introduction to functional grammar* (3rd ed.). Edward Arnold.

Hamann, E., Wortham, S., & Murillo Jr, E. G. (Eds.). (2015). *Revisiting education in the new Latino diaspora*. IAP.

Harding, S. (1991). *Whose science? Whose knowledge? Thinking from women's lives*. Cornell University Press.

Harman, R., Buxton, C., Cardozo-Gaibisso, L., Jiang, L., & Bui, K. (2020). Culturally sustaining systemic functional linguistics praxis in science classrooms. *Language and Education*. https://doi.org/10.1080/09500782.2020.1782425

Hoff, E. (2006). How social contexts support and shape language development. *Developmental Review, 26*(1), 55–88.

Jakobson, B., & Axelsson, M. (2017). Building a web in science instruction: Using multiple resources in a Swedish multilingual middle school class. *Language and Education, 31*(6), 479–494.

Kieffer, M. J., Lesaux, N. K., Rivera, M., & Francis, D. J. (2009). Accommodations for English language learners on large-scale assessments: A meta-analysis on effectiveness and validity. *Review of Educational Research, 79*(3), 1168–1201.

Kim, S., Kuak, M., Cardozo-Gaibisso, L., Buxton, C., & Cohen, A. (2017). Statistical and qualitative analyses of students' answers to a constructed response test of science inquiry knowledge. *Journal of Writing Analytics, 1*(1), 82–102.

Kirmaci, M., Allexsaht-Snider, M., & Buxton, C. (2019). "Being on the other side of the table": Lessons from a community-based science learning program with Latinx families. *Urban Education*. https://doi.org/10.1177/0042085919877934

Kusters, A., Spotti, M., Swanwick, R., & Tapio, E. (2017). Beyond languages, beyond modalities: Transforming the study of semiotic repertoires. *Translation and Translanguaging in Multilingual Contexts, 14*(3), 219–232.

Lee, O. (2017). Common core state standards for ELA/literacy and next generation science standards: Convergences and discrepancies using argument as an example. *Educational Researcher, 46*(2), 90–102.

MacDonald, R., Miller, E., & Lord, S. (2017). Doing and talking science: Engaging ELs in the discourse of the science and engineering practices. In A. Oliveira & M. Weinburgh (Eds.), *Science teacher preparation in content-based second language acquisition* (pp. 179–197). Springer.

Major, J. (2018). Bilingual identities in monolingual classrooms: Challenging the hegemony of English. *New Zealand Journal of Educational Studies, 53*(2), 193–208.

Maton, K. (2014). *Knowledge and knowers: Towards a realist sociology of education*. Routledge.

Maxwell, J. (2013). *Qualitative research design: An interactive approach* (3rd ed.). Sage publishers.

National Academies of Sciences, Engineering, and Medicine. (2018). *English learners in STEM subjects: Transforming classrooms, schools, and lives*. The National Academies Press.

National Research Council. (2012). *A framework for K-12 science education: Practices, crosscutting concepts, and core ideas*. The National Academies Press.

NGSS Lead States. (2013). *Next generation science standards: For states, by states*. The National Academies.

Otheguy, R., García, O., & Reid, W. (2015). Clarifying translanguaging and deconstructing named languages: A perspective from linguistics. *Applied Linguistics Review, 6*(3), 281–307.

Paris, D. (2012). Culturally sustaining pedagogy: A needed change in stance, terminology, and practice. *Educational Researcher, 41*(3), 93–97.

Paris, D., & Alim, H. S. (Eds.). (2017). *Culturally sustaining pedagogies: Teaching and learning for justice in a changing world*. Teachers College Press.

Poza, L. (2017). Translanguaging: Definitions, implications, and further needs in burgeoning inquiry. *Berkeley Review of Education, 6*(2), 101–128.

Poza, L. E. (2018). The language of ciencia: Translanguaging and learning in a bilingual science classroom. *International Journal of Bilingual Education and Bilingualism, 21*(1), 1–19.

Rosebery, A. S., Ogonowski, M., DiSchino, M., & Warren, B. (2010). "The coat traps all your body heat": Heterogeneity as fundamental to learning. *Journal of the Learning Sciences, 19*(3), 322–357.

Turkan, S., de Oliveira, L. C., Lee, O., & Phelps, G. (2014). Proposing a knowledge base for teaching academic content to English language learners: Disciplinary linguistic knowledge. *Teachers College Record, 116*(3), 1–30.

Chapter 3
Translanguaging for STEM Learning: Exploring Tertiary Learning Contexts

Juliet Langman, Jorge Solís, Lina Martin-Corredor, Nguyen Dao, and Karla Garza Garza

Abstract In recent years, *translanguaging* has emerged as both a theoretical and pedagogical concept through which to examine how multilingual speakers learn by leveraging their full set of linguistic resources (Vogel S, García O, Translanguaging. Publications and Research. https://academicworks.cuny.edu/gc_pubs/402, 2017). With a translanguaging lens, researchers in STEM education have examined how multilingual learners draw on both home- and school-based, informally and formally taught concepts as they develop proficiency in math and science concepts and processes. This study examines unplanned (spontaneous) translanguaging practices in tertiary engineering lectures and provides an initial analysis of how faculty and students think about language and translanguaging as part of their teaching and learning. Finally, we explore how sharing information on spontaneous students translanguaging may provide faculty with insights into pedagogical practices that can address Hispanic and multilingual Spanish English-speaking students' needs.

Keywords Translanguaging · STEM · HSI · Higher education · Lesson study · Spontaneous

J. Langman (✉) · J. Solís · N. Dao · K. G. Garza
University of Texas at San Antonio, San Antonio, TX, USA
e-mail: juliet.langman@utsa.edu; jorge.Solis@utsa.edu; nguyen.dao@utsa.edu; karla.garzagarza@utsa.edu

L. Martin-Corredor
Metropolitan State University Denver, Denver, CO, USA
e-mail: lmartin@msudenver.edu

© The Author(s), under exclusive license to Springer Nature Switzerland AG 2021

A. Jakobsson et al. (eds.), *Translanguaging in Science Education*, Sociocultural Explorations of Science Education 27,
https://doi.org/10.1007/978-3-030-82973-5_3

3.1 Introduction

This chapter presents initial exploratory results on the ways in which translanguaging manifests itself in the context of undergraduate university-level engineering classes at a university in the Southern United States designated as a Hispanic Serving Institution. The aim of this paper is to explore how the nature of unplanned translanguaging occurs in the context of large (more than 80 students) lecture classes, where multilingual learners informally and formally translanguage to appropriate STEM concepts and processes and engage in identity work. Using this as a starting point, the paper moves on to examining faculty, researchers, and students' perspectives and language ideologies on the potential value of planned translanguaging in university contexts with large numbers of multilingual speakers with varied past educational experiences both in and outside of the United States. The findings highlighted in this study also seek to contribute to the growing body of translanguaging research taking place in tertiary settings as well as STEM education.

3.2 Theoretical Background

The concept of translanguaging has become one of high interest in the field of education over the last 15 years. Translanguaging can be defined, according to García (2009) as the "multiple discursive practices in which bilinguals engage in order to make sense of their bilingual worlds" (p. 45). Multiple discursive practices theoretically draw upon all features that form part of bilingual or multilingual individuals' linguistic repertoires (García, 2009, p. 51; García & Wei, 2014, p. 22). Due to its link to education, much translanguaging research explores the connection between language use and learning (García & Wei, 2014; Mazak & Herbas-Donoso, 2015). Other research examines the relationship between translanguaging and identity (Fuller, 2012; Li & Zhu, 2013).

Certain features of translanguaging make it a robust theoretical lens with which to examine the connection between language use and learning. First, translanguaging, seen as a form of languaging, is simultaneously a social and cognitive act (Canagarajah, 2013). As such, we can examine the learning at the micro-level through an analysis of language in context. Second, the social, cultural, and institutional context may provide affordances as well as constraints on translanguaging, offering the possibility of examining learning contexts and learner practices. From a cognitive perspective, translanguaging practices can expand the range of affordances or opportunities for learning by drawing on the learner's full repertoire of semiotic learning tools and knowledge (Martin-Beltrán, 2014). Third, the multimodal nature of translanguaging offers another analytical lens for learning in practice, allowing an exploration of translanguaging across oral, written, and visual modes. García and Wei (2014) suggest that "successful multilingual interactions have always been aided by multimodalities" and cite examples of students using written materials in one language and discussing it in another (García & Wei, 2014, p. 28). Finally, translanguaging allows for the expression of social identity (Fuller, 2012; Wei, 2018).

3.3 Literature Review

Research on translanguaging at the tertiary level is not robust and extensively explored to date. However, several noteworthy examples of research focus on the connection between the local population's language profile, institutional guidelines on language use, and subsequent practices in and outside of classrooms, as well as the effects on student identity. For instance, Mazak and Carroll (2016) draw together a set of studies examining the tertiary context in South Africa, Denmark, Ukraine, Puerto Rico, China, India, and the United Arab Emirates. These contexts can be distinguished from one another in terms of the extent to which multilingualism is a central component of societal life and the extent to which a lingua franca, most often English, has had a long-term or shorter-term presence in the university environment.

At one end of the spectrum are contexts such as Puerto Rico with a highly contested orientation in society in general to the use of English in education, in addition to or in place of Spanish. In spite of this contentious relationship between English and Spanish in other contexts, "the context of higher education in Puerto Rico mirrors other higher education contexts around the world: English is the taken-for-granted *international language of academia* (Phillipson, 2009), but it coexists with the local vernacular (Spanish) in classrooms" (Mazak et al., 2016). Carroll and Mazak (2017) further discuss that English is also the language of science (see also Mazak & Herbas-Donoso, 2014). As an example, Mazak et al. (2016) explore the pedagogical practices of three professors at one such university in Puerto Rico, defining their practices as one enacting a *flexible bilingual pedagogy* (Creese & Blackledge, 2010, p. 70). They conclude that in the case of the three teachers they study, their pedagogy sets up affordances for student engagement across their full linguistic repertoire and has the effect of supporting both content development in the topic, as well as development of English and Spanish academic language. They note that providing such affordances "does not always mean that translanguaging practices are structured or planned, but, rather, that teachers have a disposition or openness towards translanguaging... that allows these practices to develop organically" (p. 88).

The other end of the spectrum features institutional discourses that highly value multilingualism and multilingual education. For example, Caruso (2018) documents translanguaging practices among a professor and multilingual students, who are local Portuguese, French, and Italian, in a Language and Communication Policy course at a Portugese university. The professor of the course encourages students to explicitly use their multilingual repertoires to co-construct a collective understanding of academic content, which is often in English, thus building a *co-learning classroom environment* (García & Wei, 2014). Some remarkable instances include (1) students using their home languages to discuss English scientific terms and (2) the professor allowing the students to take a structured final assessment in three languages to promote their multilingual competence. This study highlights the role of open institutional language policies in facilitating multilingualism, academic engagement, inclusion, and equity.

These two settings of Puerto Rico and Portugal contrast in many ways with tertiary education in the United States, a country without explicit language policies oriented towards multilingualism, but with a large and growing population of multilingual speakers. In contexts such as the United States where translanguaging pedagogy has been explored and implemented with increasing frequency in the last 10–15 years, such a pedagogy appears to be *sustainable* in K-12 settings (e.g., García & Kleyn, 2016), but more difficult to achieve in tertiary contexts with less loosely constructed policies for language use (Caruso, 2018; Mazak & Carroll, 2016).

A second body of research relevant to the current paper focuses on translanguaging in STEM settings, settings with a focused examination of English as the *language of science* (Mazak & Carroll, 2016). Research at the elementary and secondary level often refers to the importance of translanguaging as the code for "thinking" or grappling with new concepts presented in classroom environments (García & Kleyn, 2016; García & Wei, 2014). Gibbons (2006) highlights the teacher's role in facilitating the learning process as learners move from exploratory engagement with new concepts to presentational talk, such as that which would occur in presentations to the whole class. Gibbons continues describing the academic language transition from speaking to writing, seeing presentational talk as a preparation for writing. Other scholars examine how a translanguaging pedagogy can support this type of transition from exploratory to presentational talk and writing. More specifically, Probyn (2015) argues that:

> …the challenge for teachers is not only to bridge the gap between learners' understanding in their home language and the languaging of teaching and learning: English; but also… to bridge the gap between everyday discourse and academic discourse in this study, the discourse of science… (p. 233).

Recently, a vast array of studies taking place in K-12 contexts (e.g., Espinosa et al., 2016; Garza, 2017; Licona & Kelly, 2020; Poza, 2018) point out that translanguaging pedagogy helps deepen content comprehension, make cross-linguistic connections, appropriate mathematical meanings, and frame scientific reasoning and argumentation. Other research examines instances of students' translanguaging practices and interactions in science courses, which facilitates students' learning space and continuity of scientific reasoning (Garza & Arreguín-Anderson, 2018; Karlsson et al., 2019). Karlsson et al. (2020) raise a challenging issue regarding translanguaging at the classroom level, questioning the possibility of instructing multilingual students to differentiate between discursive and dominant language(s), i.e., the complexity of transferring scientific content from Arabic into Swedish and vice versa, to develop their meta-awareness of language and content. In a similar vein, while highlighting the role of teachers as local language planners in a culturally and linguistically responsive manner, Langman (2014) calls for more attention to academic and in-depth rather than informal forms of translanguaging often observed across K-12 classrooms.

In the settings of higher education, He et al. (2017) explore translanguaging in multimodal presentations in mathematics in the context of a Hong Kong university classroom with students who were speakers of Mandarin, Cantonese, and English, as well as readers of complex Chinese characters, simplified Chinese characters, and English. The article focuses on how the instructor navigated a translanguaged and trans-semiotized space, or a space where the three languages are integrated with many other semiotic patterns (e.g., visuals and gestures) during the meaning-making process (Lin, 2019), to accommodate students who did not share the same elements of the classroom repertoire of communicative forms. Their study outlines the strategic and planned use of translanguaging to engage and inform the diverse student population. He and Lin (2022) further examine the relationship between translanguaging and trans-semiotizing during the unpacking of science texts. With students' practices as a focus, Mazak and Herbas-Donoso (2015) showcase dynamic and recursive bilingual learning processes by examining students' multimodal display of scientific materials such as strategic oral discussion and note-taking. These purposeful practices promote not only the students' academic bilingualism but also their scientific knowledge.

A third body of research considers the connection between translanguaging and identity. For instance, Makalela's (2016) study is cast against the colonial and post-colonial context of South Africa which began to operate under a constitutional commitment to multilingualism in higher education (rather unsuccessfully according to the author) beginning in 1994. The specific study outlines the development of *ubuntu* (meaning "I am because you are, you are because we are") translanguaging pedagogy, which welcomes multiple languages in a second language learning classroom. The author outlines how this context resulted in positive experiences for learners in terms of personal identity transformation and an orientation to reclaiming and reflecting the former societal multilingualism of South Africa. This project contributes to an understanding of the role of higher education in validating and building identity orientations for students for interdisciplinary engagement.[1] Goodman (2016) discusses how translanguaging at the tertiary level is a political struggle for minoritized language recognition, both on the part of individuals and institutions, through an analysis of a Ukrainian university. Goodman's study provides examples of a range of multimodal and multi-semiotic translanguaging activities, i.e., multiple semiotic resources of visuals and languages, in both formal and informal learning spaces. Although linguistically responsive pedagogy has recently been adopted and legitimized, both at K-12 and higher education settings, Cole et al. (2016) note that minority students' identities shift across time and space, and they may not be ready to adopt the position as heritage language speakers in academic contexts, despite affordances created through translanguaging pedagogy.

[1] See also Hattingh et al. (2022) for a similar context in post-colonial South Africa.

In sum, most translanguaging research in higher education (as featured in Mazak & Carroll, 2016) pays explicit attention to planned translanguaging instances, such as teachers' lesson plans and instructional interactions. Those studies have yet to elucidate spontaneous translanguaging and the ways in which multilingual students use their linguistic repertoires to make appropriations of academic content. Few studies further examine how unplanned translanguaging connects to the expression of identity.[2] While previous studies do discuss ideology of monolingualism (e.g., Mazak & Herbas-Donoso, 2014) or multilingualism (e.g., Caruso, 2018; Makalela, 2016) by examining language policies both at the classroom-level or collegewide, little research has examined language ideologies from teachers' and students' perspectives through interviews or focus groups.

3.4 Research Goals and Questions

To address those gaps and contribute to a growing body of translanguaging research in tertiary settings, this paper examines (1) spontaneous translanguaging practices among engineering students during small-group activities within a lecture as well as (2) faculty and students' views on translanguaging, often defined by the speakers themselves as the use of languages other than English for learning purposes.

This study takes place in a university with a Hispanic-Serving Institution (HSI) designation, meaning a university with a Hispanic student population at or above 50%. The HSI designation does not directly correlate with any particular percentage of students who use Spanish in addition to English regularly; but it provides a context where multilinguals with varying educational experiences in and through various languages meet one another regularly.

Data for this paper are drawn from a larger four-year longitudinal curriculum redesign project aimed at improving the success rate of both Hispanic and first-generation students enrolled as STEM majors. Embracing the role of an HSI, the larger study further has as an aim to understand and harness the strengths of the multilingual experience and explore how multilingual affordances can lead to innovation and discovery enterprises in STEM. Using Lesson Study (LS) as a framework (Lewis et al., 2012), the larger project focuses on intensive work in redesigning Engineering curriculum, beginning with collaboration with participating engineering faculty to focus on student learning challenges. The LS framework involves recording and analyzing lessons, as well as conducting extensive interviews and focus groups with faculty and students enrolled in target gateway courses. The main goal of this paper is to explore the affordances and constraints on translanguaging discourse practices in tertiary STEM classrooms. More specifically, our research questions are:

[2] See Lemmi et al.'s (2022) discussion on bilingual students' meta-awareness of translanguaging in relation to social roles and academic purposes.

1. What is the nature of unplanned translanguaging among students in large lecture classes?[3]
2. What perspectives on language use do faculty and students share with respect to their own or observed translanguaging practices?
3. How may unplanned practices and discussions of translanguaging be transformative in terms of developing a translanguaging pedagogy orientation?

3.5 Data Sources

As mentioned above, the research was conducted at large public HSI university in the southwest, which aims to increase academic success for historically underrepresented students. Low passing rates for Hispanic and low-income students (57%) enrolled in gateway Engineering courses emphasize a need for seeking innovative means of supporting student learning. The participants of the project include an interdisciplinary team composed of faculty members and graduate students from the College of Education and Human Development as well as from the College of Engineering. Guided by Lesson Study principles, the team members collaboratively identified learning challenges, planned, taught, and evaluated a series of target lessons that aim at tackling potential learning challenges, using ongoing discussion and reflection.

The data for this paper is drawn from (1) transcripts of Lesson Study meetings recorded from September 2019 to May 2020; (2) transcripts of video recorded small group student interactions during Spring 2020 target lesson on Mass Spring systems; (3) interviews with students recorded during the class session; and (4) written course materials.

At the outset of each class under examination, groups of students were asked to volunteer to be recorded and later be interviewed. Data for this paper are drawn from two of the six groups that engaged in some translanguaging during the class. The focused student participants[4] in this study include six Engineering students, all of Hispanic descent and bilingual in Spanish and English. Below is the demographic information of the student participants (Table 3.1).

The faculty participants include Dr. Smith and Dr. Flores, both of whom are professors of the College of Engineering. Dr. Smith is from the United States and holds a doctoral degree in Mechanical Engineering with 19 years of teaching experience. Dr. Flores is from Mexico and speaks English and Spanish. He also holds a Ph.D. in Mechanical Engineering and has been in charge of Engineering gateway courses since 2019. In addition, the two authors, Dr. Langman and Dr. Solís, participated in Lesson Study meetings.

[3] See also Hattingh et al. (2022) for more case studies on spontaneous translanguaging.

[4] All names, with the exception of the authors, are pseudonyms.

Table 3.1 Student participants' demographics

Pseudonym	Gender	Origin	Major	Schooling
Diego	Male	Mexico	Mechanical engineering	Moved to the United States to attend college
Ramiro	Male	Costa Rica	Computer engineering	Moved to the United States to attend high school
Lucas	Male	Laredo, Texas	–	US schooling
Sofía	Female	Nuevo Laredo, Mexico	Civil engineering	Moved to the United States to attend middle school
Julio	Male	El Paso, Texas	Chemical engineering	US schooling
Hector	Male	–	–	–

3.6 Data Analysis

Analysis of data for this paper involves a sequential analysis of translanguaging practices in the classroom and reflection on those practices by faculty and students. All data from classroom interactions and interviews were transcribed. The first round of analysis involved identifying instances of translanguaging in the classroom data, focusing on when, by whom, and with what purposes translanguaging occurred. The second round of analysis involved the examination of interview data from participating students in the classroom, focusing on their interpretations of the value of translanguaging as part of their learning. As part of the regular lesson study meetings, faculty were shown video clips and transcripts of in-class sessions, as well as summaries and excerpts of the student interviews. The final round of analysis was drawn from transcripts of faculty's reflections on the potential for the intentional design of translanguaging as a pedagogical tool for multilingual students. In this way, the analyses connect to the view of translanguaging as a meaning-making tool particularly useful for collaborative sense-making and explore the potential affordances planned and unplanned translanguaging spaces may offer from the perspective of students and faculty at the tertiary level.

3.7 Findings

3.7.1 Spontaneous Translanguaging During Group Work

During a lesson on Mass Spring systems, Dr. Flores had students working in groups of three with a handout and a mass spring system with several removable weights. Students were required to manipulate the system and record their findings on the handout, as well as connect the experimental data to formulas and concepts (such as the Hooke's Law). The first portion of the assignment begins with the setup of the mass spring system and the determination of the initial elongation (see Fig. 3.1).

Activity 1. Analysis of a Mass-Spring System in Free Undamped Motion.

A mass-spring system is a device that consist of the following components: (1) a mass (including a hanger and a pointer), (2) a spring, (3) a support structure (including connecting rods) and (4) a scale.

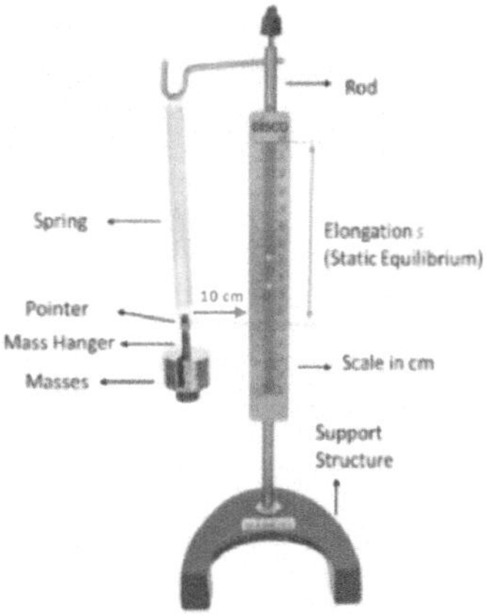

Figure SEQ Figure * ARABIC 1. Mass-Spring System in Static Conditions.

<u>**Activity 1a: Experimental Component**</u>

The first setup is analyzed in both (1) static conditions, and (2) dynamic conditions. One team member will operate the system, other member will take measurements with the pointer and scale, and the last member will record the measurements in this handout.

<u>Static Analysis:</u> Replicate the setup that is presented in Figure 1. Before installing the hanger with the masses, install the spring and make sure that one of its ends aligns with the "0 cm" marking in the scale (displace the scale up or down if needed). Then, install the hanger with the masses (9 masses of 10 grams each) and measure the elongation *s* in the spring after reaching the equilibrium position. Note that the <u>total mass</u> attached to the spring is the sum of the hanger mass (20 grams) plus the 9 individual masses (90 grams), which is <u>110 grams (or 0.11 kg)</u>.

Initial elongation *s*: ____________ [cm] = ______________ [m]
(Divide by 100 to get m)

Fig. 3.1 The mass-spring system handout

Repeat the previous experiment: take the masses with your hand and displace them <u>2 cm below the equilibrium position</u>. Then release the masses. This time, you will use your smartphone (or a clock) to measure the time it takes for the mass to complete 10 cycles of the motion (i.e., moving from the lowest point to the highest point and back to the lowest point).

Total time: ________________ [s] for 10 cycles.

Period of free vibration *T*: ______________ [s/cycle] (Divide total time by 10)   **Frequency of vibration *f* = 1/*T*:** ___________ [cycle/s]

Fig. 3.2 Handout extract

Students were also asked to measure the displacement of the mass and the amplitude and to calculate the period and frequency of free vibration (see Fig. 3.2). In the following section, we provide data from two groups who engaged in translanguaging practices throughout the lesson in their respective groups.

3.7.1.1 Seamless Translanguaging

Group 1 consists of Ramiro, a sophomore majoring in Computer Engineering from Costa Rica with a Korean father, as well as Lucas from Laredo and Diego who is from Mexico. Diego is a freshman in mechanical engineering. While Ramiro completed high school in the United States, Diego moved to the United States to attend college. Lucas did not share his schooling history with us. This group of three had not previously worked together, although later during the interviews, Lucas and

Diego both mentioned that they became friends as a result of this activity. In Excerpt 3.1,[5] we see Group 1 orienting to the activity, and using both the mass spring system as well as the handout as they discuss first steps.

Excerpt 1

Setting up the experiment

Ramiro: **¿Cómo así?** ((calibrating the spring with his pen)) **Porque alguien tiene que operar el sistema, alguien…mira aquí dice...**	Ramiro: **How so?** ((calibrating the spring with his pen)) **Because somebody has to operate the system, somebody… look here, it says…**
Diego: Uh huh	Diego: Uh huh
Ramiro: ((Reading from handout)) "one team member will operate the system, other members will take measurements with the pointer and scale, and the last member will record the measurements in this handout."	Ramiro: ((Reading from handout)) "one team member will operate the system, other members will take measurements with the pointer and scale, and the last member will record the measurements in this handout."
Lucas: **Yo, si quieres, yo hago** record the measurements.	Lucas: **If you want, I do** record the measurements.
Lucas: **pero ahorita ¿para dónde llega?** ('it' refers to the spring)	Lucas: **but now where does it stop?** ('it' refers to the spring)
Ramiro: **Ahorita de la parte de aquí,** ((pointing at the pointer)) **¿qué ves?**	Ramiro: **Now, from this part,** ((pointing at the pointer)) **what do you see?**
Diego. **Es diez** ((referring to the spot on the spring))	Diego: **It's ten** ((referring to the spot on the spring))
Ramiro: ((Writing on handout in front of the "initial elongation" item)) **Diez centímetros, lo cual es cero punto uno [metro]** ((writing '0.1' on the handout))	Ramiro: ((Writing on handout in front of the "initial elongation" item)) **Ten centimetres, which is zero point one [meter]** ((writing '0.1' on the handout))
Diego and Lucas: ((Copying what Ramiro wrote and writing on their handouts))	Diego and Lucas: ((Copying what Ramiro wrote and writing on their handouts))
Lucas: **Ajá.**	Lucas: **Uhu**
Dr. Flores: Raise your hand if you already completed this portion.	Dr. Flores: Raise your hand if you already completed this portion.
All: ((Raise their hand))	All: ((Raise their hand))

In Excerpt 3.1, Ramiro initiates the conversation in Spanish discussing the steps and the roles outlined in the handout for using the mass spring system ending with "here it says" (lines 1–4). Following Diego's hesitation marker, Ramiro begins reading verbatim in English from the handout (lines 6–10). Lucas volunteers for one of the roles first using Spanish and translanguaging to English to express the explicit task "record the measurements" (lines 11–12). From line 13 to 25, the group engages in the measuring task and recording the response "0.1" on the handout using Spanish. We note the ways in which Ramiro verbalizes the transposition of ten, i.e., from ten centimeters to "zero point one" [meter] and writes down "0.1" on the handout, which reflects Ramiro's use of academic Spanish as he records their first finding. While this small group interaction is taking place, Dr. Flores is continuing to talk to the whole class in English. When Dr. Flores elicits a comprehension check from the students, Group 1 responds as they are paying attention all along (lines 26–28).

[5] In each excerpt, the original transcription is on the left and the translation on the right. Spanish appears in **bold** in both the original and the translation.

In this example, three participants seem to move seamlessly across their linguistic repertoire which includes both Spanish and English. While the language of the lecture, the PowerPoint, and the handout are in English, most initiations take place in Spanish, with a few exceptions in which the students refer to the handout (lines 6–10, 11–12) using the written language, English. Furthermore, the excerpt reveals that Lucas, Ramiro, and Diego actively move from exploratory talk (in both English and Spanish) to presentational text (in English). While their own spoken utterances are predominantly Spanish, they are drawing on and responding to the English oral and written language throughout.

The group continues working with the handout and after 6 minutes, they move on to the second experiment (see Fig. 3.2), which requires them to measure the time it takes to complete ten cycles after pulling the spring down two centimeters.

Excerpt 3.2 highlights further evidence of the fluid movement across the linguistic repertoire of the group members. As Diego and Ramiro count the cycles of the spring out loud in unison (line 1–4), Ramiro pulls on the spring while looking at his stopwatch. As both of them verify that the mass spring has completed ten cycles of motion, they use two terms to refer to what they are measuring: "veces" ("times") (lines 6–7) and "cycles" (line 8). The students are drawing on elements of their linguistic repertoire to potentially fix the academic term "cycles" in their minds after employing the familiar term "veces" (lines 6–8) (see Otheguy et al., 2019). In this excerpt, the only English that appears are academic terms related to the activity ("cycles," "period," and "period of vibration").

Excerpt 2

Discussing the definition of a cycle

Diego and Ramiro: ((Whispering and	Diego and Ramiro: ((Whispering and
looking at the spring going up and down))	looking at the spring going up and down))
una, dos, tres, cuatro, cinco, seis, siete,	**one, two, three, four, five, six, seven,**
ocho, nueve, diez.	**eight, nine, ten.**
Ramiro: **Ya ¿verdad?**	Ramiro: **Now, right?**
Diego: **Como diez veces ¿no?**	Diego: **Like ten times, isn't it?**
Lucas: **Sí, diez veces**	Lucas: **Yes, ten times**
Ramiro: **Diez** cycles	Ramiro: **Ten** cycles
Diego: **Sí**	Diego: **Yes**
Lucas: **Seis segundos**	Lucas: **Six seconds**
Ramiro: **Seis, punto, tres, siete segundos**	Ramiro: **Six-point-three-seven seconds.**
All three: ((Writing '6.37 sec' on handout))	All three: ((Writing '6.37 sec' on handout))
Lucas: **Y luego el** period..	Lucas: **And then** the period..
Ramiro: Period of vibration	Ramiro: Period of vibration
Lucas: **XXX contrario de esto**	Lucas: **XXX contrary to that**
Diego: **Sí**	Diego: **Yes**
Lucas: **O sea, punto sobre punto, seis, tres**	Lucas: **So, point, over point six three**

Toward the end of Excerpt 3.2, Lucas expresses his thoughts on the relationship between period of vibration and frequency of vibration "contrary to that" (line 15), namely, they are in contrast to one another (as seen on the handout in Fig. 3.2 above). In other words, to measure cycles, the students are supposed to use the formula for period of vibration, which is *s/cycle*, while the contrasting formula for frequency of vibration is the opposite.

In these two excerpts, we see two types of translanguaging: one focused on reading the handout in its original form (Excerpt 3.1) and the other where the translanguaging appears to support (a) the fixing of academic terms in English and (b) cognitive support expressed through Spanish (Excerpt 3.2). Throughout Diego, Lucas, and Ramiro's entire collaboration process, which lasted the entire class period for 1 h and 15 min, they engaged in seamless translanguaging and did not refer to named languages or negotiate a preferred language.

3.7.1.2 Negotiating Translanguaging as Identity Work

This section focuses on translanguaging moments of Group 2 (in the same class), consisting of Sofía, Julio, and Hector. Sofía is a Civil Engineering student from Nuevo Laredo, Mexico, close to the United States-Mexico border. She moved to the United States at the age of 12. In middle school, she continued her studies in English although she took two advanced placement classes in Spanish. In her interview, Sofía shared that she currently feels English-dominant as she now only speaks Spanish to her mom with whom she does not have the chance to interact often and a new friend from Mexico. Julio, a Chemical Engineering major, is from El Paso, a Texas border city. He used to attend a bilingual elementary school and since then, has had his schooling in English. Hector is a bilingual Spanish-English speaker who did not share his immigration and education background with us.

In contrast to the previous group, this group's interaction begins with a negotiation of language use. At the beginning of the class, Sofía asked Hector what language he wants to use, and Hector replied he preferred Spanish. As Julio joined the group, he began participating immediately using Spanish. However, as the activity progressed, Julio used English most of the time, while Sofía and Hector used Spanish as the language for discussion. Each group member assumed a specific role outlined on the handout, namely, Hector manipulates the mass spring system; Sofía identifies the data produced by operating the system; and Julio fulfills the role of note-taker.

Excerpt 3.3 begins with Sofía asking Julio, "¿Qué es eso?" ("what is that?") while referring to the pointer on the spring system. What follows are two focused but different conversations, with Sofía acting as a mediator between Julio and Hector using English and Spanish simultaneously. Julio is trying to make sense of the concept of "period" as it relates to the action of the spring (line 3). Meanwhile, Hector and Sofía are discussing how to take the weights off the spring, but they encounter some difficulty (line 1, 5–11). While this may be interpreted as *parallel play* (Burg et al., 2013), it is clear from the video and later interactions that they are working together as a group. While they are talking, the teaching assistant (TA) walks by to pass out additional copies of the handout. Sofía, seated in the middle, turns and addresses Hector in Spanish, indicating the handout is for him and then swivels her head to the left to tell Julio the measurement to record, in English (lines 13–15).

Excerpt 3

Lesson Study Implementation

Sofía: **¿Qué es eso?** ((referring to the pointer))	Sofía: **What is that?** ((referring to the pointer))
Julio: Yeah, so the period is going to be longer.. wait, shorter	Julio: Yeah, so the period is going to be longer.. wait, shorter
Hector: **Sigue empujando, no quiere...**	Hector: **keep pushing, it doesn't want to**...
Sofía: **Si quieres, mejor quita de una vez**	Sofía: **If you want, remove it now**
Hector: **Ah no, ya, ya. Ya lo vi.**	Hector: **Ah no, already, already. I already saw it.**
Sofía: Oh Okay.	Sofía: Oh okay
Hector: **Aqui arriba, creo, creo** ((taking the weights out of the spring))...	Hector: **Over here at the top, I think, I think...**((taking the weights off the spring))...
TA: ((Passing out one handout))	TA: ((Passing out one handout))
Sofía: ((Talking to Hector)) **Yo creo que es para ti.** ((Talking to Julio)) It was like zero point zero six or eight.	Sofía: ((Talking to Hector)) **I think it's for you**. ((Talking to Julio)) like zero point zero six or eight.
Julio: ((Writes '0.06' on handout))	Julio: ((Writes '0.06' on handout))

This excerpt is indicative of the entire class period, in which Sofía, sitting between Hector (who is Spanish-dominant) and Julio (who is English-dominant) orchestrates and negotiates interactional bids. Even though there was a negotiation of language choice with Julio (Spanish) at the beginning, what follows is Sofía's consistent shifting between the two languages, which is mirrored by Hector and Julio. This could be seen, on the part of Sofía, as a one-person-one-language strategy (Brettenny & Klerk, 1995). This distribution of languages also mirrors a distribution of tasks at the beginning of the activity: Sofía is the facilitator; Hector works with the equipment and the experimental observations; and Julio interprets the data ("zero point zero six or eight") and records it on the handout.

The implicit understanding of competency leads the group dynamics to draw from both languages. That is, in the hypothetical situation that one of them was not able to understand Spanish and English, the other two would need to speak one language to ensure the participation of the "monolingual" speaker. Even though these three participants have the freedom to choose either language, two of the members consistently select one language, while the third (Sofía) draws on both. The students are not using language to translate; rather, in this example, each adopts, selects, and sticks with a language choice that aligns with their identity. In line with Cole et al.'s (2016) arguments, we may see while some bilingual students (such as Julio) are not willing to take on the position of heritage language speakers in academic settings, others (such as Sofía and Hector) are ready to assume that identity.[6]

[6] See Daniel et al. (2022) for further examples and discussion of students drawing on various languages to explain, discuss, and check one another for understanding during small group interac-

3.7.2 *Translanguaging Perceptions and Practice During Lesson Study Interviews*

The concepts of translanguaging for learning and for identity are further exemplified in the following extracts drawn from an interview with Sofía (from Excerpt 3.3) following the lesson on the mass spring system. The interview protocol has a number of general questions on study habits, class performance, and things they would like to change about the class, in addition to specific questions about the content of the lesson that had just occurred. The interview context involved two bilingual speakers, the graduate research assistant and a student who had been recorded during the target lesson. These interviews, as a crucial part of Lesson Study, were cast as informal and friendly, as well as focused on gathering information on student learning.

After the target activity on the mass spring system lesson, Lina, a graduate research assistant from Colombia, walks with Sofía from class to the interview site. During this informal transition time, the two of them converse in Spanish. At the official start of the interview, Lina asks Sofía what language she prefers. Excerpt 3.4, completely recorded in Spanish, shows Lina and Sofía negotiating as to who will decide the language of interview. Sofía makes the ultimate decision to continue in Spanish.

Excerpt 4

Negotiating the interview language

1	Lina: **Mira podemos hacer la**	Lina: **Look, we can do the interview in Spanish**
2	**entrevista en espanol on en inglés**	**or in English.**
3	Sofía: **En lo que usted quiera**	Sofía: **In the language you want**
4	Lina: **No lo que TÚ quieras**	Lina: **No, in the language YOU want.**
5	Sofía: **En español está bien**	Sofía: **In Spanish is good.**

Next, in Excerpt 3.5, Lina asks the first question in Spanish (which she translates from the English interview protocol), namely, what the purpose of the class was. In the first part of her response, Sofía inserts a single English term "spring" (line 2) as she provides a fluid and comprehensive response employing a range of academic terms in Spanish associated with mass spring systems, such as "cycles," "periods," "durations," "amplitude," and "frequency." In lines 6–7, Sofía begins to hesitate and then asks whether she can use English and Spanish, to which Lina responds "of course." Sofía then uses the academic term "Second Order Differential Equations" in English in her ongoing conversation with Lina in Spanish.

tions. See also Karlson et al. (2022) on how negotiations of meaning are increased through *multilingual and discursive loops.*

Excerpt 5

"can I say it in Spanish and in English?"

Sofía: **Eh, pues, estábamos encontrando como los ciclos del** spring, **este estamos buscando por el, el periodo, el periodo que es lo que dura en el el los ciclos, y la amplitud y la frecuencia del ciclo, entonces estamos usando también las.. uhm, ¿lo puedo decir en español y en inglés?**	Sofía: **Well, we were finding how the** spring **cycles, eh we were finding the, the period, the period that is the duration of the, the cycles and the amplitude, and the frequency of the cycle, so we were also using.. um, can I say it in Spanish and in English?**
Lina: **Claro que sí**	Lina: **Yes, of course**
Sofía: The Second Order Differential Equations, **uh, es una manera de buscar, encontrar esas cosas.**	Sofía: The Second Order Differential Equations, **um, this is a way to look for, find those things.**

Moving forward in Excerpt 3.6, Sofía and Lina both translanguage, as they move from general to more specific references to the activity handout, prompted by Lina's question beginning with the expression "Viendo el handout" ("Looking at the handout"). It is important to note that from this point on, the level of translanguaging appears to increase after Sofía realizes Lina too is using English, thus signaling the linguistic space of the interview was a safe place to draw from both languages. When Sofía explains the activities, she draws on various parts of their presumably shared linguistic repertoire. Here, in contrast to Excerpts 3.4 and 3.5, more English elements appear, potentially due to a specific focus on the handout written in English which they are reviewing.

However, Sofía is not simply reading terms off of the handout, but rather creating an explanation in Spanish of what the purpose of the activities were. Translanguaging, in this case, allows Sofía to develop an in-depth knowledge of the subject matter (Baker & Wright, 2017). Here, drawing from her linguistic repertoire allows Sofía to monitor her learning, make connections with previous concepts, and communicate her takeaways from the mass spring lesson to Lina, who shares Sofía's bilingual competencies and ideologies.

Excerpt 6

Signalling a safe translanguaging space

Lina: **Viendo el** handout, **¿puedes explicar cómo las principales, uh, actividades que te pidieron que completaran, como las cosas que te pidieron que hicieras?**	Lina: **Looking at the** handout, **can you explain the main, uh, activities that you were asked to complete, like the things they asked you to do?**
Sofía: ((Looking at the handout))...**Y también usamos todo lo que aprendimos la clase pasada del** Hooke's law **y cómo** differentiate that, that, those equations and also the acceleration we know it's like the second derivative of the initial elongation or position… **Um, aquí hicimos más lo mismo como, más este, usamos otra vez lo** the second order differential equation...	Sofía: ((Looking at the handout)) ...**And we also used all that we learned the previous class about the** Hooke's Law **and how to** differentiate that, that those equations and also the acceleration we know it's like the second derivative of the initial elongation or position…**Um, here we did more of the same like, more, we used again that of** the second order differential equation,...

Several other interviewees refer to their choice and use of languages during class. Diego from Group 1, for instance, notes that he does not ask questions in class because he feels insecure about his English language ability. In the following excerpt, Lina asks Diego "What differences did you find between this lesson and previous lessons in this course?". As Diego explains some similarities, he highlights a teaching practice that Dr. Flores normally uses in class, which is bring up questions that students would probably have, but they "no nos atrevemos a decir" ("don't feel comfortable saying"). Lina asks Diego to elaborate and relate to his own experience. Diego shares that he is not concerned about asking questions in Mexico, but here in the United States, he worries about how his accent will be perceived, even though he knows many are used to it. However, for Diego, this is a matter of how others might perceive him through the way he uses his less dominant language, English.

Excerpt 7

"this is a problem that I worry about"

Diego: ...**Explicó de la misma manera que siempre explica, o sea, muy detallado, y como que siempre.. él explica preguntas que sabe que nosotros tenemos pero que no lo vamos a decir o no sabemos cómo decirlas, no sé, preguntas que nosotros queremos hacer pero que no.**	Diego: ...**He explained in the same way he always explains; that is, with a lot of detail. It's like he always.. he explains questions he knows we have but that we are not going to say or that we don't know how to say, I don't know, questions that we want to make but we don't feel comfortable saying.**
Lina: **¿Y por qué no se atreven a preguntar? O sea, ¿tú a veces tienes preguntas y no..?**	Lina: **And why don't you feel comfortable asking? I mean, sometimes you have questions and don't...**
Diego: **Sí, claro, ...porque en México yo preguntaba sin ningún problema. Preguntaba si tenía alguna duda porque lo estás haciendo en el idioma que dominas sin ningún problema... Aunque yo veo que acá la gente ya está acostumbrada a escuchar acento así o el acento de otro país, pero yo creo que eso es un problema que me preocupa. Tal vez cómo me escuchen, que cómo está mi pronunciación.**	Diego: **Yes, of course, ... because in Mexico I asked without a problem. I asked if I had any doubts because you are doing it in the language you dominate without a problem... Even though I see that here people are already used to hearing accents like this or the accent of other countries, but I believe that this is a problem that I worry about. Maybe in relation to how they hear me, how my pronunciation is.**

These two examples from interviews mirror several others with bilingual students who seek to use Spanish when they can and express concerns about their marking of their linguistic 'status' to others. While Sofía and Diego both primarily use Spanish in their interviews, they characterize themselves as English-dominant in the case of Sofía and Spanish-dominant in the case of Diego.

3.7.3 Faculty Reflections on Translanguaging Practices

As discussed earlier, a routine practice with the Engineering lesson study group was to debrief after recorded lessons and share video and audio data of student work in class and student reflections in interviews, similar to the data presented here. Initially, for purposes of debriefing, the research assistants presented only translations of key comments as not all participants were Spanish speaking. Once the team started sharing video from class, the use of Spanish became apparent to faculty listening to and reading the meeting transcripts.

During the debrief following the mass spring system, Dr. Flores asks Lina whether she conducts "some interviews in Spanish or everything [is] in English?" Lina shared that the interview language was up to the student participants. This led to a longer discussion, one of several over the course of the academic year, on the faculty's perceptions about language use. During this meeting, Dr. Flores shares his experience with Spanish-speaking students after class:

Sometimes I notice that sometimes students are able to express better the ideas in the first language, so I speak Spanish, originally from Mexico. And sometimes when I, at the end of the class, I always have at least one or two students approaching me with technical questions of the lecture, or of the homework or the exam. And I noticed that some students detected perhaps my accent, and they started talking to me in Spanish, and were able to communicate very effectively in terms of the questions. So, I usually reply to them in Spanish in order to make these interactions a little bit friendly (Lesson Study Meeting 5; October 15, 2019).

As opposed to Diego's perspectives on accents, Dr. Flores finds students' detection of his accent as an opportunity to explain in-depth academic concepts to Spanish-dominant students, i.e., when he is addressed in Spanish, he responds in Spanish "to make these interactions a little bit friendly." Dr. Flores further notes that students who do address him in Spanish are able to "communicate very effectively in terms of the questions."

As the conversation continues, Dr. Smith, a monolingual English Engineering faculty member, wonders, for the first time: "Should we have a Spanish-only lecture or [recitation] section?".[7] What follows is a conversation in which Lina and Dr. Solís outline the difference between a fully Spanish recitation and a bilingual recitation without explicitly discussing the term *translanguaging*. All the Lesson Study team members concur that framing the section as bilingual would enhance bilingual students' academic experiences. Lina shares, "If they [students] know that it's going to be all in Spanish, and they have taken all their math classes in English, there's gonna be resistance, but if we frame it as bilingual, then they are like, oh, we're gonna use both languages so that you can easily like, make the connections between the languages...." This discussion continues with Dr. Smith adopting Lina's view,

[7] In the Engineering class structure, a lecture generally consists of 90 (or more) students and lasts 1 hour and 15 minutes, while a recitation section generally consists of 30 (or less) students and lasts 1 hour.

"So they [students] may not want to take it. They may feel reluctant to take a Spanish-only one. Yeah, maybe start the class off bilingual...."

In the subsequent meetings, conversations with regard to logistics and offering a bilingual recitation section have begun to develop. One of these is a discussion between Dr. Flores and Dr. Smith on the potential population of students who might be bilingual, as featured in the Excerpt 3.8:

Excerpt 3.8 Discussing Logistics

Dr. Flores	I think that question must be answered depending on the amount of people that are actually speaking Spanish, because I don't think the lecture is actually dominant in terms of Hispanic people, right?
Dr. Smith	Well, I know, over half my population in engineering is Hispanic doesn't mean they speak Spanish though. Maybe that's only 10%. I'm just guessing.
Dr. Flores	That's right. So maybe we could investigate how many students would be willing to take the class in Spanish and see if we can open a section in Spanish.

Following this discussion, the research team presents survey data on the number of students in the course claiming to be Spanish speakers. The data, from 501 respondents out of 749 students surveyed, shows that 52% claim their ethnic identity as Latino/Latina, Mexican, or Hispanic and 41% are multilingual (Table 3.2).

Therefore, while the faculty members' perceptions of students who self-identify as Latino is fairly accurate, a far greater percentage of students claim affiliation with the Spanish language than Dr. Smith estimated. Upon seeing these data and excerpts of Spanish language use in the classroom, the discussion on how to capitalize on this asset has made way for robust discussion in the near future.

3.8 Conclusion and Implications

In this paper, we have explored translanguaging behaviors, primarily on the part of students enrolled in engineering coursework at a Hispanic-Serving Institution. The data from students reveals varied student backgrounds in terms of Spanish and English language use, as well as varied schooling experiences, from the perspective of language(s) of instruction in and outside the United States. The data reveal that,

Table 3.2 Ethnic and linguistic background of engineering students

Percentage (n = 501/749)	Language and Education Survey Questions
33.9%	I was born outside the United States I attended one or more years of K-12 school outside the United States
41%	I speak a language other than English
39.4%	When I grew up, I spoke Spanish in the home When I grew up, I spoke another language (than English) in the home

for some students, when given the educational space to do so, Spanish and translanguaging language practices bubble up, even in tertiary settings where all educational materials are presented orally and in written form in English.

Further, interview data with students revealed that they had mixed feelings about their own language proficiency in each of their languages, when it comes to academics, and that they make choices about participation in class on the basis of those self-perceptions (particularly in the case of those who feel more comfortable in Spanish). Similarly, while the institution that served as the site for this study is designated as Hispanic-Serving, and the Engineering faculty presented here are aware of that designation and what it means, little awareness has been seen early in this study of how faculty (and students) may think about leveraging their shared multilingual linguistic repertoires. The context remains primarily viewed through a lens of monolingual English education.

The tertiary context suggests that there is a ripe ground for exploring translanguaging pedagogies. This study suggests that the first step may involve in-depth and data-supported discussions with faculty, in this case Engineering faculty, on the potential values of learning in a context that could afford translanguaging a space. It is clear from interactions with faculty that they have not considered the role of language and language development and its connection to learning in the case of Engineering education at the tertiary level. Therefore, as briefly outlined here, activities such as sharing students' practices in class and in interviews as well as survey data can serve to raise instructors' awareness of the potential for engaging in translanguaging pedagogies. Further clear is the extent to which explorations of pedagogical professional development at the tertiary level may be critical. Licona and Kelly (2022) highlight how a teacher, knowledgeable of her students' linguistic backgrounds, can use translanguaging as a scaffold. Similarly, Siry et al. (2022) demonstrate the importance of facilitating dialogic, open-ended classroom structures in order to afford spaces for culturally and linguistically diverse students to draw on their multiple semiotic resources. Without an infrastructure for professional development, particularly one that outlines how multilingual learners may employ translanguaging in planned or unplanned ways, it may be difficult to institute broad-ranging changes in pedagogical approaches at the tertiary level. The Lesson Study approach described in this article may be an important format to continue to explore in this regard. As Buxton et al. (2022) emphasizes "translanguaging research can do more to explicitly address how such an approach can support the broader epistemic practices at the heart of new science standards, such as analyzing and interpreting data, developing and using models, or designing solutions" (p. 1). We echo this view and further call for more research of STEM pedagogy at the tertiary level.

References

Baker, C., & Wright, W. E. (2017). *Foundations of bilingual education and bilingualism* (6th ed.). Multilingual Matters.

Brettenny, B., & de Klerk, V. (1995). Bilingualism: The one-person one-language bond. *Language Matters, 26*(1), 40–58.

Burg, E., Klages, M., & Sokolski, P. (2013). Beyond "parallel play": Creating a realistic model of integrative learning with community college freshmen. *Learning Communities: Research & Practice, 1*(1), 13.

Buxton, C., Harman, R., Cardozo-Gaibisso, L., & Vazquez Dominguez, M. (2022). Translanguaging within an integrated framework for multilingual science meaning-making. In A. Jakobsson, P. Nygård Larsson, & A. Karlsson (Eds.), *Translanguaging in Science Education*. Springer.

Canagarajah, A. S. (2013). *Translingual practice: Global Englishes and cosmopolitan relations*. Routledge.

Carroll, K. S., & Mazak, C. M. (2017). Language policy in Puerto Rico's higher education: Opening the door for translanguaging practices. *Anthropology & Education Quarterly, 48*(1), 4–22.

Caruso, E. (2018). Translanguaging in higher education: Using several languages for the analysis of academic content in the teaching and learning process. *Language Learning in Higher Education, 8*(1), 65–90.

Cole, M. W., David, S., & Jiménez, R. T. (2016). Collaborative translation: Negotiating student investment in culturally responsive pedagogy. *Language Arts, 93*(6), 430-443.

Creese, A., & Blackledge, A. (2010). Translanguaging in the bilingual classroom: A pedagogy for learning and teaching? *The Modern Language Journal, 94*(1), 103–115.

Daniel, S., Ryu, M., Tuvilla, M., & Wright, C. (2022). The affordances of leveraging multilingual repertoires in scientific reasoning among resettled refugee teens: Functions of translanguaging in scientific reasoning. In A. Jakobsson, P. Nygård Larsson, & A. Karlsson (Eds.), *Translanguaging in Science Education*. Springer.

Espinosa, C. M., Herrera, L. Y., & Gaudreau, C. M. (2016). Reclaiming bilingualism: Translanguaging in a science class. In O. García & T. Kleyn (Eds.), *Translanguaging with multilingual students: Learning from classroom moments* (pp. 160–177). Routledge.

Fuller, J. M. (2012). *Bilingual pre-teens: Competing ideologies and multiple identities in the U.S. and Germany*. Routledge.

García, O. (2009). Education, multilingualism and translanguaging in the 21st century. In A. Mohanty, M. Panda, R. Phillipson, & T. Skutnabb-Kangas (Eds.), *Multilingual education for social justice: Globalising the local* (pp. 128–145). Orient Blackswan.

García, O., & Kleyn, T. (2016). *Translanguaging with multilingual students: Learning from classroom moments* (1st ed.). Routledge.

García, O., & Wei, L. (2014). Translanguaging: Language, bilingualism, and education. *Bilingual Research Journal, 37*(3), 366–369.

Garza, A. (2017). "Negativo por negativo me va dar un… POSITIvo": Translanguaging as a vehicle for appropriation of mathematical meanings. In J. Langman & H. Hansen-Thomas (Eds.), *Discourse analytic perspectives on STEM education: Exploring interaction and learning in the multilingual classroom* (pp. 99–116). Springer.

Garza, E., & Arreguín-Anderson, M. G. (2018). Translanguaging: Developing scientific inquiry in a dual language classroom. *Bilingual Research Journal, 41*(2), 101–116.

Gibbons, P. (2006). *Bridging discourses in the ESL classroom: Students, teachers and researchers*. Bloomsbury Publishing.

Goodman, B. A. (2016). 4. The ecology of language and Translanguaging in a Ukrainian university. In C. M. Mazak & K. S. Carroll (Eds.), *Translanguaging in higher education: Beyond monolingual ideologies* (pp. 50–69). Multilingual Matters.

Hattingh, A., McKinney, C., Msimanga, A., Probyn, M., & Tyler, R. (2022). Translanguaging in science education in South African classrooms: Challenging constraining ideologies for science

teacher education. In A. Jakobsson, P. Nygård Larsson, & A. Karlsson (Eds.), *Translanguaging in Science Education*. Springer.
He, P., Lai, H., & Lin, A. M. Y. (2017). Translanguaging in a multimodal mathematics presentation. In C. M. Mazak & K. S. Carroll (Eds.), *Translanguaging in higher education: Beyond monolingual ideologies* (pp. 91–118). Multilingual Matters.
He, P., & Lin, A. M. Y. (2022). Translanguaging, trans-semiotizing, and trans-registering in a culturally and linguistically diverse science classroom. In A. Jakobsson, P. Nygård Larsson, & A. Karlsson (Eds.), *Translanguaging in Science Education*. Springer.
Karlsson, A., Nygård Larsson, P., & Jakobsson, A. (2019). Multilingual students' use of translanguaging in science classrooms. *International Journal of Science Education, 41*(15), 2049–2069.
Karlsson, A., Nygård Larsson, P., & Jakobsson, A. (2020). The continuity of learning in a translanguaging science classroom. *Cultural Studies of Science Education, 15*(1), 1–25.
Karlsson, A., Nygård Larsson, P., & Jakobsson, A. (2022). Students' multilingual negotiations of science in third space. In A. Jakobsson, P. Nygård Larsson, & A. Karlsson (Eds.), *Translanguaging in Science Education*. Springer.
Langman, J. (2014). Translanguaging, identity, and learning: Science teachers as engaged language planners. *Language Policy, 13*(2), 183–200.
Lemmi, C., Pérez, G., & Brown, B. A. (2022). Translanguaging in the science classroom: Drawing on students' full linguistic repertoires in bi/multilingual settings. In A. Jakobsson, P. Nygård Larsson, & A. Karlsson (Eds.), *Translanguaging in Science Education*. Springer.
Lewis, C. C., Perry, R. R., Friedkin, S., & Roth, J. R. (2012). Improving teaching does improve teachers: Evidence from lesson study. *Journal of Teacher Education, 63*(5), 368–375.
Li, W., & Zhu, H. (2013). Translanguaging identities and ideologies: Creating transnational space through flexible multilingual practices amongst Chinese University students in the UK. *Applied Linguistics, 34*(5), 516–535.
Licona, P. R., & Kelly, G. J. (2020). Translanguaging in a middle school science classroom: Constructing scientific arguments in English and Spanish. *Cultural Studies of Science Education, 15*(2), 485–510.
Licona, P., & Kelly, G. J. (2022). Translanguaging in middle school science: Written arguments about issues of biodiversity. In A. Jakobsson, P. Nygård Larsson, & A. Karlsson (Eds.), *Translanguaging in Science Education*. Springer.
Lin, A. M. Y. (2019). Theories of trans/languaging and trans-semiotizing: Implications for content-based education classrooms. *International Journal of Bilingual Education and Bilingualism, 22*(1), 5–16.
Makalela, L. (2016). Translanguaging practices in a South African Institution of Higher Learning: A case of Ubuntu multilingual return. In C. M. Mazak & K. S. Carroll (Eds.), *Translanguaging in higher education: Beyond monolingual ideologies* (pp. 11–28). Multilingual Matters.
Martin-Beltrán, M. (2014). "What do you want to say?" how adolescents use Translanguaging to expand learning opportunities. *International Multilingual Research Journal, 8*(3), 208–230.
Mazak, C., & Carroll, K. S. (Eds.). (2016). *Translanguaging in higher education: Beyond monolingual ideologies*. Multilingual Matters.
Mazak, C. M., & Herbas-Donoso, C. (2014). Translanguaging practices and language ideologies in Puerto Rican University Science Education. *Critical Inquiry in Language Studies, 11*(1), 27–49.
Mazak, C. M., & Herbas-Donoso, C. (2015). Translanguaging practices at a bilingual university: A case study of a science classroom. *International Journal of Bilingual Education and Bilingualism, 18*(6), 698–714.
Mazak, C. M., Mendoza, F., & Mangonéz, L. P. (2016). Professors translanguaging in practice: Three cases from a bilingual university. In C. M. Mazak & K. S. Carroll (Eds.), *Translanguaging in higher education: Beyond monolingual ideologies* (pp. 70–90). Multilingual Matters.
Otheguy, R., García, O., & Reid, W. (2019). A translanguaging view of the linguistic system of bilinguals. *Applied Linguistics Review, 10*(4), 625–651.
Phillipson, R. (2009). English in globalisation, a Lingua Franca or a Lingua Frankensteinia? *TESOL Quarterly, 43*(2), 335–339.

Poza, L. E. (2018). The language of ciencia: Translanguaging and learning in a bilingual science classroom. *International Journal of Bilingual Education and Bilingualism, 21*(1), 1–19.
Probyn, M. (2015). Pedagogical translanguaging: Bridging discourses in South African science classrooms. *Language and Education, 29*(3), 218–234.
Siry, C., Wilmes, S., te Heesen, K., Sportelli, D., & Heinericy, S. (2022). Young children's trans-modal participation in science investigations: Drawing on a diversity of resources for meaning-making. In A. Jakobsson, P. Nygård Larsson, & A. Karlsson (Eds.), *Translanguaging in Science Education*. Springer.
Vogel, S., & García, O. (2017). *Translanguaging*. Publications and Research. https://academicworks.cuny.edu/gc_pubs/402
Wei, L. (2018). Translanguaging as a practical theory of language. *Applied Linguistics, 39*(1), 9–30.

Chapter 4
Young Children's Transmodal Participation in Science Investigations: Drawing on a Diversity of Resources for Meaning-Making

Christina Siry, Sara Wilmes, Kerstin te Heesen, Doriana Sportelli, and Sandy Heinericy

Abstract This chapter explores children's interactions in a multilingual classroom to examine the different ways plurilingual students used communicative resources in interaction during science investigations. The aim of this research is to illuminate how these culturally and linguistically diverse young students drew on numerous resources during science investigations on the topic of worms, using a variety of approaches including digital microscopes. This ethnographic study draws upon video data collected in a mixed-age kindergarten class (4–6-year-old students) to analyze children's participation whole class dialogue and small group interactions around worm investigations and to consider the ways in which children engaged as they made and expressed meaning. We position science learning as embodied cultural enactment, and we aim to work towards new understandings of the complex processes within translanguaging spaces in science education. The claims from this research underscore the value of dialogic, open-ended classroom structures for facilitating spaces for culturally and linguistically diverse students to draw on their many resources, and to agentically participate in science investigations.

C. Siry (✉) · S. Wilmes · K. te Heesen
The University of Luxembourg, Esch-sur-Alzette, Luxembourg
e-mail: christina.siry@uni.lu; sara.wilmes@uni.lu; kerstin.teheesen@uni.lu

D. Sportelli
Pädagogische Hochschule Luzern, Luzern, Switzerland
e-mail: doriana.sportelli@stud.phbern.ch

S. Heinericy
MENJE, Roeser Primary School, Luzern, Switzerland
e-mail: sandy.heinericy@education.lu

© The Author(s), under exclusive license to Springer Nature Switzerland AG 2021

A. Jakobsson et al. (eds.), *Translanguaging in Science Education*, Sociocultural Explorations of Science Education 27,
https://doi.org/10.1007/978-3-030-82973-5_4

Keywords Translanguaging · Multimodal · Agency · Early childhood science · Plurilingual · Multilingual · Dialogic

4.1 Introduction

> Three boys stand around a table with a clear plastic container of soil in front of them. One of the boys, Eric, bends over the container, poking at the soil with a spoon, as the others, Matthieu and Lukas, observe the container intensely. Eric suddenly looks up, exclaiming loudly in Luxembourgish, *Oooooh, we have a big giant!* as he excitedly turns to face the teacher. Lukas also turns to face the teacher, smiling at her, as Matthieu takes the spoon and begins to dig in the soil. *A big giant, there!* Eric reiterates in Luxembourgish, as he tilts the container and points with the spoon at a worm. Lukas leaves the group, while Eric and Matthieu lean over the container. Matthieu takes the spoon and moves the soil around. *Oh wow! Look! Ooooooh laaaa!* Eric says excitedly in French as he points at a worm *Ohhhh laaa! There it is!* he says to Matthieu in French, who smiles enthusiastically in return, and then giggles while observing the worm with a magnifying glass. Lukas returns to join them, also looking at the container. *Heyyy, look, look!* Eric says in Luxembourgish while turning to Lukas, who leans closer over the container. *Wooo hooo!* Eric exclaims, as all three children peer excitedly into the container. Matthieu points at the container with a magnifying glass and hesitantly asks Eric in French, *There, there, another?* Eric confirms Matthieu's observation in French, *Yes, there is another. It is the baby.* Matthieu points with his magnifying lens at the container, agreeing in French, *yes, it is the baby and the mama.*
>
> (20190313 precoce 2 camera c)

The three boys in this episode are 3-year-old students in a preschool class in Luxembourg participating in a lesson about worms. Their class has joined a neighboring kindergarten class for a collaborative science project. The two classes meet weekly, and the older kindergarten children engage in science investigations with the younger preschool children. In this episode the topic of their investigation was worms. We begin with this short excerpt to introduce the communicative complexities inherent in multilingual school contexts such as in our national context of Luxembourg, a small, highly diverse, European country. In the above interaction, 3-year-old Eric moves seamlessly between Luxembourgish in interactions with Lukas and the teacher and French in interactions with Matthieu. The children in this class have varying degrees of language proficiency in the languages of instruction, and thus Eric draws on a multitude of resources in interaction with his peers and teacher around their worm investigations. He uses Luxembourgish in his interactions with Lukas, who is not proficient in French, and French in interaction with Matthieu, who is not proficient in Luxembourgish. The three boys communicate their observations and wonderings about the worms with each other and with the teacher across multiple modalities, as they draw on gestures, expressions, materials, and vocal intonations while moving in and out of languages.

Within our multilingual national context, our team conducts research into the ways in which young plurilingual[1] children such as these preschool students engage in doing science with peers and with their teachers (e.g., Gómez Fernández & Siry, 2018; Siry & Gorges, 2020; Wilmes & Siry, 2018, 2020). Guided by sociocultural theoretical frameworks, we use video-based critical ethnography (Carspecken, 1996) to examine and interpret young children's multimodal interactions as they engage in doing science together. In this chapter, we present analysis from a case study (Stake, 2003) situated in a mixed-age kindergarten class as students (4-6 years old) pursued worm investigations during five classroom sessions over a period of 3 weeks. Through detailed analysis of several episodes drawn from this video-based critical ethnography, we elaborate two overarching claims relative to children's translanguaging as they participate in science investigations. In particular, we will emphasize how the analysis reveals that dialogic classroom structures supported children's *transmodaling*, which we conceptualize as the fluid use of diverse semiotic resources, including language resources, among many others. We aim through this work to arrive at nuanced understandings of the complex processes within translanguaging spaces in science education as we reflect upon the myriad resources children draw on as they engage in science learning.

4.2 Guiding Frameworks

The study we present herein is guided by sociocultural perspectives, drawing upon two theoretical foci; the agency I structure[2] dialectic; and translanguaging in science education. Our research is grounded in cultural sociology theoretical frameworks (e.g., Bourdieu, 1992; Sewell, 1992), and we position science teaching and learning as cultural and social acts (e.g., Tobin, 2005). Science through the lenses we adopt is a cultural enactment, a practice that emerges in the doing, and one that is embodied (Roth & Hwang, 2011). Critical theoretical perspectives (e.g., Kincheloe & McLaren, 2011) support our desire to work toward equitable opportunities for culturally and linguistically diverse students to engage in science. Science presents the possibility for children to be immersed in questioning and wondering and engaging in the act of science inquiry can provide authentic ways to communicate (Bruna & Gomez, 2009), especially important for children who

---

[1] We employ the term plurilingual to describe the linguistic capabilities and abilities of the students with whom we work as discussed by the Council of Europe (2001, 2018). The purposeful use of the term plurilingual is used, instead of the use of terms that position the linguistic capabilities of actors relative to whole or native- speaker as ideals (bilingual, multilingual, language learners, etc.) Plurilingual, by contrast, valorizes the abilities and strengths of those who possess diverse communicative repertories. For a more detailed discussion of the theoretical implications of the use of a plurilingual theoretical position refer to Piccardo (2019), and in relation to our research specifically see Wilmes et al. (2018).

[2] Following Roth (2005) we adopt the Scheffer stroke (|) to indicate dialectical relationships.

might not be proficient in the languages of instruction (Wilmes et al., 2018). As students engage in pursuing questions about science phenomena, they can be supported in drawing on the diverse resources they bring to communicate their ideas and emergent understandings (e.g., Siry & Gorges, 2020). As such, we work with teachers to facilitate dialogic, open-ended opportunities to engage in doing science, (e.g., Siry et al., 2016), placing value on creating spaces for authentic communication around science observations.

Returning to the opening excerpt as an example of how students' understandings can be revealed and mediated in dialogic interaction, Lukas calls to another child, Jerome, and together they join Eric and Matthieu in observing the container of soil and worms. *There, there!* Jerome says in Luxembourgish, as he and Matthieu lean closer over the container. Eric pushes them slightly to the side as they are blocking his view. *Look, there is a big one, there is a small one,* Eric says in Luxembourgish to Lukas and Jerome, explaining what he and Matthieu had discovered earlier. As the teacher walks by the table, Eric looks up at her and exclaims in Luxembourgish, *Look, we have a big one!* while pointing with his finger at the container. She stops, bends over the container, and asks in Luxembourgish, *Ah yes? Where?* Eric points with the spoon at the worm, answering in Luxembourgish, *Th, th, that one there!* Matthieu also bends over the container, pointing with his index finger as he adds in French, *After, after the baby, it is there,* while turning to look at the teacher, Sandy. Sandy answers inquisitively in French, *There? Wow! The baby is where? There?* pointing with a finger at the soil. Matthieu bends over the container again, briefly looking through a magnifying lens, and then pointing with his index finger at the container, as Eric continues to poke at the soil with a spoon. *And then the mam…and then the papa!* Matthieu exclaims in French as he turns to face Sandy. *Aaaaah! Where is the papa?* she asks in French. *There!* Eric answers in French, pointing with a spoon at the worm. *It is there.* Matthieu confirms in French, pointing with his finger at another worm. *No, it is there, it is there!* Eric counters in French, pointing with a spoon, as Matthieu looks closely at the different worms.

We are interested in the ways in which children such as Eric and his peers are positioned to express and develop their understandings and their wonderings in science (e.g., Siry & Max, 2013) as we hope to use these understandings gleaned from research in classrooms in our work with pre- and in-service teachers through the SciTeach Center, a resource center for teachers dedicated to supporting the teaching of science at the primary school levels.[3] Our critical theoretical frameworks support us in working toward equitable classroom structures, as we seek to create transformative learning opportunities for plurilingual students and their teachers. In particular, we focus on exploring the ways in which children are positioned to take agency in classroom interactions around science investigations and the ways in which structures shape their potential possibilities for action, as explored next.

[3] For details on our collaborative teacher professional development work, see SciTeach Resource Center at https://sciteach.uni.lu.

4.2.1 Agency | Structure

We conceive of agency as the power to take action and draw on Ahearn's (2001) elaboration of agency as the "socioculturally mediated capacity to act" (p. 112). Agency is constructed in interaction (Kumpalainen et al., 2014), and is emergent within socio-cultural practices (Varelas, 2018). Through the theoretical lenses that we draw upon, agency is in a dialectical relationship with structures (e.g., Sewell, 1992), and as such, a person's ability to take agency is mediated by the structures within social systems. Such structures are schema and resources (Sewell, 1992), which shape possibilities for action. The types of structures within a classroom are vast, including material structures, such as curricula, or socially constituted structures, such as a teacher's expectations during lessons. Taking a dialectic perspective on agency and structure allows for positioning these two, seemingly opposite, constructs together theoretically and understanding that each exists with the other. Agency is mediated by the social structures at hand, but in turn structures can be impacted through agentic practices, as there is an inherent fluidity in the dialectical relationship between structure and agency. These understandings imply that agentic action can transform structures, and we hope through our work with teachers to support classroom structures that enable children to draw on a diverse array of resources in interaction around science. Aligned with the theoretical frameworks we adopt, as students are agentically positioned to draw on a diversity of semiotic and linguistic resources, they can transform existing structures through their agentic actions, and in doing so, be more equitably positioned to engage in science practices.

Several recent studies conducted by our team have examined the nuances of how young plurilingual children engage in science investigations. One, for example, explored the ways in which open-ended classroom structures mediated a plurilingual 4th grade student's agency (Siry et al., 2016) revealing that access to a range of resources such as languages, drawings, and gestures positioned him to demonstrate his understandings beyond what he explained in the target language of instruction. A related study demonstrated that a plurilingual 3rd grade student agentically moved from being a peripheral observer of science investigations to become a more central participant in his small group through playful interaction with peers and materials (Gómez Fernández & Siry, 2018). Building from these understandings, a recent case study (Siry & Gorges, 2020) explored how the structures presented in an informal classroom interaction positioned a student in a kindergarten classroom to draw on a wide range of resources in interaction, revealing that she utilised a broader array of resources to express a more complex understanding in a small group conversation than she did in a more formal part of the lesson in front of her class later. These studies layer together to, emphasize the fluid ways in which students interact across modalities, including languages. Through the understandings we have gleaned from these previous research studies, we turn next to consider the ways in which *translanguaging* can mediate students' engagement in science investigations.

4.2.2 Translanguaging and Transmodaling

Our work examines the communicative resources, meaning the material and semiotic resources, that students are able to engage as they do science. In the highly complex linguistic situation in our context, which we describe in the sections that follow, students' diverse languages can become useful resources for learning. We seek to support teachers in creating spaces to allow students to draw on their myriad resources as they do science, and we integrate theoretical perspectives on translanguaging to understand the ways in which students interact, focusing on "the multiplicity, fluidity, mobility, locality, and globality of semiotic resources" (Moore et al., 2018, p. 346) that plurilingual individuals draw upon in interaction. Translanguaging is a process of communication during which speakers draw upon resources from more than one national language (García & Wei, 2014). Translanguaging is both a communicative act, as well as a pedagogical approach that can maximize communicative potential (García, 2009) and mediate students' agentic possibilities in science as it allows for access to a wide range of resources.

The overall focus in our research is on the ways in which science is done in interaction, with a particular focus on how science concepts "are drawn out, written out, acted out, talked about" (Bezemer & Kress, 2020, p.73), We explore how science is engaged in by young plurilingual students, and how their science understandings are communicated, through highlighting the multitude of modes and modalities visible in interactions. A multimodal semiotic perspective situates our understanding that each mode contributes to meaning-making, and while each mode affords specific meaning-making potentials on its own, the meanings of combined modes and modal complexes (Norris, 2004) are interconnected. As such, we use multimodal analysis to work towards generating new understandings into the complex ways in which young children engage in science in multilingual classroom contexts.

In our team's recent research in multilingual science education contexts, we have sought to work towards an expanded notion of voice, as being embodied as well as multimodal (Wilmes et al., 2018), and we have come to use the term *transmodal* and its associated action *transmodaling* (e.g., Horner et al., 2015) to describe children's and teachers' fluid use of resources in interaction, including language resources and verbalizations embodied resources, and material semiotic resources. The use of this term grows from a focus on translanguaging and broadens our view in that it honors and encompasses the wider embodied interactional resources employed in communication beyond languages in general and in the case of an early childhood classroom that we present in this manuscript, in particular. We are interested in understanding the ways in which classroom structures mediate children's deployment of resources during science investigations, and we draw upon research examining translanguaging in the field of science education, as introduced next.

4.2.3 Translanguaging Research in Science Education

The research base exploring translanguaging in the context of science education is expanding, as this edited volume (Jakobsson et al., 2022) attests to, with scholars in diverse regions examining the ways in which translanguaging can promote engagement in science. Through the investigation of communication and meaning-making in a range of contexts, such as with Korean-English bilingual secondary students in the United States (Ryu, 2019), elementary students in South Africa (Probyn, 2019), and middle schools in Lebanon (Salloum, 2020) for example, and drawing upon differing methodological orientations, the common feature of this growing area of science education research is a move beyond ontological views of communication and meaning-making that are bounded in named-languages and structuralist views of sense-making (Lin et al., 2020). Instead, communication is viewed as dynamic, emergent, and unfolding in contextualized processes through social relations. Related to these views are renewed views of structures within these contexts that are drawn on and upon which communication occurs as "looser, more porous, blending, emergent, fluid, dynamic, complex, temporally connected, and materially mediated" (Lin et al., 2020, p.39). As Wu and Lin (2019) show in their analysis of science instruction with students and teachers in a tenth-grade Biology classroom in Hong Kong, communication and meaning-making by students involves trans-semiotization, an employment of semiotic resources that involves the "concurrent intricate entanglement of all the meaning-making resources available at that moment and space" (p.262). Their fine-grained analysis shows how sense-making through interaction draws upon combinations of semiotic resources that include the material and the embodied, that are available in the science classroom at the moment. Examining such combinations of resources that support sense making in science was the focus of Gorges' (2019) study, which analyzed the ways in which adult plurilingual students interacted while participating in lessons around the socio-scientific issue of sustainability. Situated in an alternative school, she examined how a focal student interacted within open-ended dialogic structures, finding that these allowed her to engage much of linguistic repertoire, which mediated her own, as well as her group members', understandings of concepts around sustainability (e.g., Wilmes et al., 2018). Translanguaging within open-ended classroom structures also served as a resource to middle school students in Licona and Kelly's study (2022), in which they examined how students' engagement in open-ended classroom structures provided space for deploying a range of resources, thereby facilitating students' engagement in epistemic practices of scientific argumentation. Classroom spaces that support translanguaging can serve as spaces for co-participation and the co-construction of scientific knowledge, through supporting students to appropriate scientific concepts and related terminologies (Charamba, 2022). Further examples of research grounded in translanguaging theories include Suarez's recent (2020) work examining emergent bilinguals' leveraging of semiotic resources, as he shows how students were able to draw on a range of resources when positioned within translanguaging pedagogical spaces. At the primary levels, Karlsson et al., (2019) examined a translanguaging science classroom over a three-year period to examine the ways in which language operated in interaction, revealing how the classroom constitutes a resource for science and language together. Their work has illustrated the ways in which translanguaging benefited students' contextualisation of science content to prior experiences and how this has been shown to support continuity of science learning and related student understanding (Karlsson et al., 2020).

These studies are highlighted to illustrate that the research base examining translanguaging in the context of science education is increasing recently, however we note that translanguaging in the context of early childhood science is an underexplored area. We next introduce our study and context, in order to situate our examination of how young children employed and animated a diversity of meaning-making and communication resources as they conducted investigations with worms.

4.3 The Research Study

The guiding frameworks introduced above and understandings gleaned from the research literature are layered together in our work with a critical theoretical focus, as we utilize critical ethnography as a methodological approach to guide our analysis.

4.3.1 Critical Ethnography

As a research group, we use critical theoretical lenses coupled with ethnographic approaches to guide the work that we do, and as such, this study was conducted as a critical ethnography (Carspecken, 1996). Ethnography is broadly a "form of social and educational research that emphasises the importance of studying first-hand what people do and say in particular contexts. This usually involves fairly lengthy contact, through participant observation in relevant settings, and/or through relatively open-ended interviews designed to understand people's perspectives, perhaps complemented by the study of various sorts of documents – official, publicly available, or personal" (Hammersley, 2006, p. 4). Similarly, critical ethnography works toward understanding the meanings that people ascribe to their experiences, but critical ethnography does not end with these understandings, rather using these toward examining the social, cultural, historical, and political factors that come together to structure people's experiences, with a particular focus on considering issues of equity. Critical ethnography thus layers a critical perspective onto the ethnographic question of "what is happening?" to enable asking "Why is it happening?" and work toward revealing and challenging power structures.

4.3.2 Context and Data Sources

Luxembourg is a highly diverse, multilingual country, with over 170 nationalities represented. A small country, with just over 600,000 residents, the diversity in languages and cultures is impressive, with almost half of residents having nationalities other than Luxembourgish (STATEC, 2020). Luxembourg has a multilingual school policy, meaning that students are taught in Luxembourgish, German, and French. Primary school is arranged into 2-year "cycles" and compulsory schooling begins at age 4, with cycle 1 (children ages 4–6). There exists additionally a program for 3-year-old children, précoce, that is conducted entirely in Luxembourgish. The research examined in this chapter was carried out in a cycle 1 classroom, with 16 students, and the language of instruction was Luxembourgish. All of these 16 students spoke at least one language other than Luxembourgish at home. These languages included Portuguese, French, English, Creole, Arabic, Farsi, German, Chinese, and Italian. This mirrors the demographic trends in many schools in our country, where over 50% of all students speak a language *other than* the languages of instruction in the home (MENJE, 2018). Thus, the language complexity in early childhood classrooms, such as the classroom at the focus of this study, cannot be overstated. In the research we present herein, we draw from a study in which we co-taught with the classroom teacher, Sandy, the fifth author of this chapter. As a group, we, the authors of this chapter, have been collaborating in teaching and research through our involvement in the Science Teacher Resource Center (SciTeach Center), a resource center for teachers, dedicated to supporting science at the primary school levels. Housed at the University of Luxembourg where Chris, Sara, and Kerstin work, the SciTeach[4] Center was founded in 2016 as a collaboration between the University of Luxembourg, the Luxembourg National Research Fund, the Ministry of Higher Education and Research, and the Ministry of Education, Children and Youth. The goal of the Center is to create and support a sustainable, collaborative network for primary school teacher professional development for teaching science in Luxembourg. In particular, we work at the Center to support teachers' professional growth through the insights gained from academic research in primary school science, through professional development courses, teaching resources for loan, as well as collaborative research opportunities. Our team consists of researchers, teachers, and administrative personnel, bringing a multitude of professional background and expert knowledge to the team. Sandy has been working with us as a teacher on special assignment from the beginning of the project until the end of the 2018/2019 school year in July 2019, a role in which she collaborated on the co-development and co-teaching of professional development workshops, one of the main foci of the SciTeach Center, while she also taught a kindergarten class, where this study was conducted.

Essential to our collaboration is the process of co-teaching (Roth & Tobin, 2004) that we utilized. Co-teaching supports professional development (Gallo-Fox &

[4] See www.sciteach.uni.lu for an overview of our initiatives.

Scantlebury, 2016), and it allowed us to work together with Sandy in her classroom and, through doing so, be present in the moment for observing and participating in science instruction, and to also participate in the moment-to-moment decisions and interactions that unfolded. The results of analysis we present here took place over a three-week "worms" unit we co-taught with Sandy that engaged students in posing questions and investigating aspects of worm's habitat, characteristics, and coming to know worms through scientific investigations. The science instructional approaches used were grounded in student-driven inquiry-instruction that positions students to engage in the practices of science (Siry, 2013; Wilmes, 2017). One of the essential components of an inquiry-based curriculum is that it works towards addressing students' ideas (e.g., Kawalkar, 2020), and thus the teacher's focus for instruction was on eliciting, and building from, students' ideas and questions. We collaborated on the development of activities that supported multiple modes of student expression such as drawing, whole-group, and small- group discussion, models, and other representations including photos and diagrams, and digital tools including handheld microscopes and laptops, were integrated to encourage a wide range of participation opportunities and space for students to draw on diverse resources as needed.

Data resources were collected by the second author, Sara, who participated in all lessons of the worm unit as a participant observer (Wragg, 2012), compiling approximately 15 h of video material over the five science investigation sessions, while she assisted Sandy, took photos, and interacted with students during their investigations. Further, Sara debriefed the lessons with Sandy and provided input over the course of the 3 weeks as instruction evolved. The first author, Chris, collaborated through reflective conversations with Sandy and Sara regarding the emerging focus of the data collection and participated in the second lesson as a participant observer. We layered upon the classroom data resources an informal, open-ended interview (Morehouse, 2012) with Sandy following the end of the worms unit, which provided insight into her pedagogical approaches and perspectives on the ways in which students engage in the worm investigations. The differing data sources allowed for multiple entry points for gaining perspectives of classroom interactions as they unfolded and for layered views of classroom structures. We have constructed a case of this classroom through analysis of four episodes drawn from the larger data set of these plurilingual children interacting with peers and teachers as they engaged in investigations about worms.

4.3.3 Co-researching and Co-writing as a Methodology

We, the five authors of this chapter, are members of a larger research team that utilizes theoretical approaches from the cultural studies of science education to guide our work. Our team's projects are intentionally grounded on collaborative approaches to research as well as teaching, and as such, we utilize co-writing and co-researching (Wilmes et al., 2018). As introduced above, Sara and Chris worked with Sandy to develop the classroom lessons, act as participant observers to differing degrees, and collect a variety of data resources. The third author, Kerstin, is a member of the SciTeach Center team and was involved in this study as a peripheral participant from the outset. The fourth author, Doriana, has an interesting relationship to the other authors that is worth mentioning, as she brings a further subjective understanding to this study. Doriana was previously a pre-service teacher in courses co-taught by Chris and Sara and was also a participant observer in in-service teacher education workshops at the SciTeach Center. Currently enrolled in a Masters program in Switzerland, her involvement in this chapter emerges from a research internship she participated in at the SciTeach Center, whereby she learned about qualitative video-analysis through working with the data resources for this chapter. Aside from having an interesting perspective as a student in Chris and Sara's class in the past, she brings a further unique perspective, as she herself was a student in Sandy's kindergarten class more than 20 years ago. Clearly we are each positioned with a unique view on the data resources, and we each bring different perspectives to our interpretation of the data. Our analysis has taken place collaboratively across the five authors, with each of us having differing, as well as overlapping, roles and different levels of responsibilities for the construction of the analysis and resulting manuscript.

As introduced above, Sandy was integral to the collaborative development of this study, and she participated in the data collection through engaging in numerous planning and debriefing discussions as well as a follow-up interview. After the collection of data resources was completed, Sara organized the data sets and began preliminary analysis. Doriana joined the team as a research assistant, to organize data sets and conduct first and second level video analysis together with Chris. The process of collective analysis for this chapter was conducted across the authors over time, through numerous collective as well as individual steps. Once the data sets were organized, video data resources from each of the classroom sessions on worms were viewed once at normal speed to ensure an understanding of the sequence of events over the duration of the teaching unit. After this point, the first four authors engaged in analytic discussions to begin to clarify the focus that was emerging on students' transmodal interactions as they clarified their understandings and questions to peers as well as to their teacher. After this, Chris and Doriana conducted a second round of analysis, in which videos were viewed once more at real time, from which the focus onto the differing resources the students were drawing on began to clarify. A cycle of collaborative analysis across the authors developed next, in which we combined several recursive iterative steps to emerge with claims, as well as to co-construct this chapter. Sandy integrated her perspectives into the manuscript after having viewed and reflected upon the videos. In an ongoing process of individual viewing of video excerpts, collaborative writing via a shared online document, and collective analytic discussions (face-to-face and later via video conferencing to adapt to the COVID-19 crisis in our communities), our multimodal methodological approach developed over time.

4.3.4 Methods of Analysis

Our research design was emergent (Merz, 2002) and questions were refined through a continual process of co-analysis and co-writing. From the data resources collected as introduced above, key episodes were identified using a process of event-oriented inquiry (Ritchie & Tobin, 2012). Once episodes were identified, multimodal interaction analysis (e.g., Wilmes & Siry, 2021) was employed to examine students' use of semiotic resources relative to the spaces within which they interacted. Throughout the analytical process we worked with the multimodal transcripts in their original version, meaning verbalizations were documented in the (national) language in which they were spoken. Through multi-level analysis we created multimodal transcripts depicting students in interaction, and once analysis was completed, we translated the episodes into English for publication purposes. We layer onto these narrative descriptions of the unfolding classroom episodes, and the section that follows weaves together analysis and discussion of vignettes from the worm investigations to explore the following research questions:

- *What resources do the students draw on in interaction during worm investigations and how are these resources employed?*
- *What structures mediated students' access to resources in interaction as they engaged in science investigations?*

Four episodes emerged as central for further analysis, which led to the formulation of two central claims, as elaborated next.

4.4 Analytic Discussion and Central Claims

We next present results from data analysis that leads to two overarching related claims. We posit that through the agency | structure relationship, students were able to draw on numerous resources available to participate in science through *transmodaling*. We elaborate this through two connected claims, which are:

(i) *Open-ended classroom structures supported the emergence of spaces for children's transmodaling.*
(ii) *Emergent spaces for transmodaling mediated children's students' agentic actions.*

These claims are interwoven, and although we present them separately, it is important to note that each builds off the other. Each is elaborated through data resources next individually, following which we weave both claims together as we discuss the implications of this research. The above claims emerged through a recursive, iterative process of analysis, and we have woven together analysis of classroom video data resources with ongoing individual reflection and collective dialogues, and video resources were analyzed at the meso- and micro-levels. Each layer has shaped what and how we understand the data resources and how we interpret our experiences within this research project. The claims we make emerged from this cyclical analysis, and we have drawn implications through the collective processes.

Claim 4.1: Open-Ended Classroom Structures Supported the Emergence of Spaces for Transmodaling

Analysis has revealed ways in which a variety of open-ended classroom interaction opportunities that were implemented in the worms lessons served as structures for students to co-construct explanations and understandings of worms across a variety of modalities. These spaces were structured by Sandy and the children through their interaction with each other and with the materials and classroom interaction spaces included whole-class discussions with the teacher, small group worm and soil investigation opportunities, and the use of a range of material resources including models, books, photos, magnifying lenses, and digital microscopes with laptops.

***Episode 4.1a: Worms Are Beige with a Little Pink*[5][6]**

Episode 4.1a is an excerpt from the second lesson of the worm unit, and it begins with all children sitting on benches arranged in a horseshoe shape by the blackboard with Sandy sitting in front of the blackboard, on a bench of equal height to the children, facing them. Her central objective for this lesson was to focus on the appearance of the worms, and when all are seated, she looks around at the children, puts her hands on her thighs with her arms open, looks around at the children sitting on benches, and asks the children in Luxembourgish, "What color are the worms then?", and the following interaction takes place:

Children	Pifznk!
Maria	((raising her hand quickly and pointing her finger toward Sandy to indicate the desire to speak while saying)) Pink
Charel	Or brown
Sandy	((looks around at children)) Are they all pink?
Charel	((shakes his head slightly)) or brown
Sandy	((briefly points with both hands to the compost in front of her))
Maria	Or, or, or beige
Matteo	Or dirt-brown

[5] Languages used are identified in the transcripts as follows: Luxembourgish normal font / French: *italics* / Italian: underline.

[6] The following transcription conventions have been adapted from Hwang and Roth (2011):

[and] square brackets in consecutive lines show an overlap of speech with body movements;
((draws)) words within double parentheses comment on visible body movements;
NOW capitalization marks speech that is louder than the normal speech intensity;
<<well>> words within angle brackets indicate lower speech volume;
(find out?) question mark in parentheses indicates inaudible utterance(s);
?,;. punctuation marks are used to indicate characteristics of speech production rather than grammatical units;
it's- Dash indicates sudden stop of talk

Sandy	Or beige. Or dirt-brown ((slightly lifts shoulders, pats thighs lightly with both flat hands))
Matteo	[Or dirt-brown]
Sandy	Mmmh ((Leans head briefly to the side and twists the corners of mouth down))
Maria	Or a little little beige with pink ((She rubs her two palms together, as if mixing something))
Sandy	Beige with pink. Yes ((shrugs shoulders and nods head to indicate she is considering the mentioned colors)) We will take a look

(20190312 Camera A sony +/- 00:07:38)

Episode 4.1a illustrates the ways in which the children co-constructed an explanation of the color of worms, using gestures, body movements, and Luxembourgish, guided by Sandy's openness and guiding question of *What color are the worms then?* The discussion took place with an almost square shape to the benches, thus the children and Sandy could all see each other. Sandy sat on a bench with the children, which positioned her at their level physically and also enabled her to engage in conversation while seated with them. The question served to engage the children into the topic of discussion by activating their prior knowledge on worms as well as guiding them to co-construct a description of the color of the worms. Beginning with a group suggestion of *Pink!*, Maria shares her agreement, raising her hand with a pointed finger as she says *Pink,* to which Charel adds, *or brown.* Sandy picks up the initial suggestion from the group as she looks at the children and asks *Are they all pink?* Charel shakes his head slightly, adding once more, *or brown,* to which Maria and Matteo clarify further, *or, or, or beige, or dirt-brown.* Sandy repeats what they say with shoulders slightly lifted as she pats her thighs with both flat hands, and Matteo reiterates again, *or dirt-brown. Mmmh* Sandy says while leaning her head to the side, seemingly thinking about these answers, as Maria contributes by rubbing her two palms together as if mixing things together, *or a little, little beige with pink.* Sandy nods her head with shrugged shoulders while considering these suggestions, *Beige with pink. Yes. We will take a look.* Later on, during the same whole class discussion, the conversation moves to describing the worms, and a child named Frank makes a comparison to snails, as introduced next.

Episode 4.1b: He Has No House on His Back

Frank	That's quite like a snail, but thinner, but he has no house on his back ((moves both of his hands behind his back and traces them up his back))
Sandy	((makes an astonished but pleased face as the child speaks)) Oh yes, it looks like a snail ((nodding in agreement))
Frank	and without one, like so its ((raises hands to mimic a snail's feelers by running his hands from his head along the imaginary feelers on top of his head)
Sandy	Yes, and without its ((mimics the child's gesture on her own head)) ... its feelers here above

(20190312 Camera A sony +/– 00:07:13)

Frank describes the appearance of a worm by comparing it to a snail, as he moves both of his hands to behind his back and traces a shape of shell with them while saying *that's quite like a snail, but thinner, but he has no house on his back.* Sandy appears impressed by this observation, as she nods in agreement saying, *Oh yes, it looks like a snail.* Frank continues, *and without one...* while mimicking a snail's feelers by running his hands from his head along imaginary feelers. Sandy confirms by mimicking the same gesture on her own head, *Yes, and without it's, it's feelers here above.* Frank makes his explanation of snails' body parts evident in his interaction with Sandy by using a combination of gestures and words in Luxembourgish, and he explains that a worm is shaped similar to a snail, but without a shell and with no feelers. Although he does not use these words, he communicates clearly his understanding that a worm is similar, yet different, to a snail. Sandy's open approach encourages him while also offering the word "feelers," and the interaction builds upon Frank's prior knowledge to make connections to worms while extending his Luxembourgish vocabulary through interaction with Sandy.

We consider these two episodes as one layered piece, as they both focus on a whole-class discussion that Sandy facilitated among the children with open-ended questions on the topic of worms. Sandy engages the children in the new focus of investigation by asking what they already know about worms and gathering their existing ideas. Children make their understandings known by drawing upon a variety of resources, including gestures, body movements, Luxembourgish, and facial expressions. Episodes 4.1a and 4.1b come together to demonstrate some of the ways the children transmodally made their ideas clear during circle time regarding what they already know about worms, as they co-constructed their understandings of worms in conversation. The whole-class discussions appeared to provide ample time for students to contribute to the conversation and express their emerging ideas about worms' color and shape, as they used gestures, words, sounds, and expressions to describe worms.

Episode 4.2: We Are Taking Photos Now!

Three children are standing at a table that has been set up in the classroom for using microscopes to investigate the worms. A laptop is connected to a digital microscope on the table in front of them. One of the children, Maria, tries to show the other two children, Tom and Diana, how the digital microscope works, as she demonstrates how to use the microscope by viewing her own hand with it while explaining. Maria came to kindergarten speaking Italian, but by the time of this data collection, she had developed significant proficiency in Luxembourgish. Diana recently arrived from Italy, and Maria often drew upon Italian in her interactions with her.

Maria ((holds the microscope in one hand and puts the other hand flat on the opening of the microscope. Diana looks on)) Then, then...

((Maria points towards Tom with the microscope, and Tom reaches for it. Maria pulls the microscope away))

Maria Look, go like this... ((extends her hand flat forward prompting Tom to do the same)). Look, go like this! Go like this!

((Tom shakes his head, while Diana watches carefully))

Maria Now if you do this, look like this…((looks down at the microscope and holds the microscope on (her own) flat hand)) then there ((puts her finger on the button that triggers the photo)) ... click! ((imitates pressing a button))… and now, now I took that photo ((points to the photo projected on the laptop screen))

((Tom and Diana follow her movements with their gaze))

Maria So… ((puts the microscope on Tom's palm)) Next… ((looks for the button on the microscope by rotating the microscope around))

((Tom and Diana watch her carefully))

Maria Click! ((presses the button and presses the microscope a little more firmly on the hand of Tom)) Now I took a photo! ((smiles contentedly, leans forward and looks Tom in the face and then hops briefly up and down))

((Tom also smiles and reaches for the microscope. Maria gives the microscope to Tom, while Diana stands next to them watching intently))

Tom [inaudible] ((points to something on the microscope, gesture is hidden by the laptop))

Maria Now, *if you* "click" ... ((imitates pressing the button in the air)) There is, there is a photo [inaudible]. Did you see?

((Tom follows Maria's instructions, going through the motions to take a photo))

Tom Wow! ((looks at the screen and then into the opening of the microscope))

Maria [inaudible] Photo ((Maria points at the photos displayed at the range of photos on the laptop screen. Tom leans forward and mumbles something unrecognizable))

Diana ((points to the screen)) There!

Maria Oah! Look there, that! Photo, photo, photo! ((Maria points to the photo displayed on the laptop screen. Tom and Diana look at the screen. Tom seems to be holding the microscope somewhere, all three children are watching the screen with concentration))

Maria We are taking photos now! ((Maria reaches for the microscope, but then holds back. Tom presses the button on the microscope. Everyone looks at the screen)) Bravooo! Now I will take a photo ... ((Maria takes the microscope from Tom and puts it to her mouth. Tom and Diana laugh))

Diana But yes, but now [inaudible] can I do it?

Tom ((Tom takes the microscope again and holds it against his cheek)) ((addressing Tom)) Now me!

Diana ((addressing Tom)) Now me!

Maria Now Diana!! ((Maria takes the microscope from Tom and hands it to Diana)) You do it! Then you have to, you have to push.

Diana [inaudible] ((Diana presses the microscope firmly on her hand and squeezes her eyes shut briefly))

Maria Push? No! ((Diana takes the microscope away. Yannis joins the group standing with the microscopes and the laptop at the table))

Yannis Now may I look?

Maria Look! If you do it like this... ((Maria points to the button that triggers the photo apparatus)), there is a photo, look. "Click"! ((Maria holds the microscope on her hand and presses the button))
Child 4 Now me ((Child 4 takes the microscope from Maria))

(20190313 Camera A: +/- 00:25:25)

This interaction takes place without teacher intervention, in a space that is structured by the students through interaction and structured with rules and materials by Sandy. The investigation of the worms is situated as a collective process across the students, and there is an informal type of hybrid space created in a short time frame, where the children can wrap up their interactions around the topic with each other and the materials before leaving school for lunch. Sandy had structured the space to allow for open interactions, determined by the students, and there is a wide range of modalities drawn on between the children, as Maria uses sounds, gestures, and different languages to explain how to use the digital microscope. This is a group-oriented interaction, as Maria works to assure the understandings of the others, and it takes place in an in-between space in the classroom as the lesson had ended.

Claim 4.2: Emergent Spaces for Transmodaling Mediated Children's Agentic Actions

The second claim that arises from analysis is that the open-ended approaches, in which students were able to transmodal, served as structures which mediated children's agentic actions in the context of doing science. By this we mean that in the spaces made available to them in Sandy's class, students were able to take agency with how they engaged in science. They were able to draw upon resources in interaction to express their science understandings and draw upon a diverse set of communicative resources, as illustrated next in Episodes 4.3 and 4.4.

Episode 4.3: Worms Move Big and Small

Sandy stands in front of the children who are sitting on the benches in the seating area as she begins a discussion about what the children have found out about the worms.

Sandy ((addressing all students)) yes, so how does [the worm] move then? Claude
Claude I know that!
Sandy Then show us.
Claude Like this! ((Beginning of movement not recognizable. Then Claude is on all fours on the floor))
Children ((speaking at the same time and raising fingers to indicate desire to speak)) I know! I know!
Sandy Yes, Filipe? How does, how does a worm move?
Filipe ((lays on the floor and demonstrates a crawling motion))
Sandy Hey look at Filipe!
Nico Yes, I did that once too! ((referring to Filipe's motion))
Sandy Mhmh ((as children raise their fingers to indicate they would like to share))

Matteoy	I can, I know, I can!
Sandy	Lena?
Lena	It goes like this ((Lena slides to the floor and lies down, then begins moving (not visible, below camera view))
Sandy	((points with her finger to Lena nodding head up and down)) Yeesss What does the worm do when he moves forward, what does he do? ((Sandy holds up her hands in front of her parallel to each other, palms facing each other, and brings her hands together and apart))
Matteo	IIt gets small ((Matteo pulls his arms closer to his body and lifts his knee in the direction of his belly) and then big ((Matteo then extends both arms and legs forward))
Sandy	Aaah good! ((Sandy points to Matteo)) It becomes small and then big to move ((Sandy begins to write on the chalkboard))
Claude	[unintelligible] I know why it gets small and big
Sandy	Why?
Claude	I'll show you why it gets small and big ((Claude sits down on the floor)) [Unintelligible] Look, so ... ((Claude is on all fours and pulls into himself)) short ... (Claude stretches forward)) and long.
Sandy	Yes, They show it right. ((nods, addressing the support teacher also sitting 6on the circle with the children)) Charel?
Charel	((Charel also lays down flat on the floor on his stomach. He moves with his hands forward on the floor)) It goes like this ((moving on floor)) and like this ((continuing movement))
Sandy	So what ehh ... what Matteo said, the worm gets small and big to move ((Sandy holds up her hands in front of her parallel to each other, palms facing each other, and brings her hands together and apart to indicate small and big)) Is that true? Have you seen that?

(Video 20190313 Camera A: +/- 00:38:51)

Sandy next continues the small-group conversation as she builds upon the children's ideas. She repeats a gesture she made before, holding both hands up in front of her, palms facing one another, and moves them in and out to represent "small" and "big." She asks the group, *Is that true? Have you seen that?* and Claude shares that he had in fact seen worms move that way, but he offers this with a slightly disappointed tone, perhaps because he was not called upon to demonstrate the movement. Then Sandy remembers that she may have an accordion in the classroom storage room, a musical instrument that also moves in this way through contracting to become small and expanding to become big.

Episode 4.4: How Do You Say Red Pepper in Arabic?

Later that day during a separate activity with the worms, Sandy sits at a table with a group of children and cuts a piece of bell pepper. There are multiple children in the group and as she is cutting the pepper, she discusses with the children and the support teacher what the word pepper is in other languages.

Sandy	Have you ever eaten this at home? ((Sandy looks up at Ahmad as she cuts a bell pepper into pieces and offers pieces to the children sitting at the table.
Ahmad	((shakes his head)) No
Sandy	((holding up a piece of red bell pepper)) No never eaten?
Ahmad	((shakes his head no while eating a piece of pepper))
Sandy	It's good, huh?
Ahmad	((continues shaking his head in a no motion))
Frank	((also sitting at the table with the small group)) I've eaten red pepper, but that pepper was made of chips ((referring to paprika chips, a common flavor of potato chips))
Sandy	((looking at Ahmad)) Then you have to say to your mom when you go shopping, 'Oh, I want red pepper'. Say to your mom pep-per ((looks at Ahmad while sounding out the syllables putting emphasis on each syllable)) Say it one time pep-per
Ahmad	I cannot do that in Arabic
Sandy	Ah! In Arabic .. yes .. ((Sandy looks up to Sara to explain what Ahmad said and explains in English: 'He wants to tell his mom in Arabic')) Ah...red pepper... your mother speaks a little bit of French, right? ((looking at Ahmad)) Um ((thinking))...who speaks Arabic? ((looking around the classroom)) *Name of Child*, no?
Charel	((also sitting at the table shoots his arm up in the air with finger pointed to indicate wanting to share verbally)) Me!!!
Sandy	No, you speak Farsi.
Charel	No, not yet ((while eating a piece of red pepper))
Sandy	((Holding up a piece of red pepper to the third teacher in the room)) How does one say red pepper in Arabic?
Sandy	((turns back to Ahmad sitting at the table, while holding a piece of red pepper and asks)) How do you say red pepper in Syria?
Sandy	((Sandy's gaze goes up to Bashaar standing behind Ahmad at the table and says)) and in Afghanistan, there is also red pepper, right?

(20190313 camera a: +/- 01:20:47)

As the situation unfolds in interaction, Sandy draws upon the material experience with the pepper and knowledge of Ahmed's use of Arabic as semiotic resources to discuss the pepper. In this sense, semiotic resources present in the material sense are drawn upon in connection with the use of both Luxembourgish and Arabic. The teacher's next work together with Google Translate is to translate the word "red bell pepper" from Luxembourgish into Arabic, and the small group listens to Google Translate speaking the word. The teachers practice speaking the word as well, as they work to make connections to the students' home languages and bring it into interaction with the group.

The two claims we have elaborated above are interwoven and connected in the dialectic relationship between agency and structure. The analysis has revealed the ways in which open-ended classroom structures supported the emergence of spaces for transmodaling for children, and in turn, these spaces mediated children's possibilities to take action. As the children interact with each other, with Sandy, and with the materials, they engage in the practices of sciences, asking questions, making observations and discussing their findings, as they draw upon a wide array of transmodal resources from their diverse and differing repertoires. There are two classroom settings that have emerged as central for mediating students' transmodal communication and broad access to meaning-making resources, in whole-class discussion, and in more informal in-between spaces after the science lessons have ended, and before next activities have begun. Whole class discussion is a critical part of an inquiry-based pedagogy, given that in such open-ended investigations, students may: "express their own opinions and come up with their own explanations which could be different from the canonical scientific knowledge, conclusion of the discussion is a very significant phase" (Kawalkar, 2020, p. 92). The discussions between Sandy and the children created a base for expressing a wide range of understandings and wonderings in ways that are accepting to different perspectives. Further, the spaces after the science lessons ended, yet before next class activities began, have emerged as valuable for children's continued engagement. This connects with the findings of Caiman and Lundegard (2018) who examined young children's inquiries in the in-between space between a science lesson and lunchtime. They suggest that it is in these "imagined space[s] of possibility" where "children are in charge" (p. 693). We are particularly interested in the ways in which open-ended spaces can serve as hybrid spaces (Salloum, 2022) for young children's interactions around science, interactions in which both science and languages can be negotiated (Karlsson et al., 2019), and we intend to look more closely at the different spaces cogenerated by Sandy and her students in future work.

4.5 Implications and Closing Thoughts

The episodes elaborated above serve to illustrate ways students moved between the spaces created in the classroom and, when doing so, were positioned to draw upon and move fluidly across and among modes during their investigations. Further, the episodes have been selected to demonstrate students' use of multiple intertwined resources toward meaning-making in science. We have layered theories and methods together to understand the complexities of interactions in this classroom, and each layer shapes what we see and the understandings that we gain. We have attempted to show the differing resources students drew on, how they used them in interaction, and what the use of these resources mediated. As we have introduced earlier, we have adopted the term transmodaling in our research group. Our use of the prefix "trans-" regarding modalities used in interactions refers to the copresence of language and other semiotic resources in meaning-making (Hawkins, 2018), a perspective which allows for moving beyond a focus on discrete modes, towards views of "semiotic resources as embedded and given meaning within the specific assemblage, and within trajectories of time and space, continuously shifting and re-shaping in their contexts and mobility" (Hawkins, 2018, p. 65). The findings of this work reveal how within students' interactions "the boundaries between languages become permeable" (Derry et al., 2010, p. 112), as children smoothly moved between and within modes in interactions. Through this transmodal process, science engagement became visible. Transmodaling was mediated within the agency | structure relationship, as the open-ended structures allowed for agentic interactions with peers. The findings of this study underscore the value of facilitating dialogic, open-ended classroom structures to afford possibilities for culturally and linguistically diverse students to draw on their many resources to agentically participate in science investigations. We position science learning as an embodied cultural enactment, and the implications of this research underscore the complexities of plurilingual children's classroom interactions as they participate in doing and learning science.

Using a lens of agency | structure has revealed the value of Sandy's focus on creating a culture of communication and inquiry in this particular classroom. She intentionally works toward creating spaces for student voice as well as for discovery. These structures served as translanguaging spaces, which we conceptualize as places where students are able to draw upon a diversity of material, embodied, and communicative resources in interaction and move among these flexibly and transmodally over time. The importance of these translanguaging spaces is that they "facilitate fluid practices that go beyond socially constructed language and educational systems, structures, and practices to engage students' diverse multimodal meaning-making systems and subjectivities" (García & Wei, 2014, p.3). The agency | structure relationship is central to working toward equitable opportunities to engage in meaning-making in science investigations, and echoing Maria Varelas, we believe that in order for equity to be achieved, it is critical to intentionally "create spaces and places where these students' agency is not only allowed to surface but also explicitly nurtured" (2018). Translanguaging spaces can serve as equitable classroom spaces, in that they are structures in which a range of modalities including languages can become resources to meaning-making rather than barriers (Gómez Fernández, 2019).

An interesting observation has arisen through the analysis, which is that we have noted that, when talking to the teacher, the children tend to transmodally use body language to enhance their expressions if their Luxembourgish language appears insufficient to make themselves understood. By contrast, we note that in interactions with peers, we more often observe transmodal usage including different language repertoires, for example, drawing upon resources from multiple national languages such as Maria and Diana's use of Luxembourgish and Italian. Children's use of another language from their repertoire seems to depend heavily on the interlocutor and, as far as we have seen thus far in the data, only takes place between children and not in interaction with the teacher, even when the teacher asks them to do so. For example, Sandy intervened in a dispute between several children, and asked Diana what had happened, but Diana did not reply. As Diana is a native Italian speaker, Sandy encouraged her to explain in Italian. Although Diana can be observed using Italian as a resource in multiple interactions with peers, when asked by Sandy to explain in Italian, she chose not to. We note that the children do not appear to make use of their full linguistic repertoires with Sandy, even if structures appear to welcome broad translingual participation. The question arises as to whether the children agentically choose to only draw on the languages of instruction in interactions with their teacher, given that the structures allow a wide array of transmodal participation, or whether perhaps the structures are not as flexible from the children's perspectives as might be interpreted from adult perspectives. We find this an interesting emergence from this chapter development and intend to pursue this empirical focus in a next layer of analysis.

This research has underscored the need to support teachers in creating structures that allow students to develop scientific literacy through, within, and across modes; and a transmodal perspective contributes to our understanding by illuminating how a range of accessible resources come together in fluid ways as children engage in science meaning-making. Equitable science learning environments must create opportunities for students to leverage their full semiotic repertoires for meaning-making (Suarez, 2020). and this research has highlighted how open-ended spaces for inquiry can afford such opportunities. Gaining understandings on the complex ways in which plurilingual children engage in communicating in these spaces can push back on normative approaches to pedagogies that keep languages as distinctly separate entities, which may limit students' access to the many resources they bring to science. Rather we hope to provide support for science teaching practices that honor students' plurality plurality, as we work with teachers to create structures that afford students' interaction and meaning-making in open-ended ways. This research has illustrated how such structures in this classroom allowed for young children to engage transmodally in interaction around science and how, in doing so, they drew upon a diversity of resources to make and communicate meaning in science. Open-ended structures for transmodal engagement can afford children's agency in science learning, and it is imperative to work towards such structures, especially for culturally and linguistically diverse students.

References

Ahearn, L. M. (2001). Language and agency. *Annual Review of Anthropology, 30*(1), 109–137.

Bezemer, J., & Kress, G. (2020). Semiotic work in the science classroom. *Cultural Studies of Science Education, 15*(1), 71–74.

Bourdieu, P., & Wacquant, L. (1992). *An invitation to reflexive sociology*. University of Chicago Press.

Bruna, K. R., & Gomez, K. (Eds.). (2009). *The work of language in multicultural classrooms: Talking science, writing science*. Routledge.

Caiman, C., & Lundegård, I. (2018). Young children's imagination in science education and education for sustainability. *Cultural Studies of Science Education, 13*(3), 687–705.

Carspecken, P. (1996). *Critical ethnography in educational research*. Routledge.

Charamba, E. (2022). Leveraging multilingualism to support science education through translanguaging pedagogies. In A. Jakobsson, P. Nygård Larsson, & A. Karlsson (Eds.), *Translanguaging in science education*. Springer.

Council of Europe. (2001). *Common European Framework of Reference for Languages: Learning, teaching, assessment*. Cambridge University Press. https://www.coe.int/en/web/common-european-framework-reference-languages

Council of Europe. (2018). *Common European Framework of Reference for Languages: Learning, teaching, assessment. Companion volume with new descriptors*. Council of Europe. https://www.coe.int/en/web/common-european-frameworkreference-reference-languages

Derry, S. J., Pea, R. D., Barron, B., Engle, R. A., Erickson, F., Goldman, R., Hall, R., Koschmann, T., Lemke, J. L., Sherin, M. G., & Sherin, B. L. (2010). Conducting video research in the learning sciences: Guidance on selection, analysis, technology, and ethics. *The Journal of the Learning Sciences, 19*(1), 3–53.

Gallo-Fox, J., & Scantlebury, K. (2016). Coteaching as professional development for cooperating teachers. *Teaching and Teacher Education, 60*, 191–202.

García, O. (2009). Education, multilingualism and translanguaging in the 21st century. In A. Mohanty, M. Panda, R. Phillipson, & T. Skutnabb-Kangas (Eds.), *Multilingual education for social justice: Globalising the local* (pp. 128–145). Orient Blackswan.

García, O., & Wei, L. (2014). *Translanguaging: Language, bilingualism and education*. Palgrave Macmillan.

Gómez Fernández, R., & Siry, C. (2018). 'Opening up'a science task: An exploration of shifting embodied participation of a multilingual primary student. *International Journal of Science Education, 40*(7), 771–795.

Gómez-Fernández, R. (2019). Translanguaging and equity in groupwork in the science classroom: Adding linguistic and cultural diversity to the equation. *Cultural Studies of Science Education, 14*, 383–391.

Gorges, A. (2019). *Critical perspectives on plurilingual students' identity performance in sustainability education*. Unpublished doctoral dissertation:. The University of Luxembourg.

Hammersley, M. (2006). Ethnography: Problems and prospects. *Ethnography and Education, 1*(1), 3–14.

Horner, B., Selfe, C., & Lockridge, T. (2015). Translinguality, transmodality, and difference: Exploring dispositions and change in language and learning. Faculty Scholarship, 76. https://ir.library.louisville.edu/faculty/76

Hawkins, M. R. (2018). Transmodalities and transnational encounters: Fostering critical cosmopolitan relations. *Applied Linguistics, 39*(1), 55–77.

Hwang, S. W., & Roth, W. M. (2011). *Scientific and mathematical bodies: The interface of culture and mind*. Sense publishing.

Jakobsson, A., Nygård Larsson, P., & Karlsson, A. (eds.). (2022). *Translanguaging in Science Education*. Springer Nature.

Karlsson, A., Nygård Larsson, P., & Jakobsson, A. (2020). The continuity of learning in a translanguaging science classroom. *Cultural Studies of Science Education, 15*(1), 1–25.

Karlsson, A., Nygård Larsson, P., & Jakobsson, A. (2019). Multilingual students' use of translanguaging in science classrooms. *International Journal of Science Education, 41*(15), 2049–2069.

Karlsson, A., Nygård Larsson, P., & Jakobsson, A. (2022). Students' multilingual negotiations of science in third space. In A. Jakobsson, P. Nygård Larsson & A. Karlsson (Eds.), *Translanguaging in science education*. Switzerland: Springer.

Kawalkar, A. (2020). *Transacting inquiry in middle school science classrooms: A study exploring the nature of discourse and a spectrum of outcomes*. Unpublished doctoral dissertation:. Tata Institute of Fundamental Research.

Kincheloe, J. L., & McLaren, P. (2011). Rethinking critical theory and qualitative research. In J. L. Kincheloe, K. Hayes, S. R. Steinberg, & K. G. Tobin (Eds.), *Key works in critical pedagogy: Joe L. Kincheloe* (pp. 285–326). Sense Publishers.

Kumpalainen, K., Lipponen, L., Hilppö, J., & Mikkola, A. (2014). Building on the positive in children's lives: A co-participatory study on the social construction of children's sense of agency. *Early Child Development and Care, 184*(2), 211–229.

Licona, P., & Kelly, G. J. (2022). Translanguaging in middle school science: Written arguments about issues of biodiversity. In A. Jakobsson, P. Nygård Larsson, & A. Karlsson (Eds.), *Translanguaging in science education*. Switzerland: Springer.

Lin, A. M. Y., Wu, Y. (Amy), & Lemke, J. L. (2020). 'It takes a village to research a village': Conversations between Angel Lin and Jay Lemke on contemporary issues in translanguaging. In S. Lau & S. Van Viegen (Eds.), *Plurilingual pedagogies: Critical and creative endeavors for equitable language in education* (pp. 47–74). Springer.
Merz, A. H. (2002). A journey through an emergent design and its path for understanding. *Reflective Practice, 3*(2), 141–152.
Ministère de l'éducation nationale, de l'Enfance et de la Jeunesse [MENJE] (2018). *Les chiffre clés de l'Éducation nationale: statistiques et indicateurs 2016–2017*. Retrieved from : https://men.public.lu/fr/publications/statistiques-etudes/themes-transversaux/16-17-chiffres-cles.html
Moore, E., Evnitskaya, N., & Ramos de Robles, L. (2018). Teaching and learning science in linguistically diverse classrooms. *Cultural Studies of Science Education, 13*(2), 341–352.
Morehouse, R. (2012). *Beginning interpretative inquiry: A step-by-step approach to research and evaluation*. Routledge.
Norris, S. (2004). *Analyzing multimodal interaction: A methodological framework*. Routledge.
Piccardo, E. (2019). "We are all (potential) plurilinguals": Plurilingualism as an overarching, holistic concept. *OLBI Journal, 10*.
Probyn, M. (2019). Pedagogical translanguaging and the construction of science knowledge in a multilingual South African classroom: Challenging monoglossic/post-colonial orthodoxies. *Classroom Discourse, 10*(3–4), 216–236.
Roth, W. M., & Tobin, K. (2004). Coteaching: From praxis to theory. *Teachers and Teaching: Theory and Practice, 10*(2), 161–179.
Roth, W. M. (2005). *Doing qualitative research: Praxis of method*. Brill publishers.
Ryu, M. (2019). Mixing languages for science learning and participation: An examination of Korean-English bilingual learners in an after-school science-learning programme). *International Journal of Science Education, 41(10)*, 1303–1323.
Salloum, S. (2022). Contradictions confronting hybrid spaces for translanguaging in the Lebanese context: A CHAT perspective. In A. Jakobsson, P. Nygård Larsson, & A. Karlsson (Eds.), *Translanguaging in science education*. Switzerland: Springer.
Salloum, S., & BouJaoude, S. (2020). Language in teaching and learning science in diverse Lebanese multilingual classrooms: Interactions and perspectives. *International Journal of Science Education, 42*(14), 2331–2363.
Salloum, S., Siry, C., & Espinet, M. (2020). Examining the complexities of science education in multilingual contexts: Highlighting international perspectives. *International Journal of Science Education, 42*(14), 2285–2289.
Sewell, W. H. (1992). A theory of structure. Duality, agency, and transformation. *American Journal of Sociology, 98*(1), 1–29.
Siry, C. (2011). Emphasizing collaborative practices in learning to teach: Coteaching and cogenerative dialogue in a field-based methods course. *Teaching Education, 22*(1), 91–101.
Siry, C. (2013). Exploring the complexities of children's inquiries in science: Knowledge production through participatory practices. *Research in Science Education, 43*(6), 2407–2430.
Siry, C., & Gorges, A. (2020). Young students' diverse resources for meaning making in science: Learning from multilingual contexts. *International Journal of Science Education, 42*(14), 2364–2386.
Siry, C., & Max, C. (2013). The collective construction of a science unit: Framing curricula as emergent from Kindergarteners' wonderings. *Science Education, 97*(6), 878–902.
Siry, C., Wilmes, S., & Haus, J. (2016). Examining children's agency within participatory structures in primary science investigations. *Learning, Culture and Social Interaction, 10*, 4–16.
Stake, R. E. (2003). Case studies. In N. K. Denzin & Y. S. Lincoln (Eds.), *The landscape of qualitative research. Theories and issues* (pp. 134–164). SAGE.
Suárez, E. (2020). "Estoy Explorando Science": Emergent bilingual students problematizing electrical phenomena through translanguaging. *Science Education, 104*(5), 791–826.

STATEC (National Institute of Statistics and Economic Studies of the Grand Duchy of Luxembourg) (2020). Luxembourg in figures. Retrieved from: https://luxembourg.public.lu/en/publications/statec-le-luxembourg-en-chiffres.html

Tobin, K. (2005). *Improving urban science education. New roles for teachers, students, and researchers*. Rowman & Littlefield Publishers.

Tobin, K., & Ritchie, S. M. (2012). Multi-method, multi-theoretical, multi-level research in the learning sciences. *The Asia-Pacific Education Researcher, 21*(1), 117–129.

Varelas, M. (2018). Dialectical relationships and how they shape (in)equitable science learning spaces and places. In L. Bryan & K. Tobin (Eds.), *13 questions: Reframing education's conversation: Science*. Peter Lang.

Wilmes, S. E. (2017). *Student-driven inquiry-based science in Luxembourg Primary School Contexts*. Dissertation, University of Luxembourg.

Wilmes, S. E., & Siry, C. (2018). Interaction rituals and inquiry-based science instruction: Analysis of student participation in small-group investigations in a multilingual classroom. *Science Education, 102*(5), 1107–1128.

Wilmes, S. E., & Siry, C. (2020). Science notebooks as interactional spaces in a multilingual classroom: Not just ideas on paper. *Journal of Research in Science Teaching, 57*(7), 999–1027.

Wilmes, S. E., & Siry, C. (2021). Multimodal interaction analysis: A powerful tool for examining plurilingual students' engagement in science practices. *Research in Science Education, 51*(1), 71–91.

Wilmes, S.E., Siry, C., Gomez-Fernandez, R., & Gorges, A. (2018). Reconstructing science education within the language I science relationship. *13 Questions: Reframing Education's Conversation: Science*.

Wragg, E. (2012). *An introduction to classroom observation*. Routledge.

Wu, Y., & Lin, A. M. (2019). Translanguaging and trans-semiotising in a CLIL biology class in Hong Kong: Whole-body sense-making in the flow of knowledge co-making. *Classroom Discourse, 10*(3–4), 252–273.

Chapter 5
Translanguaging in the Science Classroom: Drawing on Students' Full Linguistic Repertoires in Bi-/Multilingual Settings

Catherine Lemmi, Greses Pérez, and Bryan A. Brown

Abstract In science classrooms, teachers face the dual responsibility of teaching science and literacy. Educators are asked to deliver content knowledge while developing students' ability to use discipline-specific language found in the Next Generation of Science Standards. In theory, science should be part of the curriculum in US schools. However, in under-resourced schools, science often becomes the context for teaching literacy. Teachers are under pressure to focus on mathematics and English, running the risk of emphasizing language production over conceptual processing. With the challenges educators face in multilingual classrooms, translanguaging offers a lens to investigate interactions and promote scientific thinking. Proposed as a social justice theory and pedagogy, translanguaging offers the possibility to free the child from the limitations of the monolingual settings.

In the chapter, we analyze a vignette from an elementary science class in which a teacher and four students engage in translanguaging when talking about an experiment to determine properties of soils. The episode provides a glimpse on translingual practices within a specific context. We discuss the significance of these acts in terms of their value for the development of scientific notions and linguistic goals. In addition, we provide recommendations for science teachers on how to facilitate opportunities for students to draw on their rich language repertoires when verbalizing their scientific thinking.

Keywords Translanguaging · Science education · Bilingualism · Elementary

C. Lemmi (✉)
California State University, Chico, CA, USA
e-mail: calemmi@csuchico.edu

G. Pérez · B. A. Brown
Stanford University, Stanford, CA, USA
e-mail: greses@stanford.edu; brbrown@stanford.edu

© The Author(s), under exclusive license to Springer Nature Switzerland AG 2021

A. Jakobsson et al. (eds.), *Translanguaging in Science Education*, Sociocultural Explorations of Science Education 27,
https://doi.org/10.1007/978-3-030-82973-5_5

5.1 Introduction

Working for equity in the classroom means teachers ensure that all students get what they need in order to be able to learn (Darling-Hammond, 2015). In science classrooms, this means learning both science content and scientific communication practices (Brown & Ryoo, 2008). For instance, students who are designated as English Learners (ELs) as well as those who have the status of Redesignated Fluent English Proficiency possess a host of language abilities in two or more languages, which are often ignored in schools (García et al., 2008). Recently, we are seeing a shift away from referring to such students as English Learners, a term with a deficit focus, toward phrases like emergent bilingual or simply bilingual, which highlights students' abilities rather than labeling them perpetually deficient (García et al., 2008). Henceforth in this chapter, we will refer to these students as bi-/multilingual learners (Martínez, 2018). In traditional science classrooms, English is the dominant language of instruction. Little to no emphasis is placed on how using multiple language(s) might influence students' understanding of content and engagement in science. Rather than ignoring these realities, effective teachers of multilingual classrooms can leverage student bi-/multilingualism for the purposes of both content learning and language development (García et al., 2017; Lee et al., 2013). While traditional language-immersion programs and learning philosophies dictate that speakers use only the target-language in all classroom interactions, bilingual education has made shifts in the ways that named languages are treated. Some bilingual education scholars promote a social justice agenda to deconstruct the notion of named languages through the theory and pedagogy of translanguaging. Translanguaging is when speakers draw freely from their language repertoires without regard for traditional boundaries between languages (García & Wei, 2014; Otheguy et al., 2015). Increasingly, we are seeing a greater understanding of creating opportunities for students to draw on their language varieties as they learn (García et al., 2017; Gort & Sembiante, 2015; Lemmi et al., 2019). As a linguistic and social practice, translanguaging offers many potential possibilities for bi-/multilingual science students.

Several recent studies demonstrate the role of translanguaging specifically in the science classroom. In a study of 5th grade science classes in a bilingual education program, Poza (2018) found that allowing students to draw on their full repertoires in two languages helped to support their learning of new science content as well as new linguistic forms. Langman (2014) and Palmer et al. (2014) illustrated how bilingual science teachers themselves modeled translanguaging as a resource to help develop the academic content knowledge of middle school students. Translanguaging even promoted students' abilities to relate to the science content and contextualize it within relevant areas of knowledge (Karlsson et al., 2020). Martínez (2018) found evidence of varying ideological approaches to translanguaging among 6th grade bilingual students. Although the practice of translanguaging is gaining attention in linguistics circles, it has yet to become a part of mainstream science education conversations or professional development initiatives, and little is known about whether science teachers are aware of translanguaging or the presence of language ideologies in the classroom. In this chapter, we describe a few examples of how a 4th grade elementary teacher enables translanguaging in her science classroom.

5.2 Theoretical Framework

Humans communicate information through the means of language, whether spoken, signs, or in print form. When conveying meaning through language, individuals rarely stick to the conventions of named languages (e.g., Spanish and English) because in their heads their ways of speaking are not separate or bounded entities. In an effort to categorize the world, society constructs and reifies the boundaries between languages. Named languages such as Spanish and English are not naturally separate or bounded entities (Creese & Blackledge, 2010). For instance, dominant approaches to bilingual education view monolingualism as the norm, and bilingualism as divergent, which is measurably false on a global scale (García, 2009). Further, the traditional approach to bilingualism view the speakers as two monolinguals in one, a person who speaks both languages with native-like proficiency, no accent, and speaks each language separate from the other (García, 2009; Grosjean, 2004; Palmer et al., 2014). In contrast, heteroglossic approaches to language, drawing on the seminal work of Bakhtin, conceptualize languages and bi-/multilingualism as complex, fluid, and dynamic, most effective when speakers draw on their vast repertoire of linguistic abilities (García, 2017). When speakers draw freely from their language repertoires without regard for traditional boundaries between languages, scholars refer this practice as *translanguaging* (Otheguy et al., 2015). Distinct from code-switching, which is a more bounded temporal practice, translanguaging happens in complex ways within an exchange or utterance (García, 2009). In practical terms, some of the ways teachers use translanguaging include modeling the use of multiple languages, positioning students as bilingual even before they are, and celebrating hybrid language practices in the classroom (Palmer et al., 2014).

The education literature in translanguaging has recently provided new frameworks to center the language practices of bi-/multilingual speakers. The literature at the intersection of science education and translanguaging shows how language expectations in schools tend to limit and devalue the linguistic range of minoritized bi-/multilingual students and features attention to the range of interactions between teachers and students around literacy, relegating understanding of scientific ideas often to secondary importance. Some of these studies focus on the tensions in classroom language mixing that emerged from language policies (Probyn, 2009). Others argue in favor of translanguaging as a pedagogy or linguistically responsive approach for social meaning-making and critical awareness of scientific discourses (Infante & Licona, 2018; Poza, 2018). Most of this work has been conducted in K-12 settings, particularly in elementary and high school contexts with a focus on scientific epistemic games (e.g., argumentation, explanations) and rarely on laboratory settings. Limited attention has been given to interactions within science laboratories that highlight the role of translanguaging for both academic and social purposes in the elementary classrooms.

5.3 Research Questions

In this chapter, we answer the following research question:

1. How does translanguaging take place in one 4th grade bilingual science classroom?
2. How do acts of translanguaging serve purposes for the teacher and her students within this bilingual science classroom?

5.4 Methods

In this chapter, we analyze the ways a teacher and her 4th grade students in a bilingual classroom engage in translingual practices as they conduct a soil sample identification laboratory in a science class.

5.4.1 Context

Here, we analyze a transcript of a video that was recorded in a 4th grade classroom at a public Title I school in North Texas. The school has a demographic of predominantly Latino/a/x and Black students, representing 87% of the student body.

The transcript corresponds to a video of 1:19 min. One of the researchers, who also co-taught in the same classroom, had access to the video during the analysis process. Because the media was collected as part of a teacher's reflective practice and not for research purposes, we do not have a detailed breakdown of demographic information on the specific students participating in the experiment. The use of the video and transcript for research purposes was approved by our institutional review board after the class year had ended and limited demographic information was available to the researchers. However, we know that all students identified as Latinx/o/a, Guatamex (both Mexican and Guatemalan), and/or Mexican/Mexican American and have been labeled as Bilingual and/or English Language Learners (ELLs) at the time of the recording. Two teachers, Mrs. Sanchez and Miss Espinal,[1] worked together in the class using a co-teaching model. One of the teachers, Miss Espinal, used both Spanish and English in her interactions with students, while the other

[1] All names are pseudonyms to protect the identity of participants.

teacher, Mrs. Sanchez, only draw on her English resources without mixing languages although she was bilingual. During the clip, students are interacting with the teacher who did not mix languages, Mrs. Sanchez. She identified as Latina and had about 15 years of experience teaching in bilingual settings. Although the teacher in the video did not engage in translingual practices, translanguaging in this classroom was modeled and accepted as a common practice by the co-teacher, Miss Espinal. Both educators were intentional about modeling for their students' sense making through drawing on the linguistic resources that felt more comfortable for them as speakers while also considering their audience. Therefore, despite Mrs. Sanchez not engaging in translingual practices in the video, the teachers constructed a learning environment where messages of acceptance of different ways of speaking and knowing was verbally stated as well as modeled and encouraged.

The context of the study, a science laboratory lesson, provides an interesting look at translingual practices as rarely scholarship centers in this type of settings but on classroom discourse around argumentation and claim-evidence-reasoning. In the video, a group of four students were conducting a lab in which they were trying to identify a soil sample based on its properties, more specifically texture. This lesson is part of a bigger unit that spans for several weeks where students learned about Earth surface, including properties of soils as well as weathering and erosion. All students were randomly assigned to their groups without consideration of language. However, all students participated in a bilingual program at the school district with a focus on using students' language resources for the advancement of English as indicated in the state bilingual dispositions. Although both teachers recognized the importance of learning both languages, none of them subscribe to the idea of using Spanish in advancement of English.

In the lab, students followed a series of steps, such as increasing moisture and moldability of the sample by adding water, forming the soil into a hotdog shape with their hands, stretching it out, and making the sample into a bow shape. Students inferred the properties of the sample based on the way it responds to each of these manipulations (NRCS, n.d.; Thien, 1979). The transcript is 44 talk turns in length and takes place while the students were conducting the portion of the lab in which they are forming the sample into a hotdog shape. The group of four students are working together, and the teacher is standing alongside, supporting them to do their work. One student in the group is holding the soil sample in her hands, and the others are holding materials like a ruler, papers, and pencils.

5.4.2 Data Collection

The data that we analyzed for this chapter consisted of a transcript of the interactions between a teacher, Mrs. Sanchez, and four of her students. The teacher who did not appear in the video transcribed the video and audio recordings. She also reviewed and edited the transcriptions where necessary to match the audio file with the transcribed text and add details about tone of voice and gestures that accompanied the verbal utterances. Using the data from the cleaned transcriptions, researchers translated the Spanish portions into English and included those translations in brackets next to the original text.

5.4.3 Data Analysis

We analyzed our data using conversation analysis (Seedhouse, 2004). We identified and described social purposes accomplished by each utterance from the teacher and her students while providing insights of how speakers interact with one another. At the methodological level, conversation analysis affords us opportunities to analyze both the verbal and non-verbal interactions in connection with social practices within a given contexts (Mondada, 2013). Thus, making it an ideal lens for analyzing instances of translanguaging, especially in the context of science education where practices and discourse play a crucial role.

The authors have background in education and science teaching. Two of them are former secondary science teachers while the third one an elementary bilingual math and science teacher. For each utterance, two researchers reviewed the transcripts and discussed together the meaning and social context of what was being said at each turn of the conversation. Using the contextual knowledge of one researcher, who had participated in the data collection, as well as our own experience, as former teachers, teaching the laboratory being shown in the clip, we identified the main purpose of each turn of talk. In addition, we identified the actions accomplished through language by documenting how the hearers respond to the speakers' moves. Although our lens on translanguaging challenges the notions of boundaries between languages, we also identified the named language(s) being used in each talk turn.

5.5 Findings

Our analysis showed that participants in this classroom engaged in translanguaging for a variety of purposes (see Table 5.1). Out of 44 talk turns, most of the conversation was held in English (35 turns), and the students and teacher frequently used Spanish (9 turns) or a combination of English and Spanish within a talk turn (2 turns). Here, we characterize the ways in which participating students used translanguaging for both social and academic purposes. In the exchanges we studied, we found three main purposes that were accomplished (1) making comparisons between science and lived experiences, (2) asking and answering procedural questions, and (3) using a scientific nomenclature. In the sections, below we describe these exchanges in greater depth.

Our understanding of translanguaging encompasses the use of two or more languages for complex purposes. We do not wish to suggest that translanguaging is only the use of multiple languages within a sentence, but rather, within an utterance

Table 5.1 Analysis of talk turns during a science laboratory activity identifying soil samples. There are differences in how the excerpt is designed in width

		Speech	Action	Language
Turn	Speaker	What did they say?	What task is accomplished?	SP/EN/Both
1	Mrs. Sanchez	When it almost break down, then you know that's it	Explains to students how they will know when they are done making their "worms" out of a soil sample.	EN
2	Luis	Entonces, lo parto. [Then, I'll break it]	Explains what he is going to do to end the procedure.	SP
3	Valeria	Atrevete! [I dare you!]	Challenges previous statement.	SP
4	Angel	Se mira como una salchicha. [It looks like a sausage.]	Offers a comparison between the soil sample and a sausage.	SP
5	Mrs. Sanchez	Oh, that's kind of what you are trying to do	Responds to Angel by affirming his comparison. This statement seems to inspire more comparisons from other students.	EN

or social exchange. Here, we analyze translanguaging between teachers and students to advance specific social and academic purposes. We compare the ways in which languages were used to elucidate the similarities and differences in the social purposes accomplished by the use of each in this particular event. Finally, we also analyze the ways in which translingual interactions were within utterances.

The exchange described in the next section spanned 44 talk turns, 9 of which were said entirely in Spanish and 35 were spoken entirely in English while two were spoken using a combination of both languages within talk turn. Of the talk turns spoken in Spanish, all were said by students. Of the talk turns spoken in English, 15 were said by the teacher and 20 by students. The teacher, Mrs. Sanchez, spoke only English in this discussion, and occasionally she responded to a student's comment in Spanish with an English reply. Of the talk turns spoken in both languages, both were said by students. One of the students, Luis, only spoke Spanish in his interactions with the others, while the rest of the students interacted in both languages with different degrees of dominance between English and Spanish. For instance, Valeria mostly performed language in English with a few instances in Spanish, while Angel did the opposite.

5.5.1 Making a Comparison

One way in which students participated in translanguaging in the class was to make comparisons. Learners identified similarities between laboratory materials and commonly known objects. For example, at the beginning of the discussion, we see the following exchange.

Mrs. Sanchez: When it almost breaks down, then you know that's it.
Luis: Entonces, lo parto

Valeria: Atrevete!
Angel: Se mira como una salchica

In the above exchange, the students are making a worm out of wet soil as part of a soil identification procedure. The teacher, Mrs. Sanchez, is explaining to the students how they will know when they are done making the worm, as her directive, "then you know that's it" indicates completion of the task. The students then engage in an exchange in Spanish among themselves. First, Luis says "*entonces lo parto*" [Then, I'll break it], indicating that he plans to break the worm after this part of the procedure is completed. Valeria, who is in charge of manipulating the soil sample, responds "*Atrevete*!" [I dare you!"] as an indication of her disapproval of this plan and as a challenge to Luis. Then, Angel offers a comparison "*Se mira como una salchicha*" [It looks like a sausage]. In this exchange, the students accomplish two main purposes; they discuss what they would like to do next in the procedure (breaking the soil sample) and they make a comparison between the soil sample and a commonly recognized object. Additionally, there is an element of humor added into their discussion when Luis and Valeria banter about potentially breaking the soil sample after the procedure is completed.

These quotes suggest that even a short exchange in Spanish achieved complex social and academic purposes for the students. The students potentially show to each other that they should wait until the procedure was complete before breaking or playing with the soil sample. Additionally, Angel was able to make a useful comparison between the appearance of the soil sample and an object that the class would likely recognize. The use of humor here advances the purposes of outlining the next steps to follow and making relevant comparisons between lived experiences and science while potentially provides social coherence among the students, which may further advance their ability to work productively together as a group.

5.5.2 Asking a Question

In the clip we studied, students also engaged in translingual interactions when asking questions both for academic and social purposes. However, these hybrid interactions were not centered in the classroom. There are multiple plausible explanations for this situation. In terms of the academic purpose, English may have served as a way to focus the conversation or a mechanism to limit the expression of ideas in non-dominant languages from the perspective of the teacher and/or Valeria. In terms of the social purpose, Angel may want to provide input in a way that it is meaningful for himself and as a result contribute to the intellectual community of his classroom. For example, we see the following exchange.

Mrs. Sanchez: Are you sure you can't stretch it anymore?
Valeria: No
Angel: ¿Puedes hacer un moño?
Mrs. Sanchez: Are you sure?

Angel:	Puedes hacer un moño?
Mrs. Sanchez:	You are not sure?
Mrs. Sanchez:	If you are not sure you can [inaudible].
Valeria:	I don't know
Mrs. Sanchez:	You need to keep trying then
Valeria:	I think that's it.

During these utterances, the teacher is asking the student group whether they are able to stretch the soil sample anymore. The instructions for the laboratory activity indicate that students should stretch the sample as much as possible to measure the length of the 25 grams of soil sample before breaking. The teacher, Mrs. Sanchez, is here implying that the sample can be stretched even more. In response to the teacher's question, Valeria says "no", but Angel asks the same question twice in what may be considered Spanish, "*¿Puedes hacer un moño?*" [Can you make a bow?]. He is asking the group if they should try to make a bow out of the sample without breaking it, which is the next step in the procedure. His contribution through the question seems to be ignored by both the teacher, who responds to the group only in English, as well as the rest of the students, who do not respond to the question posed by Angel. The exchange seems to take place mostly between the teacher and Valeria, who are speaking in what may be considered English. In this instance, the use of English between the teacher and Valeria seems to serve a dual purpose, both academic and social, of keeping the conversation centered on a specific task and content while also setting up the norms of interaction in the group. The social purpose of this interaction could also be viewed as serving to silence the question being posed by Angel, spoken in Spanish. It is important to note that Angel and all students in this study interacted throughout the school year with two teachers who co-taught the lessons, Mrs. Sanchez and Miss Espinal. The former spoke mostly in English in all classroom interactions while the latter model translingual speech frequently. Sanchez typically taught language arts and social studies. Espinal usually led the science and math courses. Although Mrs. Sanchez was leading the lesson included in this chapter, Angel understood the science lesson as opportunities to draw on his linguistic repertoire due to the affordances enable by Espinal in these settings.

5.5.3 Use of a Scientific Term

A third way students engaged in translanguaging was through the use of scientific nomenclature in English, as seen in the exchange transcribed below.

Mrs. Sanchez:	Ok, next step
Valeria:	So, I need to put it there
Angel:	So, ¿no puedes hacer un moño?
Mrs. Sanchez:	Now, do you think she can measure it with the ruler? Why not?
Angel:	Si. En centimeters

During this exchange, the teacher is moving the students on to the next step in the procedure. Angel uses both Spanish and English together in one talk turn twice. In the first instance, the student asks for the fourth time about whether or not Valeria can make a bow with the soil sample. He inquires, "So, *¿no puedes hacer un mono?*" [So, aren't you able to make a bow?]. In this utterance, the only word that can be considered in English is "so". Throughout the clip, he repeats four times the question that was formerly ignored by his classmates and teacher. At the end of the exchange, there is a conversation about how to measure the soil sample. When the teacher inquiries about the possibility to measure the sample with a ruler, Angel says "*Si. En* centimeters" [Yes. In centimeters]. The first two words of this statement may be considered as linguistic chunks in Spanish while the last in English. Throughout the moment capture in the video transcripts, we noticed that Angel engaged primarily in translingual practices to communicate his ideas. He draws on his Spanish resources to make connections with his lived experiences while drawing on his English repertoire to refer to science nomenclature.

Later on, on line 36–44 (see Table 5.1), Angel continues drawing on the scientific nomenclature, centimeters, potentially signaling to others his membership in this community of practice. He may have also attempt to communicate in a way that was valued and understood by his peers. In opposition to the other instances where Angel asked about making a bow with the sample, the moments where he draws on scientific terms engaged his interlocutors in his ideas. The nature of this interaction, the fact that Angel resource to the use of scientific terms, and only accomplished attention from his peers by doing so, may be problematic and complex. Because speakers like Angel, with great bilingual dexterity, have an understanding of who their audience is, they draw on different resources to communicate their message effectively. However, it is unclear to what extent meaning may be sacrifice over effectiveness of communication.

5.6 Discussion

In this chapter, we sought to answer the research questions (1) how translanguaging take place in a 4th grade bilingual science classroom? and (2) how do acts of translanguaging serve purposes for the teacher and her students within this bilingual science classroom? We found that in this one class, translanguaging took place when students were making comparisons, asking questions, and using scientific terms. When making comparisons, the students were able to describe similarities between objects in their science lab (a soil sample) and objects that were commonly known to their peers: a mustache, a rosary, and sausage. This served the social and academic purpose of making the course content more relatable to their peers and perhaps introducing a little humor into the class as well. When asking questions, one student in particular used translanguaging to ask the same question several times, which for him may have served the purpose of attempting to be noticed by his classmates and teacher. When engaging in translanguaging for the purposes of using a scientific term, "centimeters," the student speaker may have accomplished the task of using a word that was commonly understood and used by their classmates in this course. Through the use of scientific nomenclature, the student may also want to signal membership to the practices commonly associated with this community. These varied ways of engaging in translanguaging in a short classroom segment provide a few examples of the purposes served by hybrid language practices in the classroom.

While this study demonstrates several ways that translanguaging can take place in a science classroom, there are several limitations that should be taken into account when considering our findings. Of course, as a small qualitative study our findings are not intended to be generalizable. We illustrate one example of a bi-/multilingual science class and analyze translanguaging events within it. There are many different ways translanguaging can happen that are not represented within our results. Also, we do not have extensive background information on our participants that we would like to have, which could provide additional depth to our analysis. However, our findings add to the field by providing examples of how translanguaging can happen and analysis of the value of such practices for science learning.

This chapter demonstrates the nuanced ways in which students use translanguaging for social and academic purposes in a science classroom. The data suggests that students are aware of the role context and speakers play in their interactions when they are drawing on their language resources. Often, it is assumed that science classes are English-only spaces or that English and scientific language should dominate. Educational scholars should take the responsibility to center traditionally marginalized communities' wealth of knowledge and ways of speaking in our work. The usefulness and value of languages other than English and hybrid language practices has gone largely unexamined, especially within science education research. By demonstrating three purposes accomplished by students via their participation in translanguaging, we show that these linguistic turns serve valuable academic and social purposes in the classroom.

References

Brown, B. A., & Ryoo, K. (2008). Teaching science as a language: A "content-first" approach to science teaching. *Journal of Research in Science Teaching, 45*(5), 529–553.

Creese, A., & Blackledge, A. (2010). Translanguaging in the bilingual classroom: A pedagogy for learning and teaching? *Modern Language Journal, 94*, 103–115.

Darling-Hammond, L. (2015). *The flat world and education: How America's commitment to equity will determine our future*. Teachers College Press.

García, O. (2009). Education, multilingualism and translanguaging in the 21st century. In A. Mohanty, M. Panda, R. Phillipson, & T. Skutnabb-Kangas (Eds.), *Multilingual education for social justice: Globalising the local* (pp. 128–145). Orient Blackswan.

García, O. (2017). Translanguaging in schools: Subiendo y bajando, bajando y subiendo as afterword. *Journal of Language, Identity & Education, 16*(4), 256–263.

García, O., & Wei, L. (2014). *Translanguaging: Language, bilingualism and education*. Palgrave Macmillan.

García, O., Kleifgen, J. A., & Falchi, L. (2008). *From English language learners to emergent bilinguals. Equity matters. Research review no. 1*. Campaign for Educational Equity, Teachers College, Columbia University.

García, O., Johnson, S. I., Seltzer, K., & Valdés, G. (2017). *The translanguaging classroom: Leveraging student bilingualism for learning*. Caslon.

Gort, M., & Sembiante, S. F. (2015). Navigating hybridized language learning spaces through translanguaging pedagogy: Dual language preschool teachers' languaging practices in support of emergent bilingual children's performance of academic discourse. *International Multilingual Research Journal, 9*(1), 7–25.

Grosjean, F. (2004). Studying bilinguals: Methodological and conceptual issues. In T. Bhatia & W. Ritchie (Eds.), *The handbook of bilingualism* (pp. 32–63). Blackwell.

Infante, P., & Licona, P. R. (2018). Translanguaging as pedagogy: Developing learner scientific discursive practices in a bilingual middle school science classroom. *International Journal of Bilingual Education and Bilingualism*. https://doi.org/10.1080/13670050.2018.1526885

Karlsson, A., Larsson Nygård, P., & Jakobsson, A. (2020). The continuity of learning in a translanguaging science classroom. *Cultural Studies of Science Education, 15*(1), 1–25.

Langman, J. (2014). Translanguaging, identity, and learning: Science teachers as engaged language planners. *Language Policy, 13*(2), 183–200.

Lee, O., Quinn, H., & Valdés, G. (2013). Science and language for English language learners in relation to next generation science standards and with implications for common Core state standards for English language arts and mathematics. *Educational Researcher, 42*(4), 223–233.

Lemmi, C., Brown, B. A., Wild, A., Zummo, L., & Sedlacek, Q. (2019). Language ideologies in science education. *Science Education, 103*(4), 854–874.

Martínez, R. A. (2018). Beyond the English learner label: Recognizing the richness of bi/multilingual students' linguistic repertoires. *The Reading Teacher, 71*(5), 515–522.

Mondada, L. (2013). The conversation analytic approach to data collection. In J. Sidnell & T. Stivers (Eds.), *Handbook of conversation analysis* (pp. 32–56). Wiley-Blackwell.

Natural Resources Conservation Service. (n.d.). *Guide to texture by feel*. United States Department of Agriculture. https://www.nrcs.usda.gov/wps/portal/nrcs/detail/soils/edu/?cid=nrcs142p2_054311

Otheguy, R., García, O., & Reid, W. (2015). Clarifying translanguaging and deconstructing named languages: A perspective from linguistics. *Applied Linguistics Review, 6*(3), 281–307.

Palmer, D. K., Martínez, R. A., Mateus, S. G., & Henderson, K. (2014). Reframing the debate on language separation: Toward a vision for translanguaging pedagogies in the dual language classroom. *The Modern Language Journal, 98*(3), 757–772.

Poza, L. E. (2018). The language of ciencia: Translanguaging and learning in a bilingual science classroom. *International Journal of Bilingual Education and Bilingualism, 21*(1), 1–19.

Probyn, M. (2009). 'Smuggling the vernacular into the classroom': Conflicts and tensions in classroom codeswitching in township/rural schools in South Africa. *International Journal of Bilingual Education and Bilingualism, 12*(2), 123–136.

Seedhouse, P. (2004). Different perspectives on language classroom interaction. *Language Learning, 54*(S1), 55–100.

Thien, S. J. (1979). A flow diagram for teaching texture-by-feel analysis. *Journal of Agronomic Education, 8*(1), 54–55.

Chapter 6
The Affordances of Leveraging Multilingual Repertoires in Scientific Reasoning Among Resettled Refugee Teens: Functions of Translanguaging in Scientific Reasoning

Shannon Mary Daniel, Minjung Ryu, Mavreen Rose S. Tuvilla, and Casey Elizabeth Wright

Abstract When learners' full linguistic repertoire is recognized and actively mobilized, their learning experiences can be enriched as they grow disciplinary-specific skills, expand language proficiencies, and gain expertise. Leveraging learners' multilingual repertoires is especially important when doing science, as it requires individuals to engage in oral, written, and semiotic communication to create hypotheses, develop theories, analyze or interpret data, synthesize information, and justify arguments. In this study, we illuminate the affordances of Chin-background refugee teenagers leveraging their full linguistic repertoires as they reason about scientific phenomena in a year-long after-school program focused on impacts of climate changes on human life around the globe. Based on close discourse analysis of four selected events, we show how the participants translanguaged across multiple languages to explain science ideas to one another, discuss cross-cutting concepts such as causal relationships, provide task-related social supports, and ensure that everyone in their group understood and could participate. This study provides specific

S. M. Daniel (✉)
Vanderbilt University, Nashville, TN, USA
e-mail: shannon.m.daniel@vanderbilt.edu

M. Ryu
University of Illinois at Chicago, Chicago, IL, USA
e-mail: mryu@uic.edu

M. R. S. Tuvilla
Texas State University, San Marcos, TX, USA
e-mail: mtuvilla@txstate.edu

C. E. Wright
Purdue University, West Lafayette, IA, USA
e-mail: wrigh401@purdue.edu

© The Author(s), under exclusive license to Springer Nature Switzerland AG 2021

A. Jakobsson et al. (eds.), *Translanguaging in Science Education*, Sociocultural Explorations of Science Education 27,
https://doi.org/10.1007/978-3-030-82973-5_6

examples of the functions of youths' translanguaging practices in relation to their science learning, including attention to peer-scaffolding of scientific sense-making as well as affective supports to enhance engagement. Implications are discussed.

Keywords Translanguaging · Peer scaffolding · Multilingual · Science

Mobilizing one's full linguistic repertoire can enrich learners' experiences as they grow their disciplinary-specific skills and expand their language proficiencies and expertise. In this study, we illuminate the affordances of refugee teenagers leveraging their multiple languages as they reason about scientific phenomena in a year-long after-school program focused on climate science and the potential impacts of climate changes on human life around the globe. After introducing our theoretical bases, the study, and our findings, we provide implications for educators who would like to support multilingual learners in leveraging their linguistic repertoire as they engage in learning.

6.1 Perspectives on Science Learning

In the real world, STEM disciplines are a "highly social and collaborative enterprise" (National Research Council, 2014a, p. 45). Scientists need to communicate with one another about hypotheses, theories, data, and syntheses of data through the use of formal and informal language, tools such as models and representations, and a range of technological products and applications (National Academy of Sciences, 2008, p. 4). Learning science is not just the learning of content or a discrete set of terms and facts; rather, learning science involves the development of skills that can help someone make sense of particular scientific phenomena and inquiries across their lifelong learning trajectories (Coffey et al., 2011; Hammer, 1997). Lee et al. (2013) provide some examples of practices that scientists use not only in thinking about a domain of science (e.g., biology) but also in cross-disciplinary domains, such as communication (important in both biology and English language arts) or mathematical problem-solving, which is important for many fields of science. Some of these skills include argumentation and providing evidence to support one's argument or refute other's perspectives, reason about concepts by drawing upon information from a number of sources, and use a variety of tools to analyze information and communicate ideas with others (Lee et al., 2013).

All of these skills are "language-intensive," requiring students to draw upon their academic and everyday discourses (Lee et al., 2013) and requiring educators to foster such communication in and around science by giving students authority within the learning environment (Engle & Conant, 2002). For multilingual students learning new or additional languages, "the primary process by which learning takes place is *interaction*, more specifically, engagement with other learners and teachers in joint activities that focus on matters of shared interest and that contain opportunities for learning" (Walqui, 2006, p. 159–160). In thinking about learners who are simultaneously learning science and language, then, we take a language-as-practice perspective wherein learners develop both scientific and linguistic skills through communicating with others about core concepts.

A language-as-practice perspective takes a layered view, recognizing that "language isn't a static set of words or phrases to be acquired, but a changing and adaptable way for students and teachers to participate" (Pacheco et al., 2017, p. 65). In developing an environment in which participants learn science and language, educators can afford students opportunities to leverage their full linguistic repertoires while also providing them with support in noticing and using academic language specific to the discipline (Lee et al., 2013; Pacheco et al., 2017). In this paper, we align with Lee et al. (2013) who argue that "an important role of the science teacher is to encourage and support language use *in the service of making sense of science*" (p. 231, emphasis added).

In addition to skills and practices, learning science is deeply intertwined with and informed by learners' affect toward the learning. As Jaber and Hammer (2016) argue, "affect and motivation are not just part of the dynamic of students' learning the material; they are part of the material to learn" (p. 186). This argument is supported by examples of how learners' emotions emerge in relation to the *doing* of science as well as the learning of content: when a student is excited to share a theory of why a phenomenon occurs, their affect is central to the processes of learning and doing science (Jaber & Hammer, 2016). More broadly, sociocultural views of learning that emphasize how learning occurs through interactions with others (Vygotsky, 1978) suggest that the effect of the interlocutors would impact the learning that can occur through such interactions. With this view, affective, interactive, and cognitive processes of learning science are inherently intertwined and each type of process deserves attention in researching how people engage in learning science.

6.2 Multilingual Perspectives on Learning

When people are multilingual, they constantly and strategically draw upon their full linguistic repertoires – their language systems – to accomplish communicative aims with others. This communicative process of translanguaging enables speakers to engage in "fluid practices that go between and *beyond* socially constructed language…to engage diverse students' multiple meaning-making systems and subjectivities" (García & Wei, 2014, p. 3). In the field of education, translanguaging also refers to pedagogies that encourage learners to communicate with others by using all of their linguistic resources. Pedagogically, Garcia and Wei (2015) see several potential purposes of translanguaging, such as helping students make meaning, enriching their understandings and developing their critical thinking, leveraging their background knowledge, growing metalinguistic awareness and skills across languages, and increasing engagement among learners through strengthening their identities and youth positionality.

A rich line of research examines how learners leverage their first or multiple languages while they are learning the target language, and we aim to extend this research to highlight how multilingual learners draw upon multiple languages as they engage in science learning. For example, in a study of youth speaking Bengali and English, Kenner et al. (2008) found that youth deepened their conceptual understandings when they translated poetry from English to Bengali and engaged in mathematical problem-solving. A study on a peer-tutoring program on literacy and vocabulary learning showcased how translanguaging can provide peers with cognitive support (e.g., making sense of the concepts) and social support (e.g., heightening and maintaining motivation) (Martin-Beltrán et al., 2017).

One way to investigate the affordances of translanguaging is to analyze the functions of learners' utterances, while they mobilize their full linguistic repertoires. Dicamilla and Anton (2012) showed that adult language learners drew upon their first language to achieve four functions: discussing content, determining form and meaning of the target language, managing the task, and developing interpersonal relations to "create a friendly social environment" (p. 177). More recently, Martin-Beltrán et al. (2019) showed that elementary-aged learners leveraged their first language to manage tasks, clarify language, negotiate content, build relationships, and check or confirm understanding (in that order of frequency). In a study of translanguaging practices in science learning, Karlsson et al. (2019) found that "multilingual students' use of both first and second languages often appears when they relate and contextualize the abstract content to their everyday experience" (p. 16). For multilingual students, then, the important scientific sense-making practices of connecting everyday knowledge with academic content (Tan et al., 2012) and merging colloquial and academic language (Lemke, 1990) can occur when learners mobilize all of their language resources (Karlsson et al., 2019).

A growing number of studies focus on how people translanguage when they are learning science. Lin and Lo (2017) suggest that students can develop their understandings of scientific patterns when they engage in translanguaging. In their study of a middle school, bilingual science classroom in New York, Kang et al. (2017) found that students who translanguaged were able to elaborate further on their oral and written scientific arguments, use richer persuasive language, and explain key scientific concepts (e.g., convection) by using both Spanish and English in their writing. Also studying a bilingual educational context in the United States, Poza (2018) found that fifth-grade students leveraged their full linguistic repertoire to navigate online research, discuss their inquiry project, complete a worksheet, and present their findings to classmates. Ünsal et al. (2018) studied how third-grade, Turkish-Swedish bilingual students translanguaged in a Swedish school where monolingual instruction is the norm. When students were enabled to work collaboratively in small groups, they translanguaged to discuss word meanings and brainstorm about how to make a light bulb work (Ünsal et al., 2018). In this volume, Charamba (2022) demonstrates how translanguaging helped learners to scaffold the sense-making of chemistry concepts for one another. In an after-school program focused on climate science, youth leveraged Korean and English to construct, reconstruct, and deepen accuracy of scientific understandings – going far beyond direct word-to-word translations across languages and ensuring that all participants could deepen one another's reasoning (Ryu, 2019). Other studies emphasize the role of the educator in supporting learners' scientific sense-making, such as Hattingh et al.' (2022) description of Mr. Mafunda, who guided learners in reviewing concepts and engaging in exploratory talk while translanguaging across isiXhosa and English.

To contribute to this rich and emerging line of research, we seek to understand how multilingual refugee teenagers leveraged their linguistic repertoires as they learned and communicated about ideas around climate science. In particular, we seek to build on prior research (e.g., Dicamilla & Anton, 2012; Martin-Beltrán et al., 2019; Karlsson et al., 2019) that examined the functions or purposes of translanguaging. After briefly introducing the study contexts and methods, we demonstrate the functions and affordances of translanguaging with four events of youth discussion, and conclude with a discussion of how educators can support youth in drawing upon their languages, even when educators do not share fluency in the same languages as the learners.

6.3 Purpose of the Study

Too often, marginalized communities have limited access to high-quality opportunities to engage in STEM-based literacy learning (National Research Council, 2014b). To push back against this problem, our team developed an after-school program to engage resettled refugee youth in developing critical science literacy. Drawing from notions of critical science literacy (Tan et al., 2012), critical multimodal literacy (Luke, 2012), and a view that language (and language learning) is "about social problems" (Block, 2003, p. 89), we define critical science literacy as the dialogic reading and composition of scientific ideas that can transform individuals, communities, and global societies. Luke (2012) argues that critical literacy is "focused on the use of literacy for social justice in marginalized and disenfranchised communities" (p. 5). Critical science literacy emphasizes the possibilities for transforming ways of engaging in science, spaces for doing science, and people's identities of themselves and others as scientists (Tan et al., 2012). Moreover, critical literacy can serve as an important component of English language teaching for multilingual learners by helping language learners recognize that "literacy" can be "a vehicle for social change" (Bacon, 2017, p. 427). Informed by this body of research, we designed and implemented the program with an aim to engage resettled refugee youth in critical science literacy as they learned about climate science and ultimately created videos to share their knowledge with their communities.

This framing guided our curriculum development and implementation for the yearlong after-school program focused on weather, climate, and climate change. Included in our foundational view of critical pedagogies was the importance of expanding ways of doing science in a country (the United States) where English is the dominant language of schooling and science teaching. In designing opportunities for youth to engage in learning climate science, we made intentional efforts to afford youth opportunities to draw upon their Funds of Knowledge (Moll et al., 1992) such as prior experiences living across regions of the world as well as their Funds of Strategies (Daniel, 2018) such as their multiple languages and multimodal literacy skills.

6.4 Context

This study occurred in partnership with a non-profit Burmese community organization in the Midwestern United States. The after-school program took place once each week for 90-minute sessions across the course of the 2017–2018 academic year. Rather than emphasizing scientific facts and the recitation of content among learners, we strove to engage learners in reasoning about science through interpreting and analyzing data, engaging in productive argumentation with one another, and communicating their understandings of climate change through a video they could share with their communities within and beyond the after-school program.

Program design and implementation were guided by tenets of responsive pedagogy, multilingual perspectives on learning, and aims of supporting youth in developing critical science literacy. Responsive pedagogy and curricular design emphasize the importance of attending and responding to student contributions in addition to predetermined canons of knowledge that an educator wishes students to learn (Hammer et al., 2012). In enacting responsive pedagogy, then, an educator would adapt learning goals in response to student thinking and questioning. As Hammer et al. (2012) note, educators must "first engage students in the pursuit, and then support them in their pursuit in ways that afford progress toward canonical practices and ideas" (p. 55). Related to our prioritization of student thinking, we designed the program with a goal to support students in leveraging their full linguistic repertoires.

Recognizing that multilingual youth constantly translanguage without educator prompting (Daniel, 2018), we wished to create a learning environment wherein youth felt comfortable thinking, speaking, and writing while they mobilized their multimodal and multilingual practices. This design decision was also informed by the knowledge that multiple discourses can be used for productive scientific sensemaking (Rosebery et al., 2010) and that learners make, respond to, and create new meaning from reading and composing with multiple modes and languages (Ito et al., 2013; Luke, 2012). In the sessions, youth engaged in activities such as conducting experiments, reading about climate change, researching information on the Internet to answer youth-generated questions, reasoning about science, and creating videos to communicate about climate science. Facilitators continuously emphasized that the after-school program was a multilingual space. For example, in the first session, the main facilitator (Minjung, who is fluent in Korean and English), told youth she had learned how to say a phrase in Chin, which led to a discussion of their language(s). Understanding that macro-scaffolding across a year, or repeatedly and explicitly communicating with students that translanguaging is welcomed in the learning environment (Daniel et al., 2019), facilitators frequently encouraged youth

Table 6.1 Focal participants

Participant (names are pseudonyms)	Age when data was collected	Languages Spoken (in order of comfort or frequency used)	Age upon leaving Chin state; other countries in which they lived	Length of time having lived in the United States
Lin Bo	16	English, Hakha, Zophei	5; Malaysia	~8.5 years
Efraim	17	Falam, Burmese, English	16	~1 year
Kevin		Lauto, Hakha, English, Burmese, Zophei	7 or 8; Malaysia	~3 years
Jon	19	Falam, Burmese, English	16	~1 year
Joshua	17	Zophei, Falam, Hakha, Mizo, Hindi, English	6 or 7; India	3 years

to use any language(s) that would be helpful to supporting their own and one another's engagement in learning.

The youth in this study are from the Chin state in Myanmar, wherein 20–25 languages are typically used (Center for Applied Linguistics, 2007). The teenagers, who were sophomores or juniors at the time of the study, spoke languages including Hakha, Falam, Zophei, Matu, and nationally endorsed Burmese. Focal participants are diverse in terms of their length of time living in the United States, their age, and the languages they use regularly. Though there are seven youth included in the transcripts below, we focus on five participants who are the most active in these transcripts. Of the five participants, Lin Bo had been in the United States for the longest (almost 9 years), and Efraim and Jon had been resettled for about 1 year. Another significant factor in analyzing the youth interactions is that our data in this study are from Weeks 14 and 16 of the program, and Week 14 was the first session that Efraim and Jon participated (with subsequent perfect attendance until the end of the program). This program was an optional, after-school program with 5–20 youth attending each session (Table 6.1).

6.5 Research Methods

This study is part of a larger study for which data collection included: video-recordings of each of the 22 sessions, 82 audio-recordings of small-group interactions, 36 digital recordings of youths' laptop use via screencasts, 36 semi-structured interviews with 16 participants, field notes, and copies of youth-generated artifacts, such as photographs of their posters or recordings of their groups' video productions. Iterative data analysis began during the year of data collection (2016–2017), as we watched videos of the session as a means to inform our implementation of the subsequent session.

More intensive data analysis of half of the recorded sessions (Weeks 1, 2, 3, 4, 5, 10, 11, 14, 16, 17, 18, 24) began in the 2017–2018 academic year. First, we flagged video events (Derry et al., 2010) that helped us identify youth engagement in critical science literacy and labeled events with broad keywords (e.g., multilingual competence, everyday experiences). The research team transcribed most of the key episodes to capture utterances, gesture, body posture, gaze, computer and tool use, organization of artifacts, and use of space (Norris, 2004). Next, we asked a member of the local community to translate the transcripts in which students were speaking languages other than English, and then we re-coded the translated transcripts with closer attention to features of critical science literacy (e.g., transforming scientific discourse) and scientific sense-making around climate concepts. In addition to transcribing selected events and translating utterances across languages to the extent possible, we generated thick descriptions of paraverbal and nonverbal features of interaction. We used the resulting multimodal, multilingual transcripts to identify discursive functions of translanguaging, with particular attention to how it afforded learners' critical STEM literacy practices. (A limitation of this study and area for future discussion is that no member of the research team shared fluency in the languages of Falam or Hakha, Burmese, or Zophei. Thus, we trusted the accuracy of the translated transcripts.)

Data for this study are drawn from small-group activities and discussions during Weeks 14 and 16, at a time in the year when learners grappled with causal relationships between climate change and severe weather events. Youth had read a textbook excerpt on this topic in Week 13, and they were making sense of these ideas together in Weeks 14 and 16. The events from which the data were drawn were flagged with keywords such as multilingual competence, positioning as a language broker, and facilitator norm setting.

6.6 Affordances of Translanguaging for Science Learning

Undergirded by our overlapping framing of science practices, critical science literacy, and translanguaging, our line-by-line analysis resulted in four themes emerging from the data: Translanguaging enabled youth to explain ideas to one another, discuss core skills related to scientific sense-making, provide task-related social supports, and open up the dialogue to more participants in the program. These themes highlight the conceptual and affective benefits of translanguaging. Moreover, our findings reveal that youth spoke across languages even when they did not share an in-depth knowledge of the language with which the other is most comfortable. In other words, several of the excerpts below show one person speaking in Hakha with the other responding in Falam. Because these distinct languages are closely related, youth could comprehend one another's utterances in one language while responding in another language. The complexity of how youth meshed languages – usually Hakha, Falam, and English – goes beyond views of bilingualism and highlights how they were able to draw upon their full linguistic repertoires to propel their scientific reasoning. In other words, while prior studies (e.g., Karlsson et al., 2019; Martin-Beltrán et al., 2019) focus on learners translanguaging across two languages (Arabic and Swedish or Spanish and English, respectively), our study sheds light on how youth drew upon at least three languages.

6.7 Explaining Science Ideas

At the beginning of session 14, youth were asked to review and discuss concepts from a text they read in the prior session about how climate change is impacting weather and society. Because some youth had not attended the prior session, the teenagers were both reviewing the ideas they had read the prior week and explaining what they learned to those who had not attended. In this transcript, Lin Bo and Dachen (all names are pseudonyms) had read the text in the prior week, while Efraim, the third group member, was attending the program for the first time. In this excerpt, they discuss typhoons and their impacts, with Lin Bo striving to help Efraim understand what they had learned in the prior week.

Speaker	**Actual utterance** • Languages uttered are typed in regular font with English in italics, to help make evident the translanguaging that occurred	**English translation, as needed** • Actions relevant to the utterances are denoted in brackets []. Full English translation provided for readers to make sense of learners' discussions
Lin Bo (speaking Hakha)	Si loin *typhoon* cu rili nganmi ah khin aa thok, rili ngan Tak mi ah khin, *where is that*? *Ocean* na thei maw *ocean*?	Typhoon starts from big ocean, very big ocean, where is that..Uhm do you know what ocean is?
Efraim (speaking Falam)	Thei	Yes
Lin Bo (languaging in Hakha and English)	*Ocean* ah khin a thok I khi zeiilo [Efraim nods] Ka hin ocean khi a thok le *typhoon* a ra I mah hi ka hin a kal…kahin a kal le zei pohpoh a tawngh mi kha a rawk dih...a rawk dih I mah cun *flooding* zong a tuah pah, mah bantuk hi *typhoon* a si, *typhoon* cu *tornado* bantuk he a lo ko, Ka hin rili ngantak mi in a thok I mah bantuk ah khin a hrawh dih… Mah hi *typhoon* si	It starts from ocean and that - [Efraim nods] Typhoon starts from here in the ocean [Lin Bo is showing Efraim using pictures from the prior week's text]. And it goes there, and destroyed so many things. It causes flooding, too. Typhoon is similar like tornado. So typhoon starts from big ocean and destroyed something like this [pointing something in the picture]… this is typhoon
Dachen	*Wait, um uh, you say when ice melting sea level have -*	
Lin Bo	*- yeah, sea level rise....*	
Dachen	*Oh.* [leans back, nods] *sea level.*	

(continued)

Speaker	**Actual utterance** • Languages uttered are typed in regular font with English in italics, to help make evident the translanguaging that occurred	**English translation, as needed** • Actions relevant to the utterances are denoted in brackets []. Full English translation provided for readers to make sense of learners' discussions
Lin Bo (languaging in Hakha and English)	*Melting ice* nih a tuah mi na thei ma?	Do you know what melting ice does/causes?
Efraim (speaking Falam)	Oh, Ti a ti ter I ti a luang	Oh, water melts

This group fluidly moves back and forth between Hakha, Falam, and English as they discuss typhoons. Lin Bo explains the ideas from the English-medium text in Hakha, and Efraim responds and contributes to the discussion in Falam. Then, Dachen joins the conversation, in response to Lin Bo's utterance in Hakha, to ask if the melting ice causes the sea level to rise. Lin Bo responds to Dachen in English and then shifts to Hakha to check if Efraim knew what the melting ice causes, to which Efraim responds that the water melts. In addition to the oral utterances across three languages, Lin Bo refers to the text and photographs within the text to help Efraim make sense of the ideas.

It is hard for us to know what Efraim's last utterance of "water melts" meant; he could have been making a causal connection between ice melting and water rising or he could have been repeating the English phrase that Lin Bo had used in his prior utterance in Hakha. Translation often requires interpretation and negotiation when word-for-word equivalence from one language to another may not be possible (Kenner et al., 2008). Moreover, translanguaging often involves the conscious borrowing of words (Grosjean, 1985) in ways that are relevant to the specific context and interlocutors (Canagarajah, 2006).

Of their own volition, the youth leveraged three languages – mostly Hakha – to communicate about typhoons. In this transcript, Lin Bo speaks in Hakha to support Efraim, while he also borrows English words specific to the topic (e.g., ocean, tornado, typhoon, flooding melting ice). Our findings echo Karlsson et al.' (2019) point that "subject-specific words are primarily expressed in the second language, while the more descriptive everyday phrases are expressed through the first language" (p. 11). Responses to utterances were often in a different language (e.g., Hakha from Lin Bo led to a Falam response from Efraim; Lin Bo continued in Hakha and then Dachen responded in English), thus expanding our understanding of how youth mobilize multiple languages within one meaning-making conversation.

6.8 Discussing Cross-Cutting Concepts

Later in that same session, the facilitator prompted the groups to make causal maps that explain the relationship between climate change and extreme weather. The task was for each group to represent causal relationships on a poster paper, and youth could use markers, paper, the text from the prior week, and laptops to conduct research as they worked toward this goal. The same group – Lin Bo, Dachen, and Efraim – collaborated to create a causal map on their poster, which they later shared with the larger group. While working on the task, the three teenagers began thinking about how to represent causation on their poster. After Lin Bo suggested that they draw a circle map as part of their representation, he paused and asked Efraim if he understood causal relationships.

Speaker	**Actual utterance** • Languages uttered are typed in regular font with English in italics, to help make evident the translanguaging that occurred	**English translation, as needed** • Actions relevant to the utterances are denoted in brackets []. Full English translation provided for readers to make sense of learners' discussions
Lin Bo (speaking Hakha)	*Cause and effect* timi na thei ma?	Do you know what is cause and effect?
Efraim (speaking Falam)	Thei lo	No, I do not
Lin Bo (speaking Hakha)	*Alright, alright...* *Cause and effect* timi cu mah hi si hi. Bianabia ah hin pakhatpa ka vuak, ka vuak I, zeidah aa cang lai?	[writes "cause."] This is cause and effect For example, I hit/punch someone, and what will happen to him?
Efraim	Mmm	
Lin Bo (speaking Hakha)	*Cause* cu keimah nih ka vuak, mah kha cause si *Effect* cu zei -	I hit/punch someone and that is cause. [Efraim nods] Effect is -
Thuya (speaking Hakha)	A chuakpi mi	It's the consequence
Lin Bo (speaking Hakha)	Aw a chuakpi mi khi *effect* si cuh. Na theithiam ma? Tu kan tuahmi hi cu *weather* Kong si cuh. A tu hi *climate change* timi cu a lin tuk cuh. A linh tuk ruang ah khin zeidah a chuak kho, a linhtuk ah cun zeidah a chuak kho	Yes, it is the consequence of what happened to him. Do you get it? What we are doing right now is about weather. What we are learning here of climate change means it is getting hotter. Because it is very hot, what can happened? What can happened if it is very hot?

In this transcript, Lin Bo and Thuya (a youth who was working with another group but sitting nearby) explained the concept of cause and effect to Efraim. Because identifying and hypothesizing about causal relationships are cross-cutting concepts for scientific reasoning within the Next Generation Science Standards (2013) and a core skill across the disciplines, Lin Bo's explanation was important for supporting Efraim's sense-making in this and future tasks. We note that Lin Bo opened this exchange by checking to see if Efraim knew the idea already, and explained only after Efraim indicated his need, thus Lin Bo was being responsive to Efraim's participation. Moreover, Lin Bo reviewed "cause and effect" using an everyday situation rather than a science-specific example, thereby helping his peer make meaning (Garcia & Wei,

2015). In this case, Lin Bo created a "zone of relevance" for Efraim to make sense of the crosscutting concept of "cause and effect" (Martin-Beltrán et al., 2017). As seen previously, Lin Bo leverages Hakha and English in his utterances, and Efraim is able to comprehend and respond using Falam and English. This exchange exemplifies how translanguaging affords multilingual youth the opportunity to connect everyday experiences with their science learning.

6.9 Providing Task-Related Social Supports

As the team worked on the task, the facilitator handed Efraim a laptop, and he immediately handed it to Lin Bo. Then, Lin Bo put the laptop in front of Efraim, saying, "C'mon, Bro. You got this, man. You are a (sic) engineer!" Efraim laughs and says, "Okay," and the following exchange occurs.

Speaker	**Actual utterance** • Languages uttered are typed in regular font with English in italics, to help make evident the translanguaging that occurred	**English translation, as needed** • Actions relevant to the utterances are denoted in brackets []. Full English translation provided for readers to make sense of learners' discussions
Efraim (speaking in Falam)	Ka ngan thiam lo	I don't know how to write
Lin Bo (speaking in Hakha)	Na thiam ko lai, na thiam ko lai. *Play with it, play with it.* Na thiam ko lai, zeihmanh a si lo na thiam ko lai. Mah ruang hin, hi zong hi na tuah hau. Na tuah thiam lo mi zong khi na tuah ko lai, kan nunnak si cuh. Nunnak ah thiam na duh ah cun mah hi hi na tuah hau. Na tuah lo cun na thiam kho lai lo	You can do it, you can do it. You can do it; it's not that difficult. [Efraim nods.] Because of that, you have to do this one. You will have to do things that you don't know how to do. That is life. In life, if you want to be good, you have to do it. If you don't try/do it, you will not be able to be good at that

In response to Efraim's utterance in Falam, Lin Bo used Hakha and English to build Efraim's confidence and help Efraim to view the task as a challenge that can help him learn. Lin Bo's encouragement began in English "C'mon bro!" and "You are a engineer!" and Efraim responded by expressing his uncertainty about what to write in Falam. This utterance launched Lin Bo's more elaborative support in Hakha.

Though this transcript is more focused on social support than on a specific scientific concept or skill, we see Lin Bo positioning Efraim as an "engineer" who can use the computer to engage productively in the scientific research task. As Lin Bo communicates with Efraim across languages, they engage in important identity and positioning work, which is a common purpose of translanguaging within learning environments (Garcia & Wei, 2015). This interaction shows critical science literacy at work as Lin Bo positions Efraim as an engineer, thus beginning to enact the goal of transforming identities (Tan et al., 2012). Similar to findings in a prior study of multilingual youth (Martin-Beltrán et al., 2017), Lin Bo mobilized his linguistic repertoire to "encourage further participation in the activities" (p. 6).

6.10 Making Sure Everyone Understands

In this event, which occurred in Week 16, the facilitator walked up to a group of four youth who were reviewing and discussing what they had learned in the previous session. The facilitator pointed to what Kevin had written on his poster and asked him to explain his idea to everyone in the group. Kevin responded,

> Our question was, 'Why is the earth's temperature rising?' …So we said the Earth's temperature is rising because the heat from the sun, heat hits the earth…and therefore, the remaining heats add up each day, so the Earth temperature is rising.

Next, the facilitator said, "Okay, and your job is to make sure everyone understands what you said. You can speak in English or any other language." From here, Kevin and the other three teens engage in discussion leveraging their multiple languages:

Speaker	**Actual utterance** • Languages uttered are typed in regular font with English in italics, to help make evident the translanguaging that occurred	**English translation, as needed** • Actions relevant to the utterances are denoted in brackets []. Full English translation provided for readers to make sense of learners' discussions
Kevin (speaking in Hakha)	Na theithiam pah ko maw tlawmpal cu?	Do you understand a little bit?
Jon (speaking in Hakha)	Aw tlawmpal	A little bit
Kevin (speaking in Falam)	Na theithiam ko maw?	Do you understand?
Jon (speaking in Falam)	Ka theithiam lo	I do not understand
Kevin (speaking in Falam)	Oh, *okay*. Kan hnu lam ah kan tuah mi kha	What we did last time was
Joshua (speaking in Hakha)	Lai holh in holh ko	Just say it in Hakha. [laughs]
Kevin (speaking in Hakha)	Tlawmpal te theithiam deuh seh ka ti ve cuh	I want him to understand a little better
Joshua (speaking in Hakha)	Ah *okay*. A hau lo cuh	Ah okay. [laughs] You don't need to

In this conversation, youth leveraged languaging skills in English, Falam, and Hakha. At first, Kevin asked Jon if he understood using Hakha, which is the language Kevin uses more frequently. Then, he asked again in Falam. Jon did adjust his response from "a little bit" in Hakha to "I do not understand" in Falam, which could indicate that Jon felt more comfortable sharing his understanding when Kevin made the additional effort to repeat the question and attempt to use Falam, the language Jon used most frequently. Next, Joshua laughed and told Kevin that they could comprehend Hakha easily, even though Jon more frequently uses Falam. In his attempt to help his peers, Kevin showed multilingual dexterity as he transitioned from predominantly using English in response to the facilitator to using Hakha and Falam to ensure his group members could join the conversation. Also, Joshua's suggestions that Kevin did not "need to" use Falam provides insight for us into Joshua's knowledge about the multiple languages, his peers' expertise, and the communicational needs within and for the purposes of this interaction.

In this event, we emphasize two features of the discourse. First, Kevin and Joshua show great linguistic flexibility in trying to support one another and Jon. Second, this elaboration occurred in response to the facilitator's cue to help all participants understand using any language. Noticeable is Kevin's shift from English to the translingual use of Hakha and Falam immediately after the facilitator explicitly suggested that the purpose was to make sure everyone in the group understood the ideas. In this episode, the facilitator scaffolded tasks to reiterate their emphasis that the after-school program was a multilingual environment wherein translanguaging was welcomed; such scaffolding can be necessary to help learners see that translanguaging can be a norm in a learning environment (Daniel et al., 2019). We expand upon the potential roles of facilitators in our implications.

6.11 Reflections from the Youth

Data from an interview with Lin Bo at the end of the program provides another angle of interpretation to our analysis. His comments provided insight on his language mobilization, friendship formation, support of peers, and feelings about science in school versus in this program. On the topic of language mobilization, Lin Bo recognized the complexity of being a newcomer, recalling how he had to "adapt" to learn English in the United States while he simultaneously felt, "I gotta keep my culture alive, my language alive...in order to help...I gotta speak my own language." Overall, Lin Bo felt "so smart" in this after-school program, which he said was "way different" from how he felt in his science classes at school.

In reflecting on video snippets presented in this paper, Lin Bo shared that he had chosen to befriend Dachen's because "[Dachen] didn't know anything about America at all, so I remember the first time coming to America, too, and how it is very lonely if you have no one to talk to, and I was trying to make jokes…" When Efraim joined the program, Lin Bo recalled, "I believe he spoke a little bit of my language, but he understood [Dachen], and so I work with [Dachen] so I could tell him what he was doing and so [Dachen] can translate back to him." Lin Bo reflected that "I liked explaining to them" during that session. Lin Bo's reflections corroborate our perceptions of the affordances of translanguaging in science learning – he enjoyed helping his peers understand ideas, he acknowledged the need for the group to mobilize three languages in order to encourage full participation from all group members, and he felt "smart" in science during these after-school sessions.

6.12 Discussion

Overall, the youth drew upon their full linguistic repertoire for several conversational purposes that were essential in propelling them forward in achieving the tasks and furthering their scientific understandings. The use of multiple languages enabled youth to engage one another and participate more deeply in the tasks in various ways. The teens helped one another understand topics and cross-cutting concepts by providing examples and explanations that were relevant to their everyday lives and using language practices that they use regularly outside of formal school settings.

This study provides specific examples of the functions of youths' translanguaging practices in relation to their science learning. While prior research has outlined more general purposes for translanguaging pedagogies (e.g., Garcia & Wei, 2015), our work here aims to highlight how teens' leveraging of their multilingual repertoires propelled their engagement in discussions and tasks related to climate science. For example, prior researchers (Garcia & Wei, 2015; Karlsson et al., 2019) suggested that when youth translanguage, they can leverage their background knowledge. Our data provides examples of Lin Bo supporting his peers by explaining science ideas such as typhoons and cross-cutting concepts such as causal relationships. Similar to Martin-Beltrán et al. (2017) and the findings from Ünsal et al. (2018), Lin Bo was able to create a "zone of relevance" for his group mates in ways that the facilitators alone may not have been able to achieve in a whole-group or whole-class discussion. The four excerpts here highlight four functions of translanguaging for science learning: explaining science ideas, discussing cross-cutting concepts, providing social supports, and including everyone by making sure each group member understands. While prior research discusses functions of translanguaging for language and literacy learning (Dicamilla and Anton; Martin-Beltrán et al., 2019), our study extends this body of work to discuss how the functions of translanguaging can be disciplinary-specific in science learning.

Tying back to Jaber and Hammer (2016) who argue that affect and motivation are a part of science learning, we emphasize the value of the affective dimensions achieved when youth translanguaged. Lin Bo encouraged Efraim to use the laptop by speaking in both English and Hakha. When Kevin shifted his attention and interaction from primarily focusing on the facilitator toward primarily focusing on his peers, he opened up opportunities for participation and sense-making among his group members Jon and Joshua. Lin Bo stating that feeling "smart" in this afterschool program in contrast to school-based science also provides insight into the types of identity transformation that can occur when youth engage in translanguaging, which we view as a part of critical science literacy (Tan et al., 2012) for multilingual youth. As Garcia and Wei (2015) suggest, the leveraging of teens' multilingual repertoires heightened engagement among more members of the program.

In contrast to prior studies that examined dual language learners (e.g., Arabic/Swedish speakers or Spanish/English speakers), our study adds a layer of linguistic complexity and fluidity of youth translanguaging, as they communicate across three or more languages. Youth were listening and comprehending input in one language (e.g., Hakha) while responding and producing output in another language (e.g., Falam) to discuss concepts and tasks that had been presented to them in English. Moreover, in continuing a thread from recent research (e.g., Tuvilla et al., 2018; Poza, 2018; Ünsal et al., 2018), we present findings that speak to how and why youth leverage their multiple languages in the pursuit of science learning. In this study, youth provided both affective support and scaffolded scientific tasks for one another through the explanation and discussion of key disciplinary concepts.

Three of the transcripts highlight how Lin Bo was supporting Efraim's scientific engagement. Borrowing from Lave and Wenger (1991), we see Lin Bo as the oldtimer in this particular Community of Practice, while Efraim was a newcomer. Duff (2007) particularly points to how language socialization can occur among multilingual learners, and that a Community of Practice framework can elucidate questions of how newcomers learn oral and written practices through interactions with oldtimers. In this study, we can see how Lin Bo and Efraim are an oldtimer and newcomer, respectively, in two timescales and layers of community: Lin Bo had been in the United States for almost 9 years and had participated in the after-school program for several weeks, while Efraim had been in the United States for about 1 year and was engaging in his first session of the program in the first three transcripts presented in this study. Lin Bo's reflections in the interview also indicate that he felt he could support newcomers (Dachen when he was new to the United States and Efraim when he was new to this program). Though we did not analyze the data using a Community of Practice framework, we can see opportunities for future research to investigate how translanguaging practices might be used differentially across learners who are oldtimers and/or newcomers in various learning environments.

Because youth in our study sometimes began their conversations in English and then shifted to broader linguistic repertoires in their conversations, we believe that the deepening of scientific sense-making and the social supports that furthered science learning could not have been achieved at the same level if youth were coerced into using English only. We echo Poza (2018) who argues that "fomenting spaces of collaboration" in which translanguaging is a "meaning-making process" and "normative behavior" is essential to multilingual students' support of one another's learning through the leveraging of all the communicative practices they use in their everyday lives (p. 15). Though prior studies have demonstrated how youth leverage their multilingualism to accomplish academic tasks across the disciplines (Daniel, 2018; Daniel & Zybina, 2019), this study emphasizes how translingual practices enabled participants to engage in scientific sense-making specifically in small-group interactions focused on climate science.

Our data mostly highlight how the teenagers' translanguaging afforded them opportunities to deepen their engagement and participation in their scientific pursuits in this after-school program. However, we note that educators have a key role in creating learning environments wherein youth feel comfortable drawing upon languages other than English. As Duran argues, "multilingual spaces reinforce agency and multilingualism" (2015, p. 78), and it is time for educators to "*proactively* include a multilingual approach in our classroom practice" (2016, p. 226). In the next section, we draw from our data and prior literature to provide suggestions for how educators can make such efforts.

6.13 Educational Implications for Productive Translanguaging

The last transcript in this paper provides an example of how the facilitator shifted the directionality of discourse from occurring between her and Kevin toward encouraging all teens in the group to discuss concepts together. This move of emphasizing youth ownership of learning is essential for them to engage in scientific reasoning (Lee et al., 2013). In addition to emphasizing the goal of making sure "everyone understands," the facilitator explicitly suggested that learners use any language they choose.

Our findings suggest that consistently and explicitly encouraging students to draw upon multiple languages helped to develop the after-school program as a multilingual learning environment where drawing upon individual and collective multilingual repertoires was a norm.

In their discussion of mathematics learning, Yackel and Cobb (1996) suggest that educators must develop norms that create a "taken-as-shared sense of when [and how] it is appropriate to contribute to a discussion" and "what counts as acceptable" (p. 461). Similarly, we found that explicit suggestions and subsequent affirmations of translanguaging helped youth deepen their scientific reasoning by leveraging their full linguistic repertoires.

Our findings are consistent with prior research on developing multilingual learning environments. Daniel et al. (2019) suggest that consistently scaffolding tasks with attention to translanguaging, helping learners to recognize translanguaging as a learning strategy, and asking open-ended questions that require youth to negotiate for meaning can help youth recognize that translanguaging practices are welcomed in classrooms or other learning environments. Questions without specifically correct answers also make room for responsive science teaching (Hammer, 1997) and the types of rich discourse that are necessary for scientific inquiry and reasoning (Lee et al., 2013). In their study of translanguaging, Martin-Beltrán et al. (2019) pointed out that "tasks that [are] highly structured, didactic, and [following] initiate-respond-evaluate discourse patterns...constrained students' productive engagement with language" (p. 15). In contrast, the tasks that learners in this study were accomplishing were more open-ended, thus allowing for students to discuss complex ideas and mobilize their full linguistic repertoire.

Poza (2018) emphasized that teachers' "explicit support of students' linguistic flexibility" (p. 8) enabled youth to translanguage productively in science. The teacher in Poza's (2018) study modeled code switching when leading discussions and engaged students' in metalinguistic analyses of one another's video productions. Ünsal et al. (2018) point out that students rarely engaged in scientific reasoning or translanguaging during whole-group discussions employing Initiate-Respond-Evaluate discourse patterns. In contrast, when the "teacher was not in control of the dialogue," they observed "constant and fluid change between Turkish and Swedish" during small-group activities (p. 1046). Similarly, our research shows that having a shared understanding that each person is multilingual and we are living in a multilingual world helped in encouraging youth to collaborate on tasks that require scientific sense-making (e.g., interpreting data, explaining causal relationships, making predictions). Though detailed analysis of the learning environment and curriculum of the program are beyond the scope of this paper, we recognize that the principles of responsiveness, critical science literacy, and a value of multilingualism led to a youth-centered learning environment that afforded youth to translanguage as they engaged in science learning.

6.14 Conclusion

Findings in this study show that opportunities for learners to leverage their full linguistic repertoire can broaden their engagement in scientific reasoning and emphasize the strategies multilingual learners bring to their scientific sense-making. Thus, we recommend that science educators in a variety of contexts (schools, after-school and community programs, museums, and other learning environments) can encourage learners explicitly to leverage multilingual practices as they make sense of scientific concepts and phenomena.

References

Bacon, C. K. (2017). Multilanguage, multipurpose: A literature review, synthesis, and framework for critical literacies in English language teaching. *Journal of Literacy Research, 49*(3), 424–453. https://doi.org/10.1177/1086296X17718324

Block, D. (2003). *The social turn in second language acquisition*. Georgetown University Press.

Canagarajah, S. (2006). Negotiating the local in English as a lingua franca. *Annual Review of Applied Linguistics, 26*, 197–218.

Center for Applied Linguistics. (2007). *Refugees from Burma: Their backgrounds and refugee experiences*. Cultural Orientation Resource Center.

Charamba, E. (2022). Leveraging multilingualism to support science education through translanguaging pedagogies. In A. Jakobsson, P. Nygård Larsson, & A. Karlsson (Eds.), *Translanguaging in science education*. Springer.

Coffey, J., Hammer, D., Levin, D., & Grant, T. (2011). The missing disciplinary substance of formative assessment. *Journal of Research in Science Teaching, 48*(10), 1109–1136.

Daniel, S. (2018). How resettled refugee youth leverage their out-of-school literacy practices to accomplish schoolwork. *Mind, Culture, and Activity, 25*(3), 263–277.

Daniel, S. M., Jiménez, R. T., Pray, L., & Pacheco, M. B. (2019). Scaffolding to make translanguaging a classroom norm. *TESOL Journal, 10*(1), e00361.

Daniel, S. M., & Zybina, M. (2019). Resettled refugee teens' perspectives: Identifying a need to centralize youths'"funds of strategies" in future efforts to enact culturally responsive pedagogy. *The Urban Review, 51*(3), 345–368.

Derry, S., Pea, R., Barron, B., Engle, R., Erickson, F., Goldman, R., Hall, R., Koschmann, T., Lemke, J., Sherin, M., & Sherin, B. (2010). Conducting video research in the learning sciences: Guidance on selection, analysis, technology, and ethics. *The Journal of the Learning Sciences, 19*(1), 3–53.

DiCamilla, F. J., & Antón, M. (2012). Functions of L1 in the collaborative interaction of beginning and advanced second language learners. *International Journal of Applied Linguistics, 22*(2), 160–188.

Duff, P. (2007). Second language socialization as sociocultural theory: Insights and issues. *Language Teaching, 40*(4), 309–319.

Duran, C. (2015). Theorizing young language learner agency through the lens of multilingual repertoire: A sociocultural perspective. In P. Deters, X. Gao, E. Miller, & G. Vitanova (Eds.), *Theorizing and analyzing agency in second language learning: Interdisciplinary approaches* (pp. 73–90). Multilingual Matters.

Duran, C. S. (2016). "I want to do things with languages": A male Karenni refugee's reconstructing multilingual capital. *Journal of Language, Identity & Education, 15*(4), 216–229.

Engle, R. A., & Conant, F. R. (2002). Guiding principles for fostering productive disciplinary engagement: Explaining an emergent argument in a community of learners classroom. *Cognition and Instruction, 20*(4), 399–483.

García, O., & Wei, L. (2014). *Translanguaging: Language, bilingualism and education*. Palgrave Macmillan.

García, O., & Wei, L. (2015). Translanguaging, bilingualism, and bilingual education. In W. E. Wright, S. Boun, & O. García (Eds.), *The handbook of bilingual and multilingual education* (pp. 223–240). Wiley Blackwell.

Grosjean, F. (1985). The bilingual as a competent but specific speaker-hearer. *Journal of Multilingual and Multicultural Development, 6*(6), 467–477.

Hammer, D. (1997). Discovery learning and discovery teaching. *Cognition and Instruction, 15*(4), 485–529.

Hammer, D., Goldberg, F., & Fargason, S. (2012). Responsive teaching and the beginnings of energy in a third grade classroom. *Review of Science, Mathematics and ICT Education, 6*(1), 51–72.

Hattingh, A., McKinney, C., Msimanga, A., Probyn, M., & Tyler, R. (2022). Translanguaging in science education in South African classrooms: Challenging constraining ideologies for science teacher education. In A. Jakobsson, P. Nygård Larsson, & A. Karlsson (Eds.), *Translanguaging in science education*. Springer.

Jaber, L. Z., & Hammer, D. (2016). Learning to feel like a scientist. *Science Education, 100*(2), 189–220. https://doi.org/10.1002/sce.21202

Kang, E. J., Swanson, L. H., & Bauler, C. V. (2017). "Explicame": Examining emergent bilinguals' ability to construct arguments and explanations during a unit on plate tectonics. *Electronic Journal of Science Education, 21*(6), 12–45.

Karlsson, A., Nygård Larsson, P., & Jakobsson, A. (2019). Multilingual students' use of translanguaging in science classrooms. *International Journal of Science Education, 41*(15), 2049–2069.

Kenner, C., Gregory, E., Ruby, M., & Al-Azami, S. (2008). Bilingual learning for second and third generation children. *Language, Culture and Curriculum, 21*(2), 120–137.

Lave, J., & Wenger, E. (1991). *Situated learning: Legitimate peripheral participation*. Cambridge, UK: Cambridge University Press.

Lee, O., Quinn, H., & Valdés, G. (2013). Science and language for English language learners in relation to next generation science standards and with implications for common core state standards for English language arts and mathematics. *Educational Researcher, 42*(4), 223–233.

Lemke, J. (1990). *Talking Science: Language, Learning, and Values*. Norwood, NJ: Ablex.

Lin, A. M., & Lo, Y. Y. (2017). Trans/languaging and the triadic dialogue in content and language integrated learning (CLIL) classrooms. *Language and Education, 31*(1), 26–45.
Luke, A. (2012). Critical literacy: Foundational notes. *Theory Into Practice, 51*(1), 4–11.
Ito, M., Gutiérrez, K., Livingstone, S., Penuel, B., Rhodes, J., Salen, K., Schor, J., Sefton-Green, J., & Watkins, S. (2013). *Connected learning: An agenda for research and design.* Digital Media and Learning Research Hub.
Martin-Beltrán, M., Daniel, S., Peercy, M., & Silverman, R. (2017). Developing a zone of relevance: Emergent bilinguals' use of social, linguistic, and cognitive support in peer-led literacy discussions. *International Multilingual Research Journal, 11*(3), 152–166.
Martin-Beltrán, M., Montoya-Ávila, A., García, A. A., Peercy, M. M., & Silverman, R. (2019). 'Time for una pregunta': Understanding Spanish use and interlocutor response among young English learners in cross-age peer interactions while reading and discussing text. *International Journal of Bilingual Education and Bilingualism, 22*(1), 17–34.
Moll, L., Amanti, C., Neff, D., & Gonzalez, N. (1992). Funds of knowledge for teaching: Using a qualitative approach to connect homes and classrooms. *Theory Into Practice, 31*(2), 132–141.
National Research Council. (2014a). *STEM integration in K-12 education: Status, prospects, and an agenda for research.* The National Academies Press.
National Research Council. (2014b). *STEM learning is everywhere: Summary of a convocation on building learning systems.* The National Academies Press.
NGSS Lead States. (2013). *Next generation science standards: For States, by States.* The National Academies Press.
Norris, S. (2004). *Analyzing multimodal interaction: A methodological framework.* Routledge.
Pacheco, M. B., Daniel, S. M., & Pray, L. C. (2017). Scaffolding practice: Supporting emerging bilinguals' academic language use in two classroom communities. *Language Arts, 95*(2), 63–76.
Poza, L. E. (2018). The language of ciencia: Translanguaging and learning in a bilingual science classroom. *International Journal of Bilingual Education and Bilingualism, 21*(1), 1–19.
Rosebery, A. S., Ogonowski, M., DiSchino, M., & Warren, B. (2010). "The coat traps all your body heat": Heterogeneity as fundamental to learning. *Journal of the Learning Sciences, 19*(3), 322–357.
Ryu, M. (2019). Mixing languages for science learning and participation: An examination of Korean-English bilingual learners in an after-school science-learning programme. *International Journal of Science Education, 41*(10), 1303–1323.
Tan, E., Barton, A. C., Turner, E., & Gutiérrez, M. V. (2012). *Empowering science and mathematics education in urban schools.* University of Chicago Press.
Tuvilla, M., Wright, C., Ryu, M., & Daniel, S. (2018). *How do multilingual learnners support one another's science learning and participation?* In The 2018 proceedings of the international conference of the learning sciences, pp. 1761–1762. London, UK.
Ünsal, Z., Jakobson, B., Molander, B., & Wickman, P. (2018). Language use in a multilingual class: A study of the relation between bilingual students' languages and their meaning-making in science. *Research in Science Education, 48*(5), 1027–1048.
Vygotsky, L. S. (1978). *Mind in society: The development of higher psychological processes.* Harvard university press.
Walqui, A. (2006). Scaffolding instruction for English language learners: A conceptual framework. *International Journal of Bilingual Education and Bilingualism, 9*(2), 159–180.
Yackel, E., & Cobb, P. (1996). Sociomathematical norms, argumentation, and autonomy in mathematics. *Journal of Research in Mathematics Education, 27*(4), 458–477.

Chapter 7
Students' Multilingual Negotiations of Science in Third Space

Annika Karlsson, Pia Nygård Larsson, and Anders Jakobsson

Abstract This chapter focuses on the complexity of learning processes in a translanguaging science classroom (TSC). We explore multilingual students' use of their first and second languages in authentic meaning-making in translanguaging situations in a middle school in Sweden. In the analysis, we interpret these classroom situations as multilingual hybrid spaces, in which both content and languages are simultaneously negotiated in order to create meaning. The aim is to investigate how these situations may contribute to the development of students' conceptual knowledge and language use in science. The negotiations are illustrated as movements in *multilingual discursive loops*, which includes immediate and dynamic movements between "national languages" (Swedish and Arabic) and different discourses (every-day and scientific) with support from multimodal resources. We describe these movements in a model and conclude that the students' language and conceptual development is largely enabled by opening up the multilingual negotiation spaces that constitute a TSC.

Keywords Language and conceptual development · Multilingual discursive loops · Science education · Subject-specific language · Translanguaging

7.1 Introduction

[1]Increased globalization has led to new encounters occurring when students from different language, cultural, and social backgrounds converge in educational contexts. In recent years, this development has contributed to European schools involving

---

[1] In this chapter, the term first language is used to define the languages that the student learns from infancy in the child's immediate environment. Thus, multilingual students can have several first languages. The term second language defines the languages that the student learns after the first

---

A. Karlsson (✉) · P. Nygård Larsson · A. Jakobsson
Malmö University, Malmö, Sweden
e-mail: annika.karlsson@mau.se; pia.nygard-larsson@mau.se; anders.jakobsson@mau.se

© The Author(s), under exclusive license to Springer Nature Switzerland AG 2021

A. Jakobsson et al. (eds.), *Translanguaging in Science Education*, Sociocultural Explorations of Science Education 27,
https://doi.org/10.1007/978-3-030-82973-5_7

a variety of languages and cultures, which we argue places particular demands on science education as well as on the education systems in general. For example, several scholars (e.g., Cummins, 2008; Turkan & Liu, 2012) argue that students' acquisition of subject-specific language in science is crucial for everyone, but especially important for multilingual students, as they face the dual task of learning a new language while learning the subject matter through this language. Others have emphasized that language usage in school science contexts may be characterized by high lexical density, abstraction, and technicality that further tends to increase the complexity for students who do not have the language of instruction as their first language (Martin & Veel, 1998; Seah et al., 2014). However, research in the area has pointed out that science teaching is less frequently designed to assume multilingual or intercultural perspectives, which risks a relatively large group of students experiencing exclusion from teaching (Buxton & Lee, 2014; Van Laere et al., 2014).

Schleppegrell (2016) and Cummins (2008) also emphasize the new opportunities for cultural and language exchange between students, which may be understood as important educational assets and resources in science classrooms (see also Buxton et al., 2022). Gutiérrez (2008), Stevenson (2013) and Lin and Wu (2015) point to the positive effects on students' performance when cultural and language experiences are taken into consideration (see also Siry et al., 2022). These and other studies (Karlsson et al., 2019, 2020) clearly indicate that if students' backgrounds, all their language resources, and individual experiences are used, the commitment and empowerment tend to increase dramatically. These authors also argue that important ways of enhancing science education for all students is to make hybrid language and discourses explicit, which implies merging and comparing everyday worlds with the languages of science.

In recent years, educators have increasingly focused on learning in science as a process of appropriation of a subject-specific language (Fang & Wei, 2010; González-Howard et al., 2017; Seah et al., 2014). This aspect is, of course, important for all students regardless of their language background. In this perspective, science teaching may be understood as a conscious and intentional encounter between different language usage, discourses, and epistemological worldviews. This implies a practice that aims to move between and make explicit the students' everyday expressions and scientific ways of talking, understanding, and interpreting the world. Several studies within science education have found that students are often unaware of these discursive distinctions, which may lead to limitations in the process of learning and understanding the content (e.g., Kambrelis & Wehunt, 2012; Nygård Larsson & Jakobsson, 2020; Serder & Jakobsson, 2016). All of these studies also show that an important prerequisite for learning science is to increase the students' discursive awareness and mobility in relation to content and language (Nygård Larsson, 2011, 2018). This also means engaging students in dialogical negotiations about the meaning of the content and the specific language use.

language in an environment where the second language spoken, for example, in school. A student who has attended school in several different countries can thus have several second languages. The terms first and second languages are used in the Swedish school system.

However, in multilingual classrooms, this also includes giving students the opportunity to use all their languages and other semiotic resources in these negotiations (García & Wei, 2014; Karlsson et al., 2019, 2020).

Therefore, the focus of the present chapter is on multilingual students' use of "national" and discursive languages in translanguaging situations in science classrooms. The aim is to explore and interpret the students' negotiations about the scientific content, words, and expressions and to analyze whether these situations can facilitate and contribute to the development of conceptual knowledge and language usage. To be able to analyze these complex situations in authentic classroom environments, we have developed a model that aims to investigate and interpret the simultaneous movements between discourses and languages. Our hope is that the model can contribute to describe and clarify the use of different resources in translanguaging situations in science education, thereby increasing the awareness of science teaching as a specific language activity.

7.2 Translanguaging Science Classrooms

A translanguaging science classroom (TSC) can be defined as a classroom in which students are allowed and consciously encouraged to use all their language resources in order to develop an understanding of the subject content and the language used in teaching (Karlsson et al., 2019, 2020). The pedagogical purpose of such a strategy is that multilingual students do not become limited in their acquisition of knowledge exclusively through the language of instruction (monolingualism) but are able to benefit from their entire language repertoire to create meaning. However, several studies have shown that translanguaging situations in science education do not consist only of movements between different languages. For example, in Karlsson et al. (2019, 2020), the students used language resources in both Swedish and Arabic when discussing the science content, but the conversations also indicated movements between everyday and subject-specific discourses in both languages. The analysis uncovered that the students' experiences were mainly expressed through the first language (Arabic) and that the second language (Swedish) was primarily used for the scientific concepts and subject-specific expressions. The analysis also showed that this type of multidimensional movements was especially common in situations where the students were able to negotiate and relate the subject content to their everyday language use and experiences. This means that, in the negotiations about the scientific content, the new subject-specific words were often expressed in Swedish, while the descriptive, clarifying, and interconnecting words and phrases were commonly expressed in Arabic. From a semantic perspective, it becomes important for all students, and especially for multilingual students, to have access to this kind of interconnecting words in their meaning-making processes (as, for example, part/whole relations).

These results clearly indicate that the pedagogical use of translanguaging in subject teaching, such as science, is much more complex than just encouraging students to use their different national languages. It also places specific demands on the teaching when it comes to developing an awareness of language use in the specific subject area. Nevertheless, it is important to emphasize that the analysis displayed many situations where it become obvious that the use of both first and second languages, as well as other semiotic resources, facilitated and clarified important semantic and taxonomic relations between words and concepts.

7.3 The Science Classroom as a Hybrid Discursive Practice

In this chapter, we assume that learning scientific concepts and theories imply socializing into a specific social practice, a discourse and language use, all of which are acquired by participating in and including knowledge and experiences within this practice (Gee, 2015; Jakobsson, 2012; Wertsch, 1998). In this way, a science classroom may be understood as a venue for different ways of expressing oneself, for separate worldviews, where students' understanding of the world and scientific ideas may be negotiated and transformed in relation to the content in the education (Gee, 2015; Warren et al., 2001). Several scholars have stressed this *hybrid* (Bakhtin, 1981) nature of learning spaces in science education. For example, Kamberelis (2001) uses the concept of *hybrid discourse practices* to describe the micro-cultures that arise when different discourses encounter in science classrooms. Other researchers have noted that hybrid activities usually take place on an implicit level and with an unconscious use of the language of teaching, which may result in increased complexity and risks of misunderstandings (Kambrelis & Wehunt, 2012; Ünsal et al., 2018b).

However, such a development may be prevented by making these hybrid spaces explicit to the students by comparing colloquial and scientific discourses and by creating opportunities for increased awareness about the use of language (Nygård Larsson & Jakobsson, 2020; Poza, 2018; Serder & Jakobsson, 2016). Wallace (2004) also notes that hybrid discourses can be interpreted as the result of dialogic teacher–student or student–student negotiations in the *third space* (Bhabha, 1994). This constitutes the space in between scientific and everyday ways of acting and expressing oneself. The aim of opening up this space in educational settings is to invite students to negotiations about how to understand the subject content and to jointly construct interpretations and create meaning. For students with a multilingual background, this means that the teaching is organized so that the negotiations can be carried out with the help of several different languages and other available resources.

Thus, in a translanguaging classroom practice, the opening of the third space can be understood as a *hybrid translanguaging space* (Wei, 2011) or as the space of *a hybrid language use* (Gutiérrez et al., 2001). Even in this perspective, however, school science practice can be regarded as a disciplinary discourse with specific ways of thinking, arguing, interpreting, and reporting that differ in many ways from daily perspectives on the world (e.g., Halliday & Martin, 1993). By opening up science teaching to all students' individual language resources, these more elusive goals will be more available to multilingual students.

In all educational contexts, discursive movements occasionally arise in relation to the content of knowledge and the individuals' different experiences. Nygård Larsson (2011, 2018) uses the notion of *discursive mobility* to describe teachers' and students' multimodal movements between and within different discourses and the associated ways of thinking and acting. In education, this mobility involves movements between daily and subject-specific meanings, as well as concrete, abstract, specific, and general meanings, expressed both in language and through other semiotic resources. Discursive mobility can also be explained from the perspective that teachers have developed awareness of and pedagogical abilities when it comes to explicitly moving between discourses. A high level of discursive mobility implies that a teacher – or a student – is consciously able to move between different expressions in order to maximize learning opportunities. Nygård Larsson and Jakobsson (2020) found that students' capabilities regarding discursive mobility are crucial for their performance in science.

In science education, a range of various semiotic resources is intertwined with and inseparable from the content, such as graphs, formulas, signs, symbols (Kelly & Licona, 2018), visual representations (Evagorou et al., 2015), argumentation (Özdem Yilmaz et al., 2017), and even gestures (Zhang, 2016). Consequently, several recent studies have explored the specific demands that are put on students' multilingual and multimodal meaning-making in relation to these kind of resources (Lin, 2019; Siry & Gorges, 2020; Ünsal, Jakobson, Wickman, & Molander, 2018a; Wu & Lin, 2019). Other studies have reported that students' ability to discuss, argue, and define subject-specific words and concepts in science seems to increase when students are allowed to use translanguaging (e.g., Karlsson et al., 2019; Lin & Wu, 2015; Msimanga & Lelliott, 2014; and the studies in this anthology[2]).

According to a social-semiotic perspective and the systemic-functional theory of language (SFL), languages and other semiotic resources are used and developed functionally for communicative purposes in specific contexts and discourses (Halliday & Martin, 1993; Halliday & Matthiessen, 2004; Lemke, 2012). According to this perspective, language and content are intertwined and inseparable and thereby constitute each other. Thus, SFL scholars make an important distinction between language use in colloquial contexts and language in school practices. This distinction implies that language proficiency in science can be regarded as *register-specific* and that the notion of register concerns expressions and ways of expressing oneself in specific contexts, rather than focusing on the use of single words. This means that the notion of register emphasizes the highly contextualized usage of language and highlights the importance of language awareness when it comes to teachers' and students' language usage in science classrooms.

[2] Buxton et al. (2022); Charamba (2022); Daniel et al. (2022); Hattingh et al. (2022); He and Lin (2022); Langman et al. (2022); Lemmi et al. (2022); Licona and Kelly (2022); Salloum (2022); Siry et al. (2022)

7.4 Risk of Oversimplification

Poza (2018) emphasizes the need for precaution when it comes to oversimplifying translanguaging pedagogies to a linguistic free-for-all, stressing the importance of teachers being aware of socially oriented theories of second language development. The author argues that such awareness is an important prerequisite for creating increased opportunities for multilingual students to interact in meaningful ways. It is about eliminating the risks of reducing translanguaging processes to simple codeswitching or translation without socialization into the subject culture. Other researchers have demonstrated the overall challenges in multilingual classrooms and the need for attention to avoid simplifications of content and language use, which tends to greatly disadvantage this group of students (Karlsson et al., 2020; Ünsal et al., 2018b). For example, Nygård Larsson (2011) identified the obvious risk that teachers will use overly simplified language when unconsciously moving between daily and scientific discourses. Karlsson et al. (2020) show that even when semiotic resources are fully utilized, interruptions in classroom activities occur relatively often when teachers and students move between discursive and national languages. These situations may be consequences of the challenge of transforming complex scientific ideas into a language use that the students understand. These results raise serious questions about how students' dynamic language usage in such a translanguaging practice may be further explored and described in order to help increase knowledge about the affordances of a more conscious approach in science teaching and learning in multilingual classrooms.

7.5 Multilingual Discursive Loops

As described above, the aim of this chapter is to explore the ways in which students' multilingual negotiations in a translanguaging science classroom may contribute to a development of their subject knowledge and language use. This includes focusing on whether – and, if so, how – the movements between different national and discursive languages have an impact on this development. As mentioned, we interpret translanguaging situations in a third space as multilingual hybrids, in which both content and language are negotiated in joint meaning-making processes. We have referred in several previous publications to these situations where students move between national and discursive languages as *linguistic loops* (Karlsson et al., 2019, 2020).

However, since this chapter focuses on simultaneous movements between different languages, discourses and other available resources, we decided to define them as *multilingual discursive loops*. In other words, this means that we interpret the students' negotiations as movements in multilingual discursive loops, which includes their movements between discourses, different national languages, as well as the use of other multimodal resources. Figure 7.1 illustrates the direction of these

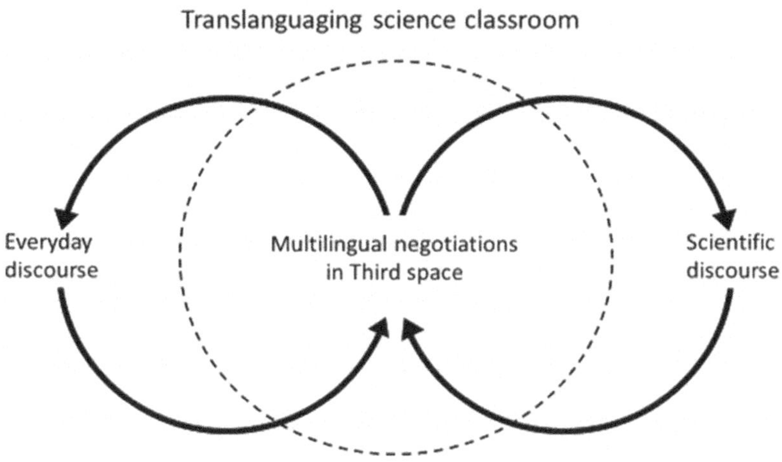

Fig. 7.1 Multilingual discursive loops in a translanguaging science classroom

loops and includes a description over the translanguaging science classroom (TSC). The center of the model, within the dashed line, represents the students' multilingual negotiations in the Third Space. The large arc-like arrows, respectively, directed to and from every-day and scientific discourses, are intended to describe how these negotiations can be oriented toward or from different language use and experiences. In the analysis, the four arrows can be used to describe the direction in which the negotiations are oriented and what resources are used to create understanding. This means that utterances of individual students or the teacher can lead to the conversation moving in a certain direction. The model can also be used to highlight the complexity that multilingual students encounter in a science classroom and we will return to a discussion on this. In the analysis, we will also relate the students' negotiations to the concept of discursive movements (Nygård Larsson, 2018).

7.6 The Context of the Study

The material used in the analyses in this chapter consists of data that were originally collected in a 3-year longitudinal and ethnographic research project that focused on Swedish students' discursive language usage in multilingual science classroom (grades 4–6) contexts (Karlsson, 2019). All students in the observed classroom are both Arabic and Swedish speakers and most were born outside Sweden. Most of the class had earlier received multilingual teaching during lower primary school, which in this case implies that a Swedish-speaking teacher and an Arabic-speaking teacher were teaching the class together and the students were given the opportunity to use both Arabic and Swedish in the classroom. However, the multilingual teaching ceased in Grade 4, and the Arabic-speaking teacher was only participating temporarily to support newly arrived students. Encouraged and supported by the teacher, the students continued to use both languages throughout middle school and were often invited to relate their prior experience and knowledge to the subject content they encountered in the science classroom. Thus, in this perspective, the situation may be defined as a typical TSC, where the students were allowed and encouraged to use all their language resources to create meaning in science. In the next section, we present analyses of four excerpts from different teaching situations (A–D) in this middle school science classroom. Our aim is for the excerpts to help to exemplify and clarify the various movements between languages, discourses, and multimodal resources in the students' meaning-making processes. The approach in the analysis has been described in more detail in two previous articles (Karlsson et al., 2019, 2020) and is further clarified in the results description below.

7.7 Results from the Analysis

7.7.1 Situation A: What Is a Stalk?

In Situation A (Fig. 7.2), the students in Grade 4 have just been watching an instructive film about *photosynthesis* that introduces several new Swedish words and expressions. The teacher has previously selected some of these words and the students are expected to discuss the meaning of the words in small groups. The everyday terms include "leaves," "flower," "stalk," "needle," "tree trunk," and "branch." The students are also expected to discuss scientific terms such as "chlorophyll," "solar radiation," "nutrients," "carbon dioxide," "oxygen," "glucose," and "stomatas." The conversation in Excerpt 1 is an example of such a discussion and starts when Rayan points to the word "stalk" and begin to discuss the meaning of the word in Arabic.

Rayan starts the conversation by reasoning in Arabic about the meaning of the word **stalk**, arguing that the word … **says something about the flower** and asking herself how **a flower** actually is built (1). This utterance indicates that she understands the word "stalk" is related to the construction of or to parts of the plant. In this sense, the meaning of the word "flower" starts to move from an everyday notion and a daily use of the word towards more of a school science discourse (see Fig. 7.3). This means that the daily meaning of the word "flower" (as a plant) seems to be transformed into a more scientific meaning and use of words (the flower and stalk as specific components of the plant). In this way, Rayan's use of Arabic enables her to discuss the central semantic relation (the part/whole-relation) between the stalk and the plant, which may contribute to increased conditions to develop further understanding of the flower as a special part and as the reproductive structure of plants. Thus, the students' negotiation about the words "flower" and "stalk" could potentially create the conditions for understanding the meaning of and the semantic relation between the words. However, although the students are able to use both

Turn	Person	Transcription and translation	Arabic	Contextual description
1.	Rayan:	**Keft tkoun elwarde / echi ^an elwarde / hay kelme echi ^an elwarde** **(How is the flower built? / something about the flower / this word says something about the flower)**	كيف تكون الوردة/ إشي عن الوردة/ هاي كلمة إشي عن الوردة	Rayan ratiocinates loudly in Arabic, "**How is the flower built** [constructed]? / [The stalk describes] **something about the flower / This word** ['stalk'] **says something about the flower**".
2.	Hanan:	Flowers **ha hiya** flowers (Flowers **here is** flowers)	هاهي	Hanan points to the word 'flowers' and says, "Flowers **here is** [the word] flowers".
3.	Amer:	Flowers…		Amer says, "Flowers…"
4.	Rayan:	… Stalk / what is stalk called / flower **khalas / kalasna minha / bas** stalk **cho hiya** (…Stalk / what is stalk called?/ flower **enough / we are already finish with it / but** stalk **what is that)**	خلاص خلصنا منها باس شو هي	Amer is interrupted by Rayan, who declares that they had already figured out what flowers are and must now describe what a stalk is.
5.	Rayan:	**Hay lkelme same^tha / hay ichi bil** trees **(This word I have heard before / it is something on** trees)	هاي الكلمة سمعتها هاي إشي بال	Rayan had heard the word 'stalk' before, and relates it to 'trees'.
6.	Hanan:	…		Hanan points with a pen at the potted plastic plant on the table.
7.	Rayan:	…		Rayan then raises the whole potted plant and looks at the petioles.
8.	Hanan:	**Hadoul** stalk (**These ones** stalk)	هادول	Hanan points to the petioles and says, "[Are] **these ones** stalk [stalks]?"
9.	Hanan:	**La moch** stalk (**No it is not** stalk)	لا موش	She then corrects herself and says, "**No, it is not** [a] 'stalk'".
10.	Amer:	…		Rayan is sitting with the plastic potted plant in her hands and Amer looks at her in a questioning manner.
11.	Rayan:	Investigate		Jourmana explains to Amer, "[I will] investigate [if there is any stalk]".
12.	Amer:	What are you doing?		Amer mumbles and asks Rayan what she is doing.
13.	Rayan:	Maybe I will find something		Rayan giggles and says, "Maybe I will find something".
14.	Amer:	It says stalk		Amer emphasizes, "It says 'stalk'".
15.	Hanan:	**Khalas** / we pass it (**Enough** / we pass it)	خلاص	Hanan suggests that they should not go on discussing the word 'stalk': "[It is] **enough**; we pass it".

Fig. 7.2 Excerpt 1: Group discussion between Rayan, Hanan and Amer. Column 3 shows the transcription of Arabic and Swedish speech (Swedish has been translated into English). The whole utterance follows in English, within parentheses

Swedish and Arabic, they do not fully develop the meaning of the word "stalk" in this situation. Instead, Rayan turns to the idea that stalks may be related to trees (5). However, the conversation takes another direction when Hanan points at the potted plastic plant on the table (6) and Rayan raises the plant and looks at the petioles (7).

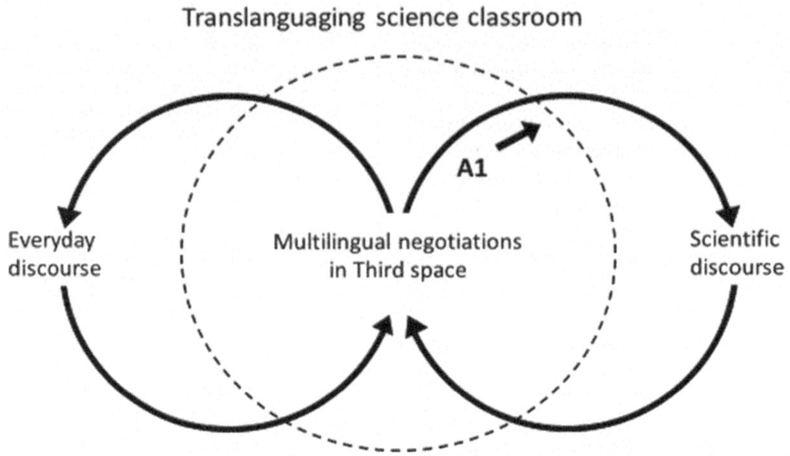

Fig. 7.3 Students' multilingual discursive loops in a translanguaging science classroom. In Situation A1, the meaning and usage of words move from an every-day use towards a school science discourse

These actions can be interpreted as the students using the plastic potted plant to help them contextualize and concretize the word "stalk" in their negotiations about its relation to the whole plant. However, this specific object has no observable and distinct stalk and therefore does not function as an effective resource for the students in their efforts to understand the word. Finally, Hanan suggests that they should move on to the next word.

Consequently, Situation A constitutes an example in which the students are not able to reach a productive solution in their meaning-making process, despite the use of both multilingual and multimodal (visual) resources. It turns out that there is a particular difficulty with these words in relation to their use in the Arabic language. We will return to this in the next situation. However, Situation A also indicates that when Rayan reasons for herself about the meaning of the word "stalk," she relates the word "stalk" to the word "flower" and thus seeks to clarify the semantic relationship between the words. In doing so, she potentially begins to transform the meaning of the word "flower" from an everyday meaning and usage (the whole plant) into a partial/whole relationship where the words flower and stalk constitute specific part of a plant. Therefore, Situation A1 can be interpreted as the conversation moving toward the scientific discourse (Fig. 7.3).

7.7.2 *Situation B: Multilingual Negotiation About Stalk and Tree Trunk*

The whole-class conversation in Situation B (Fig. 7.4) takes place in Grade 5 without any connection to the previous situation. The science teacher and the students take turns reading a text in Swedish about the structure of a tree and its growth. When they come to the terms "tree trunk" and "stalk," the teacher asks one of the students to draw a tree trunk and a stalk on the whiteboard. One of the students draws a tree with a tree trunk and a plant with a stalk to illustrate the words. The teacher then points to the illustrations and asks the students about the Arabic word for "stalk." None of the students respond to the question, which leads the Arabic-speaking teacher to express an expectation that the students should be able to answer this question (the Arabic mother tongue teacher attends the lesson to assist the students as an extra resource).

When the science teacher and the students encounter the terms "tree trunk" and "stalk" in the text, the teacher chooses to concretize these words for the students by asking one of them to draw a tree trunk and a stalk on the whiteboard. The illustrations on the whiteboard clearly display a tree with its trunk and a plant with a stalk and, with the support of this image, the teacher asks what a "stalk" is called in Arabic. None of the students seem to know, but Adnan finally answers, **tree trunk ("Gid^")** (17); however, he is immediately contradicted by Zein who states, **that is [on a] tree** (18). The Arabic-speaking teacher then explains the word stalk by using everyday expressions in Arabic, **the tree trunk to the tree and the tree trunk to the flower ("gid^ chagara wa gid^ elwarde")** (19). By relating the Swedish words for "tree trunk" (trädstam) and "stalk" (stjälk) to everyday expressions in Arabic, the crucial meanings and the semantic part/whole relations of these words are made available and visible in the conversation (see Fig. 7.5).

Thus, the use of an image as a visual resource contributes to contextualize and concretize the words and leads to a productive negotiation between the teachers and students. Moreover, this visual concretization enables increased use of the subject-related terms "tree trunk" and "stalk," which has significance for the students' opportunities to further develop crucial semantic relationships. Thus, the analysis indicates that this type of simultaneous movement between everyday and scientific discourse in different languages (Swedish and Arabic) constitutes important meaning-making resources in a TSC.

However, these types of situations seem to pose a particular complexity as the same expression may have different connotations and discursive meanings in different languages. In this situation, the negotiations are enabled by the two teachers and by the interplay between multimodal and multilingual resources; this obviously leads to the use of productive multilingual discursive loops, which may support students' language development in both languages and their discursive awareness. A particular difficulty in Situations A and B is that the Arabic word "Gid^" may be used for both tree trunk and stalk; this constitute an example of how words in one language do not always have simple equivalents in other languages. As mentioned,

Turn	Person	Transcription and translation	Arabic	Contextual description
16.	Science teacher:	Stalk / is there any word in Arabic / that on the flower show …?		The teacher repeats the question and asks Hanan (who is sitting next to Haydar) to point out the stalk in the picture in front of them.
17.	Adnan:	**Gid^** **(Tree trunk)**	جدع	Adnan says, "tree trunk" in Arabic.
18.	Zein:	That is [on a] tree		Zein remarks that this is incorrect.
[14:35-14:57]		(…..)		Hanan quietly whispers something to Haydar and then turns to the mother-tongue teacher to ask her something.
19.	Arabic-speaking teacher:	It is called the same / **Zay maqolt / gid^ chagara wa gid^ elwarde** (It is called the same / **as I said / the tree trunk to the tree and the tree trunk to the flower)**	زي ماقلت، /جدع الشجرة وجدع الوردة	The Arabic-speaking teacher explains to the class that 'tree trunk' and 'stalk' corresponds to the same Arabic word 'gid^'.
20.	Hanan:	Do they have the same name?		Hanan (the student from Situation A, who suggests that they should pass over the word 'stalk') asks, "Do they ['tree trunk' and 'stalk'] have the same name [gid^ in Arabic]?"
21.	Arabic-speaking teacher:	Yes, **gid^ cha…** (Yes **the trunk to the thr…)**	يا، جدع الش	The mother-tongue teacher confirms Hanan's question
22.	Science teacher:	…Does it have the same name on a flower as on…?		but is interrupted by the Swedish-speaking teacher, who seems surprised.
23.	Arabic-speaking teacher:	…Yes it has the same name / **gid^ chagara wa gid^ elwarde** (…Yes it has the same name / **the tree trunk to the tree and the tree trunk to the flower)**	يا، جدع الشجرة وجدع الوردة	Once again, the mother-tongue teacher confirms.
24.	Furkan:	Listen / the tree's and the flower's		Furkan explains eagerly to the science teacher, "Listen, the tree's [tree trunk] and the flower's [tree trunk]".
25.	Hanan:	Then you just add tree and flower		Also, Hanan is eager and asks the Arabic-speaking teacher once again.
26.	Zein:	The tree's trunk and the flower's trunk		Zein also becomes agitated and translates to the science teacher.
27.	Arabic-speaking teacher:	Yes we add / we just add the word tree or flower		The Arabic-speaking teacher answers Hanan's question.

Fig. 7.4 Excerpt 2: A whole-class conversation among the students, the science teacher, and the Arabic-speaking teacher, who attends the lesson to support the newly arrived students in the class. Column 3 in the excerpts shows the transcription of Arabic and Swedish speech (Swedish is translated to English). The whole utterance follows in English, within brackets

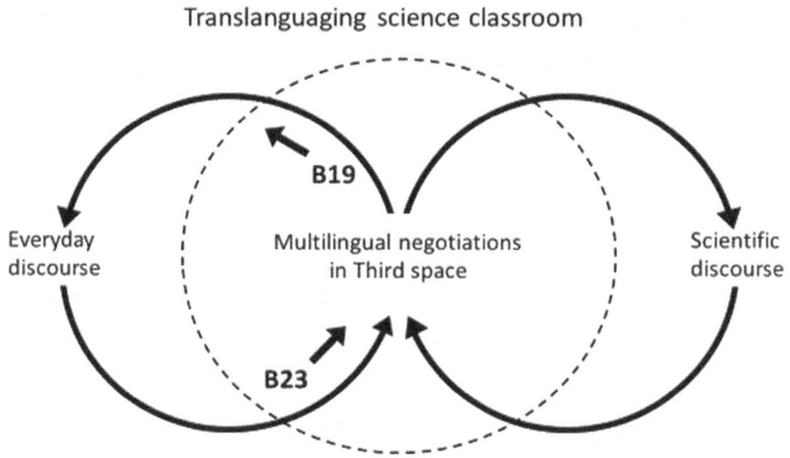

Fig. 7.5 Students' multilingual discursive loops in a TSC. In situation B19 the meaning and usage of words move from the negotiation space towards an everyday discourse in Arabic. In B23, the teacher and students bring new meanings of the words back to the negotiating space

the Arabic-speaking teacher clarifies the meaning of these words by using everyday Arabic expressions. Without this support, it may have been a far greater challenge for the students to become aware of this difference between the two languages. In Arabic, the word "saq" is an academic word for "stalk" that is rarely used in everyday language, while the Swedish counterpart may be used in both contexts.

Figure 7.5 aims to illustrate some of this complexity. In the figure, B19 represent the situation in which the teacher clarifies the meaning of the word "tree trunk" and "stalk" by relating them to everyday expressions in Arabic. In this way, the conversation moves toward an everyday discourse. In situation B23, the teacher and the students bring the new meanings of the words (from everyday discourse and language use) into the negotiating space by expressing them as "the trunk of the tree" and "the trunk of the flower." This leads to a clarified understanding of the semantic relationships between the words and a development of an awareness about the complexity of translating from one language to another.

7.7.3 Situation C: If the Sun Was Not There

Situation C (Fig. 7.6) concerns the introduction of a new field of work regarding *photosynthesis*, *combustion*, and *ecological connections* in Grade 4. The teacher introduces the new area by reconnecting to the students' prior experiences and previous knowledge, by relating to the significance of solar radiation for life on earth. She does this by asking the students a hypothetical question: "What would happen on Earth if the Sun disappeared?" The students are expected to discuss this question in small groups and write down their ideas. By this point, one of the students, Ali,

Turn	Person	Transcription and translation	Arabic	Contextual description
28.	Yousef:	I think we would have died / because we need to ...		Yousef formulates the text they are going to write.
29.	Furkan:	.. Because we need / a tree / because a tree needs …		
30.	Yousef:	Energy / Energy / and that's the Sun / trees give us oxygen / we give them carbon dioxide		Yousef fills in and takes charge of the conversation.
31.	Mariam:	What are we going to write …		Mariam is the student whose role in the group is to take notes.
32.	Yousef:	... Aha teacher, teacher, we wouldn't be able to breathe without the Sun because the Sun helps the trees / and the trees help us and we help the trees // that's why we would have died		The teacher passes their desk and Yousef seeks the teacher's attention to explain his conclusion.
33.	Ali:	If the Sun wasn't there / could not breathe so well / why because we could not breathe well / because…		Ali is eager and requires the teacher's attention.
34.	Yousef:	…I know		Yousef interrupts Ali.
35.	Ali:	**Sss wesel / Waqef bes eshajer bins...** **(we understand ... stop / only the trees...)**	وسس.. وصل وقف! بس الشجر بنس... الأشجار	Ali raises his voice and speaks in Arabic.
36.	Musa:	… Are you going to collaborate with us or not		Musa interrupts Ali.

Fig. 7.6 Excerpt 3: Group discussion among Yousef, Furkan, Mariam, Ali, and Musa

had recently left the introductory class for newly arrived students to participate in the regular science class and the other four students have been in Sweden since preschool. Excerpt 3 begins with Yousef expressing his ideas.

The group conversation in this situation is mainly conducted in Swedish except for some single statements. Yousef starts to reason about the importance of solar radiation for life on Earth by emphasizing that without the Sun **we would have died** (28). When doing so, Yousef uses subject-specific expressions, such as **oxygen** and **carbon dioxide** (30). He also contextualizes the content by using everyday expressions, such as **the Sun helps the trees, and the trees help us** (32). In his reasoning, Yousef connects concrete expressions to more abstracted reasoning of the subject content and wording in his second language. We interpret this situation as Yousef displaying an ability of discursive mobility (Nygård Larsson, 2018), which offers the group an initial and overall understanding of the photosynthesis. Figure 7.7 visualizes these discursive movements.

In the next sequence of the conversation, Ali becomes eager and seemingly attempts to focus the discussion on the importance of solar radiation by using everyday expressions: **If the Sun wasn't there … we could not breathe well, because …** (33). However, Yousef, who does not seem to encourage this attempt, interrupts Ali (34) by claiming that he is already aware of this. This leads Ali to raise his voice when he tries to explain his ideas in Arabic (35), but he is again interrupted, this time by Musa (36). We interpret this situation as Ali being unable to fully utilize his language resources in his second language and therefore having difficulty

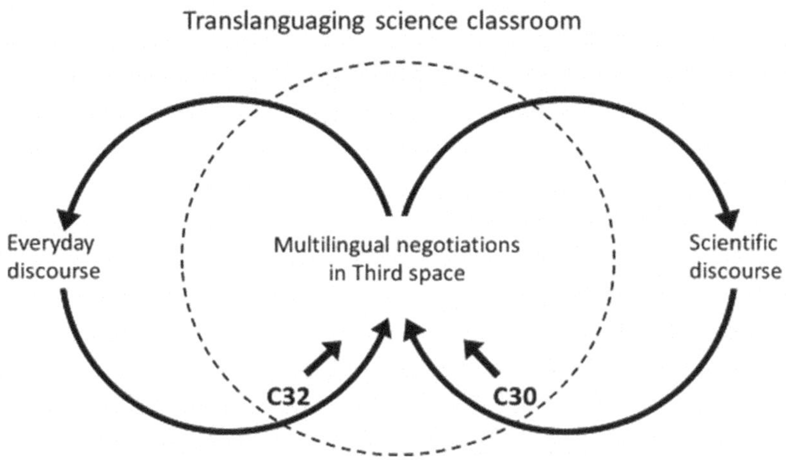

Fig. 7.7 Students' multilingual discursive loops in a TSC. In Situation C30 the students bring concepts and expressions from the scientific discourse into the negotiating space, while in Situation C32, the students bring expressions from everyday discourse

contributing to the conversation in a way that could have led to further deepening of the discussion. However, in the next situation (Situation D), Ali is allowed to expand his reasoning about the photosynthesis in Arabic, with the support of the Arabic-speaking teacher. We will return to this situation.

Figure 7.7, C30 illustrates the situation when Yousef brings the more subject-specific concepts (energy, oxygen, and carbon dioxide) into the negotiating space in order to concretize the abstract processes that take place in photosynthesis. C32 represents when Yousef uses the everyday expression "the Sun helps the trees, and the trees help us" in order to explain his understanding and conclusion of the processes in photosynthesis for the teacher. In this way, Fig. 7.7 illustrates two movements into the negotiating space, one from scientific discourse (30) and one from the every-day discourse (32), which indicates a functional discursive movement in the conversation.

7.7.4 Situation D: Expanding Reasoning About the Photosynthesis

In Situation D (Fig. 7.8), Ali and Furkan collaborate by writing a text about photosynthesis. In this situation, the Arabic-speaking mother tongue teacher also participates and scaffolds Ali's reasoning in Arabic. When the teacher asks Ali about what he knows about photosynthesis, Ali starts to explain his ideas about the photosynthetic process in Arabic.

In this situation, Ali explains the photosynthesis in a dialogic exchange with the Arabic-speaking teacher. He describes in Arabic how green plants, through their

Turn	Person	Transcription and translation	Arabic	Contextual description
37.	Ali:	**Ashams tabât lqowa liwaraq eshagar / asheâto / shams todfie almae Tamtaso eshagara almae** **(The Sun sends power to the leaves on the trees / sunbeams / the water reaches the trees)**	الشمس تبعث القوة للورق الشجر, أشعة الشمس تدفئ الماء تمتص الشجرة الماء	Ali is eager and talks rapidly and gesticulates while he is speaking.
38.	The Arabic-speaking teacher:	**Men win btetlae** **(Where does it come from?)**	من تأتي لا أنها حيث	The Arabic-speaking teacher asks Ali in Arabic.
39.	Ali:	**Min** roo... **agza3** (**From** roo… **roots**)	والمي تطلع للشجرة من أج... أجذاع	Ali first says the Swedish word for roots but switches into Arabic.
40.	The Arabic-speaking teacher:	**Shatour** **(Good)**	شطور	
41.	Ali:	**Aw el**…carbon dioxide **byetla^** … (**Or**…carbon dioxide **that comes from…**)	أو ...التركيب الضوئي	Ali uses the Swedish word for carbon dioxide. (koldioxid)
42.	The Arabic-speaking teacher:	…**Sho howa** carbon dioxide? (…**What is** carbon dioxide?)	ثاني أكسيد الكاربون	The Arabic teacher supports Ali with the Arabic word for carbon dioxide.
43.	Ali:	**Anna anna…** **(That that…)**	أن،أن	
44.	The Arabic-speaking teacher:	**Tanioksid alkarboun** **(Carbon dioxide)**	ثاني أكسيد الكاربون	The teacher pronounces the word clearly.
45.	Ali:	**Tanioksid alkarboun yaeti min assayarat wa min alensan** **(Carbon dioxide comes from cars or from people)**	ثاني أكسيد الكاربون يأتي من السيارات و من الأنسان	
46.	Ali:	**Waqtama yetle^** oxygen **we sockar / sockar bedal biqaklb eshajara we** oxygen **biyetla** (**When it goes out** oxygen **and sugar / the sugar remains in the tree and** oxygen **goes out**)	وقتما يطلع (سيره) والسكر يظل بالشجرةو.... تطلع برا	Ali uses Arabic when explaining.
47.	The Arabic-speaking teacher:	Very good, Ali **oktob lan bilârabi** (Very good, Ali, **you can now write it in Arabic**)	اكتب الأن هذا بالعربي	Ali looks satisfied and Furkan and Ali continue writing.

Fig. 7.8 Excerpt 4: A dialogue between Ali and the Arabic-speaking teacher during students' group work

leaves, use energy from sunlight to convert carbon dioxide and water to oxygen **and sugar** (46), and, in answer to a direct question from the teacher, he also emphasizes that the tree absorbs water through the roots (37/39). Furthermore, he stresses that carbon dioxide originates from **cars and people** (45) and relates this to the processes and products of the photosynthesis by expressing that **the sugar remains in the tree and** oxygen **goes out** (46).

When Ali explains, he almost entirely uses Arabic as meaning-making language resource. However, at one point he does use his second language (Swedish), when he expresses the subject-specific terms carbon dioxide [**koldioxid**, (41)] and oxygen ([**syre**, (46)]. This leads the Arabic-speaking teacher to scaffold Ali by giving him the Arabic word for carbon dioxide [**Tanioksid alkarboun** (44)]. This situation constitutes an example where the option to use both his first and second languages considerably increases Ali's opportunities to express prior knowledge and expands the opportunities for him to build on the subject content. In his reasoning, he displays both specific and general knowledge about the processes and products of photosynthesis, which becomes evident when he uses the specific words and expressions belonging to this scientific field (39, 41, 45, 46). Further, we argue, Ali's use of Arabic in a TSC allows him to express himself within a school science discourse in a way that would most likely have been unattainable if he had only been allowed to use his second language. Furthermore, when Ali concretizes the processes of photosynthesis by relating these to a more specific level through every-day expressions, the semantic relationships between the words and concepts is strengthened.

This situation clearly indicates that opening up the opportunity for translanguaging in science classrooms creates increased conditions for multilingual negotiations about the subject content, languages, and discourses, similar to the ones that arise between Ali and the Arabic-speaking teacher in this situation. Thus, the analysis shows that Ali's use of multiple language resources becomes a prerequisite for him to move in productive *multilingual discursive loops*, which is an important condition to develop further knowledge in the field. Figure 7.9 illustrates Ali's movements during the negotiation of the meanings of words and concepts.

In Situation D37 (Fig. 7.9) Ali brings the everyday expression that **the Sun sends power to the leaves** (37) into the negotiation space. We interpret this situation as an attempt to use everyday expressions in order to concretize; to focus on the importance of solar radiation for photosynthesis and to the water absorption of the roots. In Situation D46, however, Ali uses expressions and words in both Arabic and

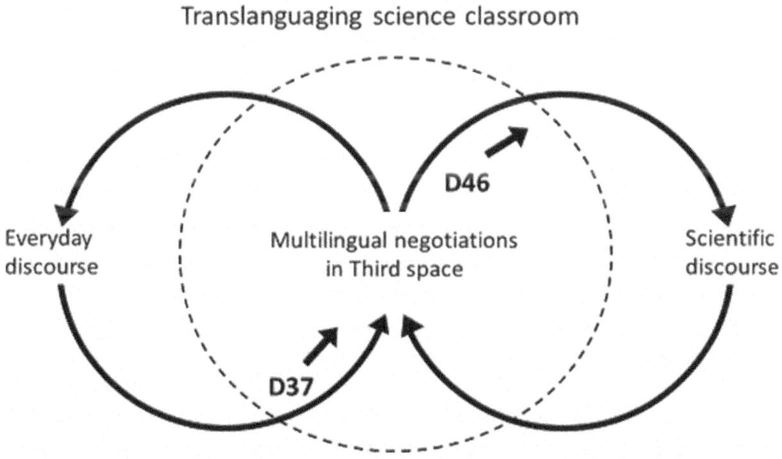

Fig. 7.9 Students' multilingual discursive loops in a TSC. Situation D37 represents a movement from an every-day discourse towards the negotiation space, and in Situation D46, the movement is directed towards a scientific discourse

Swedish – "**When it goes out** oxygen **and sugar/the sugar remains in the tree and** oxygen **goes out**" – to reason about the conversion to oxygen and sugar (glucose) and by doing so moves toward the science discourse.

7.8 Discussion

We would first like to emphasize that we understand the term *translanguaging* as a concept that constitutes a general description of a natural process where humans, in encounters with others, use all available language resources and other multimodal resources to create shared meaning. However, we also understand the concept as a conscious educational strategy or method that can be used to increase multilingual students' content learning and language development related to various subject areas. Therefore, our focus in this chapter has been on multilingual students' use of national and discursive languages with the support of multimodal resources in translanguaging situations in a middle-school science classroom (Karlsson et al., 2019, 2020).

The research literature sometimes uses the term translanguaging exclusively on the basis that multilingual students use their different languages in educational contexts and that these situations imply increased learning opportunities. We can also detect several examples of such positive developments in our empirical material. However, our intention in this chapter has been to highlight and focus on the students' negotiations about the content as simultaneous and immediate movements between both national and discursive languages with support from multimodal resources. Accordingly, our analyses have focused mainly on the students' *multilingual negotiations in the third space.* We argue that this space (Bhabha, 1994; Gutiérrez et al., 2001; Wei, 2011) constitutes an important negotiating space in science education, where both everyday and scientific means of expression are used to create meaning and understanding of the subject content. The main purpose of opening this space is to create conditions to increase all students' content knowledge, and their language and discursive awareness in science. In the analysis in this study, however, it becomes clear that it would not have been possible to carry out most of the conversations if the students had only been allowed to use their second language (Swedish). This is mainly due to the fact that most of the students in the study need to use their entire linguistic repertoire (Otheguy et al., 2015) because they have not yet developed their second language to the extent necessary to be able to conduct a complete discussion about the specific subject content in just Swedish. At the same time, we can also detect several situations where the students' first language use (Arabic) has a clarifying and positive impact on the possibilities of appropriating expressions in their second language. A number of other studies have emphasized these positive relationships between knowledge development and first and second language development (e.g., Cummins, 2014).

Nonetheless, by developing a model that aims to describe and analyze students' language and discursive movements, we wanted to put focus on a more subtle phenomenon that, we argue, is often overlooked. The analysis shows that the students' negotiations take place with the help of everyday and scientific expressions in both Arabic and Swedish, which seems to increase the level of complexity and difficulty. Across the entire data material, there are very few occasions when the teachers or the students seem to pay any attention to the fact that different language use exists in different subject areas and in colloquial contexts. Thus, we would like to emphasize that increasing teachers' and students' awareness of these conditions implies expanded opportunities for understanding of the subject content. We also see that an understanding of these processes is of the utmost importance for multilingual students as it may reduce some of the complexity that surrounds the school science discourse.

In the model, we describe the students' language movements as *multilingual discursive loops*. This implies that their utterances may be characterized as movements that have a direction towards or from discourses and that these movements simultaneously take place using different languages. These complex conditions are exemplified in Situation A, where the students negotiate the meanings of the words "flower" (elwarde) and "stalk." In the negotiation (conducted both in Arabic and Swedish) the significance of the word "flower" successively moves from an everyday notion (as a plant) toward more of a school science discourse (the flower as a specific component of the plant). However, the word "stalk" poses certain difficulties when it is translated between the two languages, as is obvious in Situation B. The word "Gid^" may be used in everyday Arabic for both "tree trunk" and "stalk," which creates confusion among the students and initially prevents them from arriving at the solution. The situation is an example of how words in one language do not always have a simple equivalent in another language. In addition, the discursive meanings and the belonging of words can also be different in different national languages, which make it even more complex. In this case, the situation can be solved because there is an Arabic-speaking teacher in the classroom. In the Arabic language, the word "saq" is an academic equivalent for "stalk" and is rarely used in everyday contexts. If the Arabic-speaking teacher had introduced this word, it may have created increased conditions for the students to move towards a more scientific discourse and strengthened the semantic relationship in both languages.

An obvious risk here is that simplified translations between national languages may lead to an unconscious transfer of the meanings of words between discourses or language usage, which does not facilitate the language and knowledge development of these students. This example also points to the need for multilingual students to, in addition to their second language development, be offered opportunities to develop their academic vocabulary in their first language. Research in the area of multilingualism in education (e.g., Cummins, 2014) has stressed the importance of increased opportunities for multilingual students to continue developing their first language alongside the new language in order to achieve the greatest possible school success.

The third situation (C) constitutes an example of how a newly arrived student experiences difficulties expressing his knowledge of the photosynthesis in the language of instruction. In this example, the discussion is conducted almost exclusively in Swedish, which means that Ali's knowledge of photosynthesis risk not becoming visible or clarified. We see opportunities here for an engaged science teacher to interact with students in similar situations to encourage them to use their first language to make themselves understood. However, in a later situation (D), the student is offered this opportunity in a conversation with the Arabic-speaking teacher. As mentioned, in this dialogue, he displays both general and specific knowledge about the processes and products of the photosynthesis, which allowed him to express himself within a school science discourse. Further, the analysis shows that when he concretizes the processes by relating to a more specific level through everyday expressions, the semantic relationship between words and concepts is strengthened. This also indicates that his discursive mobility (Nygård Larsson & Jakobsson, 2020) has increased.

Finally, we would like to emphasize that the model of a translanguaging science classroom has offered us the opportunity to illustrative analyze the students' discursive language usage. This means that we have been able to simultaneously focus on students' movements between discursive and national languages and to interpret what direction the movements have at different occasions. Further, the analysis shows that this type of movement between daily and more scientific discourses, between different languages (Swedish and Arabic), with the support of visual and other multimodal resources, constitutes important meaning-making resources in a TSC. We argue that these negotiation events are important for all students in developing a subject-specific language in science, but of utmost importance for multilingual students.

In conclusion, we argue that the model offers opportunities for science teachers at different levels as well as teacher students to deepen their understanding of the complex situation that multilingual students experience when creating meaning and understanding of a science subject content. The model may also be used in teacher education, as an analytical tool to analyze how students' conversations move between different ways of expressing themselves and thereby contribute to an awareness that science education is a communicative language activity.FundingThis work was financially supported by the research program, Disciplinary Literacy and Inclusive Teaching (LIT), Malmö University.

References

Bakhtin, M. (1981). *The dialogic imagination: Four essays*. University of Texas Press.

Bhabha, H. K. (1994). *The location of culture*. Routledge.

Buxton, C. A., & Lee, O. (2014). English learners in science education. In N. G. Lederman & S. K. Abell (Eds.), *Handbook of research on science education* (pp. 204–222). Routledge.

Buxton, C., Harman, R., Cardozo-Gaibisso, L., & Vazquez Dominguez, M. (2022). Translanguaging within an integrated framework for multilingual science meaning-making. In A. Jakobsson, P. Nygård Larsson, & A. Karlsson (Eds.), *Translanguaging in science education*. Springer.

Charamba, E. (2022). Leveraging multilingualism to support science education through translanguaging pedagogies. In A. Jakobsson, P. Nygård Larsson, & A. Karlsson (Eds.), *Translanguaging in science education*. Springer.

Cummins, J. (2008). Teaching for transfer: Challenging the two solitudes assumption in bilingual education. In N. Hornberger (Ed.), *Encyclopedia of language and education* (pp. 1528–1538). Springer.

Cummins, J. (2014). Beyond language: Academic communication and student success. *Linguistics and Education, 26*, 145–154.

Daniel, S., Ryu, M., Tuvilla, M., & Wright, C. (2022). The affordances of leveraging multilingual repertoires in scientific reasoning among resettled refugee teens: Functions of translanguaging in scientific reasoning. In A. Jakobsson, P. Nygård Larsson, & A. Karlsson (Eds.), *Translanguaging in science education*. Springer.

Evagorou, M., Erduran, S., & Mäntylä, T. (2015). The role of visual representations in scientific practices: From conceptual understanding and knowledge generation to 'seeing' how science works. *International Journal of STEM Education, 2*(1), 11.

Fang, Z., & Wei, Y. (2010). Improving middle school students' science literacy through reading infusion. *The Journal of Educational Research, 103*(4), 262–273.

García, O., & Wei, L. (2014). *Translanguaging: Language, bilingualism and education*. Palgrave Macmillan.

Gee, J. P. (2015). *Social linguistics and literacies: Ideology in discourses* (5th ed.). Routledge.

González-Howard, M., McNeill, K. L., Marco-Bujosa, L., & Proctor, C. P. (2017). 'Does it answer the question or is it French fries?': An exploration of language supports for scientific argumentation. *International Journal of Science Education, 39*(5), 528–547.

Gutiérrez, K. (2008). Developing a sociocritical literacy in the third space. *Reading Research Quarterly, 43*(2), 148–164.

Gutiérrez, K., Baquedano-Lopez, P., & Alvarez, H. (2001). Literacy as hybridity: Moving beyond bilingualism in urban classrooms. In M. de la Luz Reyes & J. J. Halcon (Eds.), *The best for our children: Critical perspectives on literacy for Latino students* (pp. 122–141). Teacher College Press.

Halliday, M., & Martin, J. R. (1993). *Writing science: Literacy and discursive power*. Falmer.

Halliday, M., & Matthiessen, C. (2004). *An introduction to functional grammar* (3rd ed.). Arnold.

Hattingh, A., McKinney, C., Msimanga, A., Probyn, M., & Tyler, R. (2022). Translanguaging in science education in south African classrooms: Challenging constraining ideologies for science teacher education. In A. Jakobsson, P. Nygård Larsson, & A. Karlsson (Eds.), *Translanguaging in science education*. Springer.

He, P., & Lin, A. M. Y. (2022). Translanguaging, trans-semiotizing, and trans-registering in a culturally and linguistically diverse science classroom. In A. Jakobsson, P. Nygård Larsson, & A. Karlsson (Eds.), *Translanguaging in science education*. Springer.

Jakobsson, A. (2012). Sociokulturella perspektiv på lärande och utveckling: Lärande som begreppsmässig precisering och koordinering [Sociocultural perspectives on learning and development: Learning as conceptual clarifcation and coordination]. *Pedagogisk Forskning i Sverige, 17*(3–4), 152–170.

Kamberelis, G. (2001). Producing of heteroglossic classroom (micro) cultures through hybrid discourse practice. *Linguistics and Education, 12*(1), 85–125.

Kambrelis, G., & Wehunt, M. (2012). Hybrid discourse practice and science learning. *Cultural Studies of Science Education, 7*(3), 505–534.

Karlsson, A. (2019). *Det flerspråkiga NO-klassrummet – En studie om translanguaging som läranderesurs i ett NO-klassrum [The multilingual science classroom – A study about translanguaging as a learning resource in a science classroom]*. Malmö University.

Karlsson, A., Nygård Larsson, P., & Jakobsson, A. (2019). Multilingual students' use of translanguaging in science classrooms. *International Journal of Science Education, 41*(15), 2049–2069.

Karlsson, A., Nygård Larsson, P., & Jakobsson, A. (2020). The continuity of learning in a translanguaging science classroom. *Cultural Studies of Science Education, 15*(1), 1–25.

Kelly, G. J., & Licona, P. (2018). Epistemic practices and science education. In M. Matthews (Ed.), *History, philosophy and science teaching: New research perspectives* (pp. 139–165). Springer.

Langman, J., Solís, J., Martin-Corredor, L., Dao, N., & Garza Garza, K. (2022). Translanguaging for STEM learning: Exploring tertiary learning contexts. In A. Jakobsson, P. Nygård Larsson, & A. Karlsson (Eds.), *Translanguaging in science education*. Springer.

Lemke, J. (2012). Analyzing verbal data: Principles, methods, and problems. In B. Fraser, K. Tobin, & C. McRobbie (Eds.), *Second international handbook of science education* (pp. 1471–1484). Springer.

Lemmi, C., Pérez, G., & Brown, B. A. (2022). Translanguaging in the science classroom: Drawing on students' full linguistic repertoires in bi/multilingual settings. In A. Jakobsson, P. Nygård Larsson, & A. Karlsson (Eds.), *Translanguaging in science education*. Springer.

Licona, P., & Kelly, G. J. (2022). Translanguaging in middle school science: Written arguments about issues of biodiversity. In A. Jakobsson, P. Nygård Larsson, & A. Karlsson (Eds.), *Translanguaging in science education*. Springer.

Lin, A. (2019). Theories of trans/languaging and trans-semiotizing: Implications for content-based education classrooms. *International Journal of Bilingual Education and Bilingualism, 22*(1), 5–16.

Lin, A., & Wu, Y. (2015). 'May I speak Cantonese?'– Co-constructing a scientific proof in an EFL junior secondary science classroom. *International Journal of Bilingual Education and Bilingualism, 18*(3), 289–305.

Martin, J. R., & Veel, R. (1998). *Reading science: Critical and functional perspectives on discourses of science*. Routledge.

Msimanga, A., & Lelliott, A. (2014). Talking science in multilingual contexts in South Africa: Possibilities and challenges for engagement in learners home languages in high school classrooms. *International Journal of Science Education, 36*(7), 1159–1183.

Nygård Larsson, P. (2011). *Biologiämnets texter: Text, språk och lärande i en språkligt heterogen gymnasieklass [The biology subject's texts: Text, language and learning in a linguistic heterogeneous upper secondary class]*. Malmö University.

Nygård Larsson, P. (2018). 'We're talking about mobility': Discourse strategies for promoting disciplinary knowledge and language in educational contexts. *Linguistics and Education, 48*, 61–75.

Nygård Larsson, P., & Jakobsson, A. (2020). Meaning-making in science from the perspective of students' hybrid language use. *International Journal of Science and Mathematics Education, 18*(5), 811–830.

Otheguy, R., García, O., & Reid, W. (2015). Clarifying translanguaging and deconstructing named languages: A perspective from linguistics. *Applied Linguistics Review, 6*(3), 281–307.

Özdem Yilmaz, Y., Cakiroglu, J., Ertepinar, H., & Erduran, S. (2017). The pedagogy of argumentation in science education: Science teachers' instructional practices. *International Journal of Science Education, 39*(11), 1443–1464.

Poza, L. E. (2018). The language of ciencia: Translanguaging and learning in a bilingual science classroom. *International Journal of Bilingual Education and Bilingualism, 21*(1), 1–19.

Salloum, S. (2022). Contradictions confronting hybrid spaces for translanguaging in the Lebanese context: A CHAT perspective. In A. Jakobsson, P. Nygård Larsson, & A. Karlsson (Eds.), *Translanguaging in science education*. Springer.

Schleppegrell, M. J. (2016). Content-based language teaching with functional grammar in the elementary school. *Language Teaching, 49*(1), 116–128.

Seah, L. H., Clarke, D. J., & Hart, C. E. (2014). Understanding the language demands on science students from an integrated science and language perspective. *International Journal of Science Education, 36*(6), 952–973.

Serder, M., & Jakobsson, A. (2016). Language games: The meaning potentials of scientific literacy surveys. *Science Education, 100*(2), 321–343.

Siry, C., & Gorges, A. (2020). Young students' diverse resources for meaning making in science: Learning from multilingual contexts. *International Journal of Science Education, 42*(14), 2364–2386.

Siry, C., Wilmes, S., te Heesen, K., Sportelli, D., & Heinericy, S. (2022). Young children's trans-modal participation in science investigations: Drawing on a diversity of resources for meaning-making. In A. Jakobsson, P. Nygård Larsson, & A. Karlsson (Eds.), *Translanguaging in science education*. Springer.

Stevenson, A. R. (2013). How fifth grade Latino/a bilingual students use their linguistic resources in the classroom and laboratory during science instruction. *Cultural Studies of Science Education, 8*(4), 973–989.

Turkan, S., & Liu, O. L. (2012). Differential performance by English language learners on an inquiry-based science assessment. *International Journal of Science Education, 34*(15), 2343–2369.

Ünsal, Z., Jakobson, B., Wickman, P., & Molander, B. (2018a). Gesticulating science: Emergent bilingual students' use of gestures. *Journal of Research in Science Teaching, 55*(1), 121–144.

Ünsal, Z., Jakobson, B., Molander, B., & Wickman, P. (2018b). Science education in a bilingual class: Problematising a translational practice. *Cultural Studies of Science Education, 13*(2), 317–340.

Van Laere, E., Aesaert, K., & van Braak, J. (2014). The role of students' home language in science achievement: A multilevel approach. *International Journal of Science Education, 36*(16), 2772–2794.

Wallace, C. S. (2004). Framing new research in science literacy and language use: Authenticity, multiple discourses, and the 'third space'. *Science Education, 88*(6), 901–914.

Warren, B., Ballenger, C., Ogonowski, M., Rosebery, A. S., & Hudicourt-Barnes, J. (2001). Rethinking diversity in learning science: The logic of every-day sense-making. *Journal of Research in Science Teaching, 38*(5), 529–552.

Wei, L. (2011). Moment analysis and translanguaging space: Discursive construction of identities by multilingual Chinese youth in Britain. *Journal of Pragmatics, 43*(5), 1222–1235.

Wertsch, J. V. (1998). *Mind as action*. Oxford University Press.

Wu, Y., & Lin, A. M. Y. (2019). Translanguaging and trans-semiotising in a CLIL biology class in Hong Kong: Whole-body sense-making in the flow of knowledge co-making. *Classroom Discourse, 10*(3–4), 252–273.

Zhang, Y. (2016). Multimodal teacher input and science learning in a middle school sheltered classroom. *Journal of Research in Science Teaching, 53*(1), 7–30.

Chapter 8
Translanguaging, Trans-semiotizing, and Trans-registering in a Culturally and Linguistically Diverse Science Classroom

Peichang (Emily) He and Angel M. Y. Lin

Abstract This study analyzes translanguaging and trans-semiotizing as meaning-making practices in a science class with student diversity in linguistic and cultural backgrounds as well as cognitive abilities. Observations of the science lessons demonstrated that spontaneous translanguaging and trans-semiotizing in the lessons provided useful interactional scaffolding, which complemented well the "Concept + Language Mapping" (CLM) teaching materials that provided designed scaffolding to facilitate meaning-making in the science lessons through seamless bridging between multiple communicative resources. Although helping students to engage in the science literacy activities by shifting between everyday Cantonese, academic Cantonese, everyday Putonghua, academic Putonghua, as well as written academic Chinese, the teacher was also engaged in shifting seamlessly between every-day and academic science registers, a process called "trans-registering". During this process, the complex science texts were unpacked and repacked to help students to both grasp the key conceptual meanings of the science topics and comprehend and produce science texts to express these meanings.

Keywords Translanguaging · Trans-semiotizing · Trans-registering · Thematic patterns · "Concept + Language Mapping" (CLM) · Designed scaffolding · Interactional scaffolding

---

P. He (✉)
Faculty of Education, The University of Hong Kong, Hong Kong (SAR), China

A. M. Y. Lin
Faculty of Education, Simon Fraser University, Vancouver, BC, Canada
e-mail: angellin_2018@sfu.ca

© The Author(s), under exclusive license to Springer Nature Switzerland AG 2021

A. Jakobsson et al. (eds.), *Translanguaging in Science Education*, Sociocultural Explorations of Science Education 27,
https://doi.org/10.1007/978-3-030-82973-5_8

8.1 Research Background

This research explores science education in a multicultural and multilingual secondary school in Hong Kong. In the Integrated Science class, students consisted of both Cantonese-speaking local Hong Kong students and Putonghua-speaking students from nearby cities in mainland China (Pérez-Milans, 2016). Owing to the differences in historical backgrounds, social systems, as well as educational policies between Hong Kong and Mainland China, there are also differences in the language policies and mediums of instruction (MOIs) between schools in the two loci. For instance, the teacher and the local students in this study spoke Cantonese and read and wrote in the more complex traditional Chinese characters (繁體字), whereas the students from mainland China spoke Putonghua and used to read and write in simplified Chinese characters (简体字) before they were enrolled in the school.

As shown in Table 8.1, there are conspicuous discrepancies between the two languages in terms of the pronunciation and form of vocabulary and grammar when they are used in oral or written communication (He et al., 2017; Li & Zhu, 2013). For the students from the Mainland, although the grammar and vocabulary in the teaching materials are relatively close to those of Putonghua, they have difficulty reading the academic texts, as teaching resources in Hong Kong are written in Traditional Chinese Characters, which look very complicated to students from the Mainland (e.g., Traditional Chinese Characters "體" vs. Simplified Chinese Characters "体"). As for oral communication, the discrepancies between Cantonese and Putonghua are even greater. They are not only different in pronunciation but there are also huge differences in grammar and vocabulary. This has formed barriers for students of mainland background, both in daily life (e.g., buying food in shops) and at school; especially when colloquial Cantonese (e.g., "細[sai3]蓉[yung2]行[haang4]街[gaai1]" instead of "外賣/wài mài/小碗/xiǎo wǎn/雲吞麵/yún tūn miàn/" is used to mean "take-away dumpling in small portions") and English are frequently used in local Hong Kongers' everyday conversation (e.g., students call their teachers in Hong Kong style English such as "Wong Sir" rather than the Chinese appellation "黃老師"). It thus takes time for the non-local students to get accustomed to both Cantonese as the MOI and the traditional Chinese characters in the academic literacy practice at school.

Apart from the sociocultural and linguistic differences in students' backgrounds, there is also huge diversity in students' academic performances. For example, the students in this research were of mixed abilities who did worksheets of different difficulty levels and finished different sets of exam papers. Although many students appeared to be talkative in class, they lacked confidence in reading and writing. This was explained by the teacher, "they are afraid of reading longer sentences or paragraphs," and admitted by the students, "any tasks without much writing will be okay for us." The diversity in learning ability was even widened owing to the enrolment of special educational needs (SEN) students who had difficulty in academic learning owing to linguistic or cognitive specificities, such as a slight degree of speech and language impairment, dyslexia or intellectual disability.

Table 8.1 Examples of differences in vocabulary and grammar of the oral and written communication of students from Hong Kong and Mainland China

English	Traditional Chinese characters 繁體字	Cantonese Romanization	Simplified Chinese characters 简体字	Putonghua Pinyin
Umbilical cord	臍帶	臍[chi4]帶[daai3]	脐带	/qí dài/
Sex hormone	性腺體	性[sing3]腺[sin3]體[tai2]	性腺体	/xìng xiàn tǐ/
Give birth to a baby	生BB/ 生仔	生[sang1/saang1] BB 生[saang1]仔[jai2]	生宝宝/ 生孩子	/shēng bǎo bǎo/ /shēng hái zi/
Indecent, crude	粗鄙	粗[chou1]鄙[pei4]	粗俗	/cū sú/
Hung Hom	紅磡	紅[hung4]磡[ham3]	红磡	/hóng kàn/
Wonton/dumpling noodle	雲吞麵 /蓉	雲[wan4]吞[tan1]麵[min6] /蓉[yung2]	云吞面	/yún tūn miàn /
Take-away	行街	行[haang4]街[gaai1]	外卖	/wài mài/
Have lessons	上堂	上[seung5]堂[tong4]	上课	/shàng kè/
Mr Wong; Miss Wong	黃老師	Wong[wong2] sir[səI4] Miss[mis6] Wong[wong1]	黄老师	/huáng lǎo shī/
3rd person plural pronoun "they"	佢哋	佢[keui5]哋[dei6]	他们	/tā men/
Present progressive "-ing"	緊	緊[gan2]	正在	/zhèng zài/
Present perfect "Have done"	咗	咗[jo2]	了	/le/

Science education is challenging for school learners not only because of the abstractness of concepts in the subject but also because of discursive practices associated with science learning, which may well be seen as "foreign" to students (Lin, 2016; Wellington & Osborne, 2001). Given the complicated background of the learning community, in this study we attempt to investigate: first, how the science teacher engaged students of diverse cultural and linguistic backgrounds and cognitive abilities to achieve meaning-making in the science lessons; and second, how the teacher guided the students to improve their academic literacy through the Integrated Science lessons.

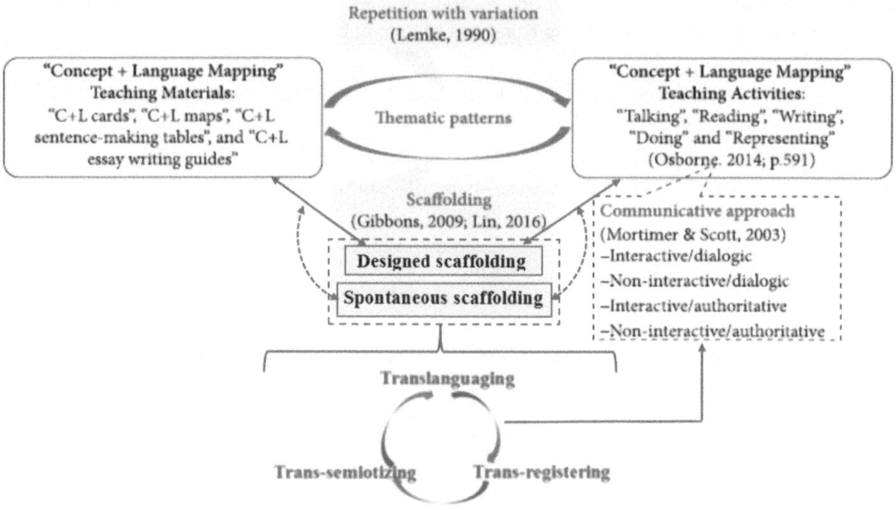

Fig. 8.1 The framework of Concept + Language Mapping pedagogy facilitated by translanguaging, trans-semiotizing, and trans-registering. (Adapted from He & Lin, 2019)

8.2 Literature Review

The centrality of language in mediating learning (Halliday, 1993; Tytler and Tytler & Prain, 2010) and meaning-making in science education has been emphasized in the literature (Danielsson, 2016; Hodgson-Drysdale, 2014; Wellington & Osborne, 2001). This study proposes a conceptual framework of "Concept + Language Mapping" (CLM) pedagogy facilitated by translanguaging, trans-semiotizing, and trans-registering (Fig. 8.1). Based on this framework, this section first introduces the development of the thematic pattern-based CLM pedagogy, then reviews the research on translanguaging and trans-semiotizing and elaborates on how the research findings enrich the CLM pedagogy by incorporating translanguaging and trans-semiotizing as spontaneous scaffolding during science lessons. It further highlights the importance of academic literacies and the necessity to incorporate trans-registering into the meaning-making process of science education.

8.2.1 *Thematic-Pattern-Based "Concept + Language Mapping"*

In school education, concepts are fundamental for content learning, but can be difficult owing to their abstractness and complexity. In science education, "concepts" are emphasized, whereas the role of "language" is generally neglected. Lemke (1990) drew science educators' attention to the notion of *thematic patterns* and emphasized that concepts are thematic patterns, which are patterns of connections among the meanings of words in a particular field of science or human activities. In Lemke's words, thematic patterns are "a network of relationships among the scientific concepts in a field, but described semantically, in terms of how language is used in that field" (p.12). The thematic pattern theory thus highlights the relationship between "concept" and "language"; namely, concepts are represented by language in networks of meaning relationships about a particular subject. According to Lemke (1990), thematic patterns constitute the core of the curriculum; in different lessons the same thematic patterns are presented and re-presented in different language expressions through a series of classroom communication activities in a strategy of "repetition with variation"; for example, they are represented in textbooks, lesson plans, and blackboard notes, talked about during discussions, illustrated in experiments, and expressed in worksheets and test papers (Osborne, 2014). Grounded in the thematic pattern theory, CLM pedagogy has been developed and adopted in content and language integrated learning (CLIL) classrooms (Lin & He, 2017). By mapping concepts and the corresponding language expressions representing the concepts, the CLM pedagogy visualizes the patterns of interrelationship between the meanings of thematic items through language, hence facilitating students' understanding of the abstract concepts and the complex inter-relationship among the concepts. Research into CLM pedagogy showed that multimodal animated "concept + language mapping" with e-copy and hard-copy CLM materials (e.g., "C + L cards," "C + L maps," "C+L sentence-making tables," and "C+L writing guides"[1]) as well as CLM activities as designed and spontaneous scaffoldings helped to facilitate students' development of both content and academic language knowledge in biology lessons (He & Lin, 2019).

8.2.2 *Trans-languaging and Trans-semiotizing as Spontaneous Scaffolding*

Although thematic patterns as parts of the thematic–conceptual system of a curriculum constitute fundamental knowledge for students, they are not pre-existing entities or accomplished acts waiting to be acquired; rather, they need to be co-constructed through a dynamic process of recognizing, thinking, and participating in negotiations and co-reflections in the learning community. Recent studies (Li, 2018; Lin, 2019; Swain & Lapkin, 2013) in CLIL and multilingual education focused on the notion of *languaging*, which was first proposed by Maturana & Varela (1980) and later appropriated and elaborated in language and education scholarship (e.g., Thibault, 2011; Swain & Lapkin, 2013). Simply put, *languaging* can be seen as the process by which "one uses language, among other purposes, to focus attention,

[1] "C + L" is the abbreviation for "Concept + Language"; e.g., "C + L cards" means "Concept + Language cards."

solve problems and create affect" (Swain & Lapkin, 2013, p.105). This recent attention given to *languaging* further accentuates the role of language in a dynamic and dialogic learning process. Pedagogically, *languaging* in collaborative dialogue is essential for meaning-making in content-based classrooms (Lin & He, 2019), as deep learning requires *languaging*, which allows individuals to talk about what they think they have learned and understood (Coyle's presentation in Working CLIL Digital, 2018). Theoretically, *languaging* is related to the view of distributed cognition and is referred to as "an assemblage of diverse material, biological, semiotic and cognitive properties and capacities which languaging agents orchestrate in real-time and across a diversity of timescales" (Thibault, 2017, p.82). Accordingly, knowledge construction as *languaging* cannot just rely on the functioning of an abstract code system of an isolated language (i.e., our common notion of language A, B, or C) but goes beyond boundaries of codified languages or modalities through a dynamic and dialogic meaning-making process entangling with a collection of material, biological, semiotic, cognitive resources, and affordances deployed by the knowledge-building participants in real time or across timescales. In this sense, *languaging* is a first-order activity, whereas language is a second-order construct (Thibault, 2011). This distributed cognition view theoretically underpins the dynamic, dialogic, fluid, and distributed view of translanguaging as *flows* (Lemke, 2016; Lin et al., 2020), on which we shall elaborate below.

In a multicultural and multilingual science classroom, knowledge construction and meaning-making become more complicated owing to the diverse linguistic and cultural backgrounds of the teacher and students. Translanguaging (García & Li, 2014) and trans-semiotizing (Lin, 2015) have been reported to be pedagogical and identity affirmation strategies (Lin & He, 2017) in a multilingual and multicultural education context. According to García and Li (2014), translanguaging refers to "the dynamic process whereby multilingual language users mediate complex social and cognitive activities through strategic employment of multiple semiotic resources to act, to know and to be" (p.42). They proposed a holistic view of semiotic repertoire, which "signals a trans-semiotic system with many meaning-making signs, primarily linguistic ones that combine to make up a person's semiotic repertoire." This echoes previous studies that explored human learning from social semiotic perspectives and emphasized the role of multimodality in meaning-making (Danielsson, 2016; Kress et al., 2001; Lemke, 1998); for example, Kress et al. (2001) explicated the amplification of meaning-making through dynamic shifting of modes and orchestration of meanings in science classrooms; namely, the backgrounding and foregrounding of verbal, visual, and actional modes, each of which illustrates specific aspects of the meaning represented and provides affordances for meaning-making from different angles, hence maximizing meaning-making and facilitating comprehension during communication. Echoing Halliday's (2013) "trans-semiotic" view, Lin (2015) developed the notion of "trans-semiotizing" to conceptualize the well-coordinated use of diverse semiotic resource systems (traditionally called multimodalities) in meaning-making. She also emphasized "the need to draw on multiple resources in the communicative repertoire of the students to provide language and semiotic support to them when they are learning content using a second or foreign language (an L2)" (p.84). Such a semiotic repertoire lens expands the multimodal nature of meaning-making and enriches the notion of translanguaging by extending the inventory of the semiotic repertoire to include communicative resources beyond the boundaries of "linguistic resources" during meaning-making in multilingual and multicultural contexts (Kusters et al., 2017).

Grounded in the distributed language view (Thibault, 2011), Lemke (2016) explicated translanguaging and *flows*, which further transformed the concept of translanguaging by expanding meaning-making and the semiotic repertoire into an eco-social system perspective (Lemke, 2018) that evolves in both temporal and spatial dimensions. According to Lemke (2016), human communication involves "material *flows* of matter energy and information." The different parts of the body (e.g., the brain) of a participating human and those of the other participants are mediums in which the *flows* occur. Hence, meaning is made through different mediums including the bodies of one or more participants, and other artifacts and material media involved. Past events and experiences may leave traces in a series of material media within the eco-social system, just like "envelopes" of prior events leaving traces for the new events to fit in. The entanglement of material *flows* may include human bodies, artifacts, features of the landscape in the communication setting, etc., and the histories of past communications may be related to the potential outcomes in future communications. Drawing on both the translanguaging and trans-semiotizing strategies, He et al. (2017) reported the meaning-making processes of a multicultural and multilingual mathematics education seminar that consisted of different languages, PowerPoint (PPT) demonstration, coordinating gestures, facial expressions, sounds, and visual images in a dynamic *flow* of meaning-making. In a recent study, Wu and Lin (2019) conducted fine-grained multimodal analysis of a CLIL biology classroom drawing on Lemke's (2016) fluid, distributed, and dynamic process view of human meaning-making and reported a positive impact of translanguaging and trans-semiotizing on the students' biology learning and their development of a positive attitude toward multilingualism.

The theorization and research findings on translanguaging and trans-semiotizing shed light on the re-conceptualization of the framework of the CLM pedagogy, which emphasizes the role of scaffolding in facilitating students' CLIL development. Lin and He (2017) proposed the CLM pedagogy drawing on Gibbons' (2009) differentiation between designed/planned scaffolding and interactional/spontaneous scaffolding, with the former referring to "the support teachers consciously plan in advance," whereas the latter referring to "the support teachers provide contingently through dialogue during instruction or other interaction" (Lin, 2016, p.158). CLM pedagogy was adopted in their later research, which showed that the thematic pattern-based designed scaffolding and spontaneous scaffolding intertwined to support the students' development of their content and language knowledge (He & Lin, 2019). Based on these research findings, we propose that, in multilingual and multicultural classroom contexts, teachers may incorporate translanguaging and trans-semiotizing into their spontaneous scaffolding strategies. By adopting appropriate communicative approaches (Mortimer & Scott, 2003), the thematic patterns of the curriculum re/presented and embodied by both the designed scaffolding (i.e., the

CLM materials and activities prepared before lessons) and the spontaneous scaffolding (i.e., the ongoing translanguaging and trans-semiotizing during classroom interaction) can be activated and mediated during the dynamic, dialogic, fluid, and distributed process of meaning-making and co-construction of knowledge in the science classroom.

8.2.3 Academic Literacy Development and Trans-registering in Science Education

Given the role of CLM materials and activities in both designed scaffolding and spontaneous scaffolding facilitated by translanguaging and trans-semiotizing practices during the lessons, the material *flow* of information and resources unfolds as the thematic patterns in the curriculum are re/presented by various semiotic artifacts lesson after lesson following the principle of "repetition with variation." The co-constructed semiotic repertoire shared between the teacher and students also expands, consisting of various representations that can be more formal and academic such as those written in the syllabus, textbooks, CLM materials, and blackboard notes; or more typical of the colloquial and everyday conversations such as the examples provided by the teacher or group-work discussions among students. Namely, both features of academic registers and colloquial registers emerge and entangle with one another during meaning-making in the lessons. In systemic functional linguistics (SFL), register refers to a variety of linguistic features that are typically used for a particular purpose or used in a particular communicative situation (Halliday & Hasan, 1976). Academic language is a set of linguistic registers that construe multiple and complex meanings at all levels and in all subjects of schooling (Schleppegrell, 2009). Coyle (presentation in Working CLIL Digital, 2018) commented that academic language is "nobody's language." Previous research also pointed out that academic literacy is not only challenging for students whose first language (L1) is not the medium of instruction (MOI) of the subject, it can be "foreign" to students even if their L1 is used as the MOI (Lin, 2016; Schleppegrell, 2009; Wellington & Osborne, 2001). However, the fluidity and hybridity of academic language must be pointed out and the best way to conceptualize academic registers is to think of them as a spectrum of features overlapping and entangling with everyday language features (see discussion in Lin et al., 2020).

In science education, academic language is a basic communication resource and academic literacy is a key learning outcome. According to Osborne (2014), a fundamental sense of science literacy is the ability to "construct meaning through interaction with the multiple forms of semiotic communication that are used within the discipline of science" (p.591). The language of science is typical of academic registers that consist of various modes, including words, diagrams, pictures, graphs, maps, equations, tables, charts, and other forms of visual and mathematical expressions (Lemke, 1998). To help students to bridge the gap between the academic and colloquial registers in science education, Lemke (1990) suggested,

> "Students will begin to grasp semantic and conceptual relations in colloquial language first. Then they will substitute scientific, technical terms for colloquial words … … Teachers should use all the stylistic and rhetorical means available to communicate science to students, including narrative and dramatic presentations; humor, irony, and metaphor; fiction and fantasy; reference to actual scientific activities, disputes, and persons; personal anecdotes and historical examples." (p.172–174)

Accordingly, apart from providing scaffoldings via translanguaging and trans-semiotizing, teachers may also deploy flexibly semiotic resources from an expanded communicative repertoire (Lin, 2012) to help students to bridge the gap between colloquial and academic registers in science education. Such trans-registering practice as an expanded notion of translanguaging has been proposed by Lin et al. (2020) in their exploration of the theoretical underpinnings of translanguaging. It broadens the range of designed scaffolding and spontaneous scaffolding strategies (Gibbons, 2009; Lin, 2016), and contributes to the ongoing conceptualization of the thematic pattern-based CLM pedagogy.

In this section, we introduced the framework of CLM pedagogy facilitated by translanguaging, trans-semiotizing, and trans-registering (Fig. 8.1), which has been adapted from and further developed the framework of thematic pattern-based CLM pedagogy (He & Lin, 2019) in our previous research. Based on the literature review and the updated conceptual framework, the present study addresses the following two research questions:

1. How do translanguaging, trans-semiotizing, and trans-registering as spontaneous scaffolding strategies facilitate meaning-making in the multilingual science classroom?
2. How is academic literacy development facilitated by trans-registering in Integrated Science lessons?

8.3 Research Methods

8.3.1 Research Site and Participants

In this study, the collaborative school aimed at inclusive education and had been dedicated to catering for learner diversity for decades. Apart from enrolling students whose academic performances were of high or average levels in the city, the school also offered education opportunities to students who had learning difficulties, who were unmotivated, or who lacked proper learning support from their family. Hence, there was huge learner diversity among the students with regard to cognitive ability and academic performance. To cater for student diversity, the school had classes of mixed abilities in junior grades, with a relatively small number of students in each class. The MOI in the school was Cantonese. However, owing to its proximity to the border of Mainland China, the school also enrolled students from the nearby mainland cities who communicate in a very different language (i.e., Putonghua) from that of the local students (i.e., Cantonese). Among the 19 students in the participating class, 13 were Cantonese-speaking local students in Hong Kong and 6 were Putonghua-speaking students from nearby mainland cities. Wong Sir[2] was a Cantonese-speaking science teacher in his early 40s. He developed a friendly rapport with the students and was able to create a relaxing atmosphere during the science lessons.

The participating students were in Secondary 1 (S1, grade 7) who had been upgraded from primary schools for not more than 3 months. According to Wong Sir, the students had great curiosity in asking questions and inquiring about new knowledge, but the academic literacy of most students was relatively low, which had, to some extent, led to their poor memory. Some students could not remember many words, which were the complex traditional Chinese characters, and they were afraid of reading "long" texts. Many students did not want to read the textbook because they found "too many words." Some students whose L1 was the same as the MOI (i.e., Cantonese) had the misconception that they understood the lessons, but they could not write much in their assignments or tests. For those students who were from Mainland China, as their L1 is Putonghua and they were used to reading and writing the simplified Chinese characters, they had difficulty not only in reading and writing the complex traditional Chinese characters, but also in communicating with their Cantonese-speaking teacher and classmates, as they were not good at listening to or speaking Putonghua. There were also a few SEN students who had dyslexia and who were receiving speech therapy. They could not focus on reading, even though it was just a short paragraph on a PPT slide. Owing to their young age and cognitive immaturity, some students could not concentrate on or listen patiently to the lesson if the teacher only presented what was written in the textbook. As the school emphasized that teaching must not be "離地 (off ground)," namely, "too abstract, technical or unrealistic"; Wong Sir tried to make the science lessons relevant to the students' daily life; for example, by referring to local TV programs or using the experiences of the students' friends as examples in the lesson.

8.3.2 Data Collection

This research is part of the second author's SCOLAR project[3] (Hong Kong Education Bureau), which involved a series of university–school collaborations with different institutions. Although the overall project research drew on Reeves' (2000) "development research" framework and adopted a mixed-method study (Creswell, 2003), as this study focused on the translanguaging practices in the Integrated Science lessons of a participating S1 class, only the qualitative data collected were analyzed. These data included the transcripts of lesson observations, a semi-structured

[2] All names of the teacher and students in this research are pseudonyms.

[3] The Standing Committee on Language Education and Research (SCOLAR) (Project #2015–0025)

Table 8.2 Summary of data collection

Data sources	Details
Lesson observation	Video recording of five lessons (270 min)
Semi-structured interview with teacher	Audio recording (40 min)
Focus group interview with students	Audio recording (36 min)
Documents and artifacts	CLM cards and CLM maps, PPT slides, and game worksheets
	Sample student assignments
	Pre-test and post-test papers

interview with the teacher, and a focus group interview with the students, as well as documents and artifacts such as textbooks, PPT slides, and CLM materials and CLM materials like the "C + L cards," "C + L maps," worksheets for the CLM activities, sample student assignments, and pre-test and post-test papers (Table 8.2).

Before data collection, the researchers obtained ethical approval clearances and consent from the principal, Wong Sir, the students, and their parents. The teacher selected a section "human reproductive system" from S1 Integrated Science for the collaboration, based on which the researchers designed the teaching materials. They then shared with Wong Sir the conceptualization of the thematic-pattern-based CLM pedagogy and the design and use of the CLM materials including "C + L cards," "C + L maps," "C + L sentence-making tables," "C + L writing structures," as well as the worksheets and PPT slides for the CLM activities. The teacher then selected and adapted the materials according to the students' cognitive abilities and the lesson time. During the collaboration, Wong Sir adopted the teaching materials to carry out the teaching and learning activities, but he was free to decide on the manner of interaction with the students, including the MOI, questioning, error correction, examples or feedback to the students. During data collection, the students had the pre-test a week before the intervention and then had five 55-min lessons in the following 3 weeks. Wong Sir taught the lessons according to the CLM pedagogy with the CLM materials distributed to the students as worksheets and notes. During the collaboration, Wong Sir's lessons were observed and videotaped, and the students' worksheets and assignments were collected. At the end of the intervention, students had a post-test. Six students of high, intermediate, and low learning ability were invited to a focus group interview about their perceptions of the teaching materials and learning activities adopted in the lessons. An individual semi-structured interview was conducted with Wong Sir about his perceptions and attitudes toward the CLM pedagogy, his reflections about the collaboration, and suggestions about how to improve the pedagogy. The interviews with both the students and the teacher were audiotaped, with the former lasting 36 minutes and the latter 40 minutes.

8.3.3 Data Analysis Methods

Data analysis was conducted in three rounds following an inductive coding method (Miles & Huberman, 1994). In the first round of analysis, the first author reviewed all qualitative data including lesson observation video clips, interview transcriptions, and documents and artifacts (i.e., teaching materials and student work) in a context-sensitive manner identifying the critical moments of teacher–student interactions, meanwhile taking into consideration the sociocultural backgrounds of the school community. Second, drawing on the theoretical tools for micro-ethnographic analysis of classroom language and literacy events (Bloome et al., 2005), the episodes in lesson observations involving strategies of translanguaging, trans-semiotizing, and trans-registering were analyzed in a fine-grained manner. Scenarios (e.g., those reported in this article) that engaged the majority of the students and aroused most students' interest were selected for further analysis. The selected scenarios were transcribed verbatim by the first author, who is proficient in both Cantonese and Putonghua. The transcripts were then proofread and translated into English with the source languages marked in the translated transcript. Then, a third-round analysis was carried out focusing on the interrelationship among the translanguaging, trans-semiotizing, and trans-registering in the lessons. Based on the data analysis, the preliminary theoretical framework (i.e., the framework of thematic-pattern-based CLM pedagogy (He & Lin, 2019) was re-conceptualized. A research assistant, who had obtained an MA degree in a MEd CLIL program, graded the pre-test and post-test papers. As the current article focuses on the discourse analysis of classroom interactions (Bloome et al., 2005), we employed the quantitative data analysis as supportive data for corroboration of the qualitative research analysis. To ensure trustworthiness, the analysis of the lesson observations was further validated through the focus group interview with the students and the semi-structured interview with the teacher (Lincoln & Guba, 1985). During the data analysis cycles, both the first and second authors had discussions about the interpretation of the data and the re-conceptualization of the theoretical framework. Full transcription of the classroom interactions and thick description of the situation of the excerpted lesson episodes were presented in the article to achieve transferability and allow readers to determine the reliability of the data (Merriam, 1998).

8.4 Analysis

To cater for learner diversity and engage the students, the researchers designed both the CLM teaching materials and CLM teaching activities, which allowed students to learn the science knowledge through both self-directed learning and collaborative learning; and at the same time earn the bonus – the "likes" and "stars" offered by the teacher and peers as means of encouragement. This motivated the class and attracted even the lower achievers to participate in the learning activities.

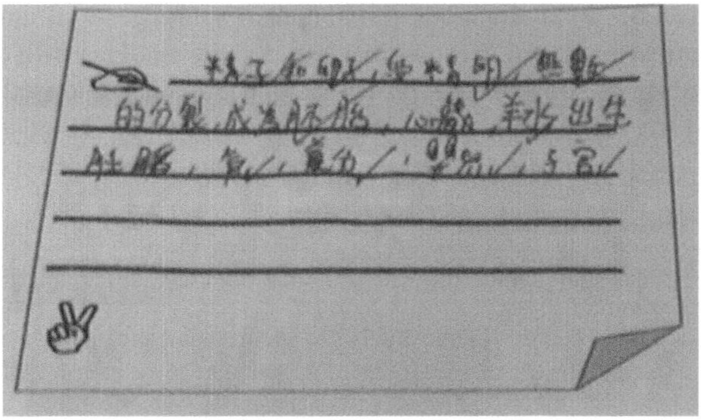

Fig. 8.2 Student worksheet with the "science words" they captured in the cartoon

8.4.1 CLM Materials as Designed Scaffolding for Meaning-Making Though Translanguaging, Trans-semiotizing, and Trans-registering

At the beginning of the unit, Wong Sir guided students to brainstorm the process of development from a cell to a baby. To facilitate brainstorming, Wong Sir gave students a "spider graphic organizer" and allowed them to use homophone or simply a cross to indicate the words they did not know. He then played an animated cartoon about "the birth of new life" to lead in the topic of the unit and gave students a worksheet (Fig. 8.2) to note down the "science words" that they could identify in the cartoon. As the animated cartoon was for primary learners, the students did not have difficulty understanding the content, but many of them did not know how to write the characters of the science vocabulary they had captured (e.g., 臍帶umbilical cord). For both activities, the teacher invited students to present their answers and offered his feedback on the students' answers. He then displayed and explained some possible answers to the worksheet exercises during which he highlighted the science vocabulary and the students noted down the "science words" beside their own answers on the worksheets.

After leading in the topic of the unit, Wong Sir started teaching the science knowledge in different lessons. This was facilitated by the demonstration of another video "The birth of a new life," which he played episode by episode according to the relevant concepts taught in different lessons. Compared with the animated cartoon, this video clip was much more technical as it is an authentic recording of the process of the birth of human life. Before playing the video, guiding questions about the science knowledge were listed on a PPT slide which Wong Sir elaborated to the students and reminded them to think about the questions while watching the video. Then, he played the video so that students were able to focus on the specific knowledge points of the science concepts. As the video was mute, during video playing, Wong Sir also elaborated on the science knowledge by defining, describing features and processes, or explaining the functions and properties of the corresponding concepts. The teacher's simultaneous verbal narration and gestural simulation of the science phenomena displayed in the video provided the students with extra information about the abstract concepts and processes in the science knowledge, which not only facilitated the comprehension of the science knowledge but also engaged the students who raised interesting questions, which Wong Sir answered one by one.

Excerpt 8.1: "Go! Find the Egg! Let's Go!"

[*The scene of ejaculation appeared in the video. Wong Sir started describing the scene in* ***Cantonese***]

T: Look. Ejaculation begins. A large number of sperms come out. And then ... Go! Find the egg! Let's go!

[*the video showed the scene of thousands of sperms swimming swiftly.*]

Sam: The sperms are swimming like crazy.

Jack: I wonder if there are a hundred of them, or two hundred.

T: Shh! Can you see? Their tails are wagging. The sperms are swimming at their best. Look! Fighting each other! So you all are the best! You won the battle. If you lost, you wouldn't have come to this world. You are the winner. You beat tens of thousands of your brothers and sisters to come to this world.

[*The video showed the scene of sperms entering the oviduct.*]

T: Look. Now the sperms have entered the body of a woman, at the uterus, above the wall of the oviduct. Search and search! Here and there! But there are limitations. There are limitations. The lifespan of a sperm is only several hours. If they can't find the egg, or at improper time, then they get nothing.

Ss: Will they die?

Lucy: Wong Sir. What if they all die?

T: What?

Lucy: What if they all die?

T: Then no baby will be born.

[*The video showed the scene of sperms swimming at an egg*]

T: Look here! The egg! We got it! The treasure! Come on!

[*The video showed the scene of the inner wall of the oviduct, which was full of cilia*]

Sam: It looks like a towel.

T: Shh! Yeah.

[*The video showed the scene of an egg.*]

Jack: Like a football!

Sam: Sir. What's this grass-like stuff?

[*The video showed the scene of a sperm swimming in the oviduct*]

Jack: Like a parasite!

T: Wow, smart boy. Shh! Those are not grass. They're cilia at the inner wall of the oviduct in a woman's body. Why? [*holding up his left fist to represent the fertilized egg, wavering the fingers of his right hand beside the left fist to imitate the swaying of cilia to pass the fertilized egg*] Because the fertilized egg can't move, it's passed down to the uterus by the swaying of the cilia, so that the woman is able to get pregnant. So, it is not just by chance that all of you managed to come to this world. First, you beat millions of your brothers and sisters, and found the egg; second, you entered the egg successfully which was then fertilized; third, the fertilized egg was successfully sent to the uterus by the cilia in the inner wall and then adhered to the endometrium …

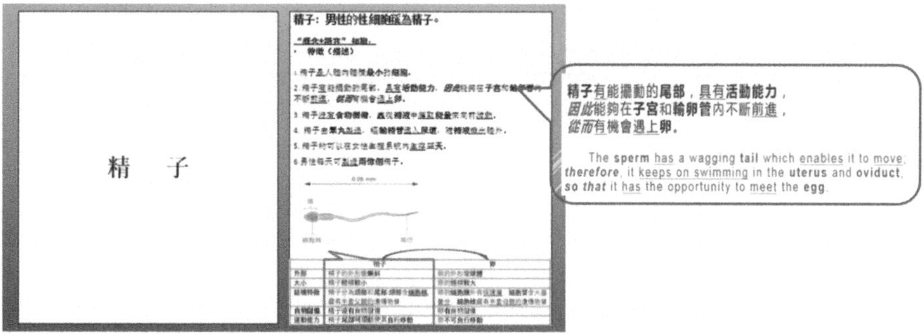

Fig. 8.3 A C + L card of "sperm"

Apart from the video demonstration and the teacher's explanation, the students' science learning was also supported by the CLM materials. A set of C + L cards were distributed to each student, so that they had relevant resources to refer to during their learning of the science lessons. The C + L cards included all relevant concepts as well as the basic information about them.

As Fig. 8.3 shows, the C + L card of "sperm" provided a definition of the concept, its appearance, functions, and properties. The card included a diagram of the concept so that students could have the visual information related to the ideas. It also provided a table comparing the concepts of "sperm" and "egg" in terms of their shape, size, structure, food storage, and movement. It should be noted that different symbols were used to indicate the thematic items (concepts) and the semantic relations between the items (Lemke, 1990) in the verbal information on the C + L card. For example, the bold words are subject-specific words (i.e., thematic items: e.g., 精子sperm, 卵egg, 子宮uterus, and 輸卵管oviduct); the main verbs (i.e., Process: e.g., (游動)前進 swim, 遇上meet) are underlined; and the logical connectors (i.e., cause-and-effect; e.g., 因此therefore, 從而so that) are bold and italic. During the science lessons, Wong Sir also encouraged the students to use academic language to answer the questions, as he reminded them, "記得用翻啲相關專業嘅字眼嚟回答喔!(Bear in mind the relevant professional words and use them to answer the questions!)." As the students generally lacked the ability to write complete sentences and failed to answer the open-ended questions in proper sentence patterns and text-type structures, Wong Sir also provided the students with sentence-making tables which were incorporated into simple writing templates to answer open-ended questions (Fig. 8.4).

The sentence-making table as shown in Fig. 8.4 provided students with a writing guide for answering an open-ended question about comparing two concepts. A comparison text-type structure was designed in a table form with the features of phenomena A and B (e.g., sperm vs egg) listed on the left and right columns in the table, which are connected by a logical connector "而while." A general introduction was provided that summarizes the "five differences (外形shape, 大小size, 結構特

比較點次序	比較對象 A (描述不同比較項特點和功能)	而 (對比)	比較對象 B (描述不同比較項特點和功能)
總體介紹	精子和卵在**外形**、**大小**、**結構特徵**、食物儲備及**運動能力**等五方面不同：		
首先，	精子的外形像蝌蚪	而	卵的外形像球體
其次，	精子的體積較小	而	卵的體積較大
再者，	精子分為頭部和尾部，頭部含有細胞核，藏有**半套父親**的遺傳物質	而	卵的細胞膜外有保護層，細胞質含有大量養分，細胞核藏有**半套母親**的遺傳物質
此外，	精子沒有食物儲備	而	卵有食物儲備
最後，	精子可自行移動	而	卵不可自行移動

Sequence of comparison	Object A compared (describing features/functions being compared)	while (comparing)	Object B compared (describing features/functions being compared)
General introduction	Sperms and eggs *are different in five aspects including* **shape, size, structure, food storage** and **movement.**		
First,	the sperm looks like a tadpole;	*while*	the egg looks like a sphere.
Second,	the sperm is smaller;	*while*	the egg is larger.
Third,	the sperm includes a head and a tail, its head has the nucleus which carries **half** of the genetic materials of the **male parent**;	*while*	the egg is surrounded by a protective layer, its cytoplasm contains food substances, and its nucleus carries **half** of the genetic materials of the **female parent.**
Fourth,	the sperm does not have food substances;	*while*	the egg has plenty of food substances.
Fifth,	the sperm can move by swimming;	*while*	the egg cannot move by itself.

Fig. 8.4 A sentence-making table incorporated into a simple writing template of comparison text-type

徵structure, 食物儲備food storage and運動能力 movement).” These different aspects are compared one by one with sequential logical connectors “首先first, 其次second, 再者third, 此外fourth, and 最後fifth” listed at the beginning of each sentence. Wong Sir tried to use examples from daily life that students were familiar with to help them to understand the importance of following the genre structure to answer questions about comparison.

T: Well, it’s just like you want to choose a suitable husband. Say, now you have options A and B. You need to compare their characteristics one by one rather than just have a general impression of each. So, you compare them item by item, like their height, appearance, family and economic status, etc.”

8.4.2 Translanguaging, Trans-semiotizing, and Trans-registering as Spontaneous Scaffolding During CLM Activities

The CLM materials were distributed to the students in the first lesson. The teacher and the students referred to these instructional materials as scaffolding resources to facilitate science teaching and learning. As the students were afraid of writing,

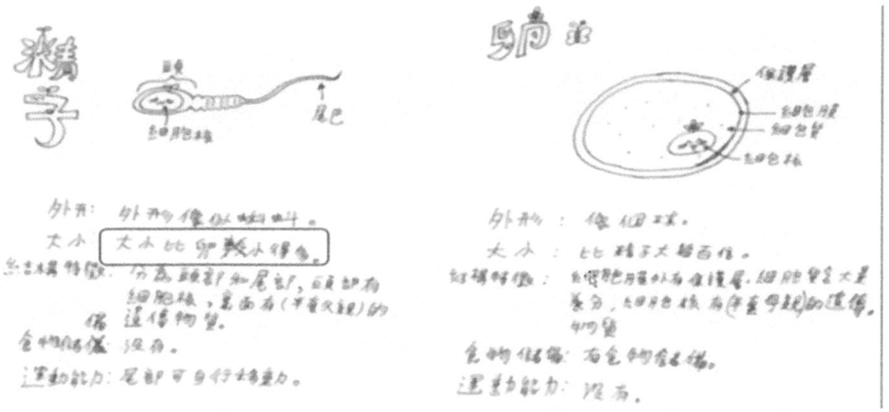

Fig. 8.5 A personalized C + L card about "sperm and egg" designed by a student

Wong Sir allowed them to make their personalized C + L cards about the concepts learned in this unit.

With the help of the CLM materials, students did not seem to find the assignment difficult. In fact, students all produced multimodal C + L cards, which consisted of both verbal texts about the concepts and visual information such as the relevant diagrams. It seemed that students were willing to write some sentences on their personalized cards although some of them could only copy some sentences on the teacher-version C + L cards or the textbook. However, Wong Sir was surprised to find some interesting designs in the students' cards. For example, Fig. 8.5 shows a C + L card about "sperm vs egg" that impressed the teacher most. Although the diagrams of both sex cells were drawn based on the scientific diagrams in the C + L cards shared by Wong Sir, the student actually re-designed the diagram by creatively adding a few "features"; namely, like the cartoons many students enjoyed reading, the diagrams of the sperm and the egg were personalized with "eyes" and "mouth"; to indicate the specific sex of the cells, the drawing became more artistic with some tiny leaves added to the Chinese characters of "精子sperm," whereas some symbols that might be identified as being more feminine, such as a flower and a butterfly being added to the Chinese character of "卵 egg." As for the verbal text, it can be seen that, although the designer of the card copied the information of the teacher-version of the C + L card originally, some information was updated to make the characteristics of the concept more precise. For example, in the description of the size of sperm, the original sentence "(精子)大小比卵較小The size of (a sperm) is smaller than an egg" was later revised to a more precise expression "大小比卵小得多 The size of (a sperm) is much smaller than an egg" – the comparison between sperm and egg that Wong Sir had emphasized during the lesson.

The CLM materials were also used as self-directed learning materials. For example, Wong Sir encouraged the students to use the C + L cards to preview (i.e., self-learn the concepts before the lesson) the male and female reproductive systems. In the next lesson, he invited Lily to present the concept she had previewed. Although

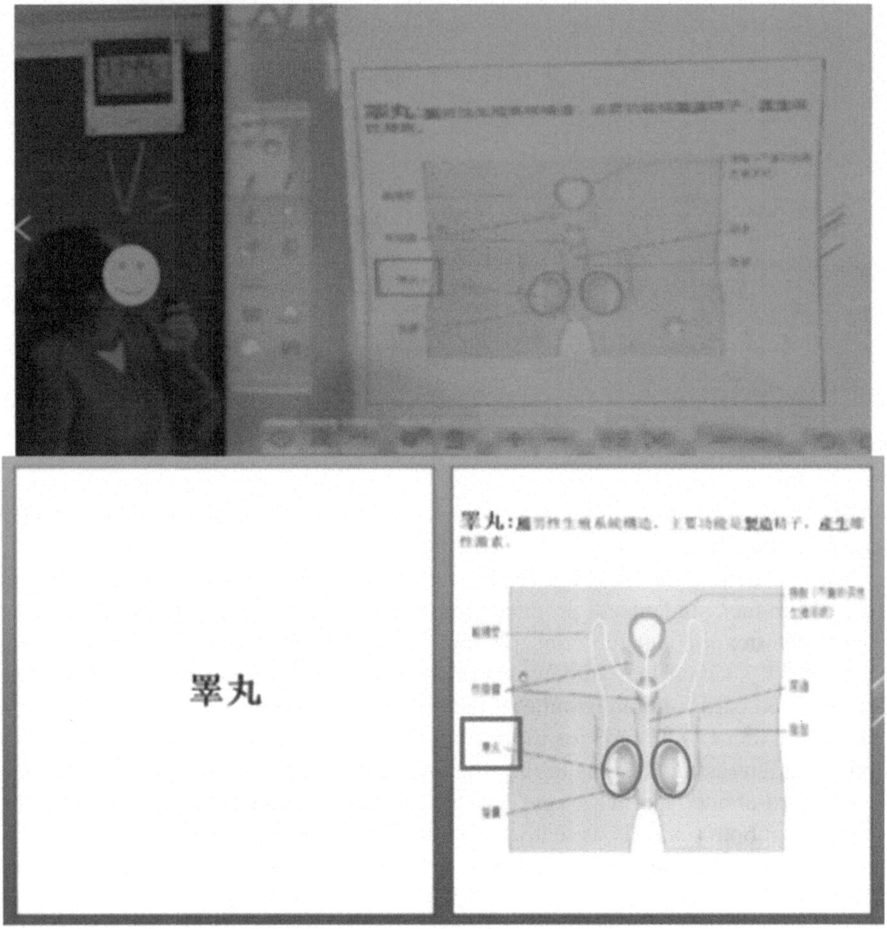

Fig. 8.6 A student introducing the concept "testes" according to the "C + L card"

Lily could only make her presentation in Putonghua – her L1, which was not the MOI of the class, as the C + L card about the concept was projected on the projector screen, both the teacher and the peers were able to understand what she was saying. Lily introduced the concept very fluently, as could be seen by the responses of the teacher and her peers (Fig. 8.6).

Excerpt 8.2: Individual Presentation of Concepts with CLM Cards

Lily: [*Standing in front of the classroom facing the C + L card "testes" on the screen and introducing the concept in* ***Putonghua***]
I think, hmm, actually testes is a male reproductive organ and it's similar to the ovary of a female. They are almost the same; both are in pairs. And it is, hmm, its main function is also similar to that of a female's ovary; that is, it produces sperms and male sex hormones.

Ss: [*Applauding and shouting in* ***Cantonese***] Bravo!!

T: [*Commenting on Lily's presentation in* ***Cantonese***]
Well, do you know what I like most in Lily's presentation? Have you noticed that: first, she's so confident; and second, she didn't just read out the words on the C + L card. What did she try to do? She tried to make some comparison. She tried to place the two reproductive systems side by side. Okay? She tried to compare which part of the female reproductive system is parallel to testes. And she did it VERY well!! She deserves a star!

According to Wong Sir, he appreciated Lily's introduction of the concept because she was able to not only interpret it in her own words but also make a proper comparison between testes and ovary, a pair of concepts that have parallel functions in the male and female reproduction systems. This was related to a worksheet in the CLM materials that asked students to make a comparison between the reproduction system of males and females following the writing template of comparison text-type similar to the one used to compare the characteristics of sperm and egg (Fig. 8.4). As Lily had previewed the lesson according to the related CLM materials, she presented the concept "testes" by making a comparison with the parallel concept "ovary."

The CLM materials also provided collaborative learning activities in the science lessons. For example, after the presentation of the reproductive systems, the students were provided with an A3-sized worksheet with an enlarged diagram of a male or female reproductive system. Students were assigned a group-work to first label the key components in the reproductive systems and then match the different components with the corresponding description of their functions, which were printed on different paper slips. After the labelling and matching work, Wong Sir invited some of the groups to present their work in front of the class (Fig. 8.5). Lily, Lucy, and Maggie were arranged in one group during this activity. As they spoke different languages, their presentation sounded multilingual (Fig. 8.7).

Fig. 8.7 Students labelling diagrams, matching concepts with functions, and presenting their group work

Excerpt 8.3: Labelling and Identifying Functions of Reproductive Organs

	[*Wong Sir helped the students to hold up the diagram of the male reproductive system, then announced the beginning of the presentation in* ***Cantonese***]
T:	The first one.
Lucy:	[*Trying her best to speak out the functions in* ***emerging Cantonese***] Urethra.
T:	[***Cantonese***] The function of the urethra. Louder please. Pull it (the paper slip) up.
Lucy:	[***emerging Cantonese***] It carries semen and urine into the body ... Oh sorry, out of the body.
T:	[***Cantonese***] Okay. That's right. Good. Next function.
Maggie:	[***Cantonese***] Hmm, testes produce sperms and male sex hormones.
Lucy:	[*Signaling to Lily and changing to* ***Putonghua*** *to remind her peer to speak*] Your turn.
Lily:	[*Reading the sentences on the paper slip in* ***Putonghua***] It enters and carries semen to the vagina of a female during sexual intercourse.
T:	[*Reminding the audience in* ***Cantonese*** *that Lily was introducing the functions of the penis*] Okay? Right. Lily was talking about the functions of the penis.
Lucy:	[***emerging Cantonese***] Sex glands. They produce sex hormones and a liquid which provides nutrients for the sperms and allows sperms to swim.
Maggie:	[***Cantonese***] Hmm. Sperm ducts. They carry sperms from the testes to the urethra.
T:	[***Cantonese***] Okay. Good. All done. Thank you girls.

Although the multilingual presentation might not sound as straightforward to the audience as one presented monolingually, it did not seem to have affected their understanding of the functions of the components of the reproductive systems. Maggie was a local student and had no difficulty presenting in Cantonese. Lucy and Lily were both from Mainland China and their L1 was Putonghua. Lily had been studying in Hong Kong for only a few months and had weaker listening to Cantonese and therefore had a slower response and needed to be reminded by peers (e.g., Lucy reminded that it was her turn to speak). Lucy had been learning Cantonese and had relatively better listening to the language. She tried to present in Cantonese, but she could not speak the language fluently and some words were pronounced incorrectly. To help the Cantonese-speaking students to understand the presentation by Lucy and Lily, and the Putonghua-speaking students to understand the part by Maggie, Wong Sir reminded Lucy to speak louder and asked the presenters to hold up the printed paper slip so that the whole audience was able to capture both audio and visual information related to the presentation. The teacher also reminded the audi-

Fig. 8.8 Students building and presenting their C + L maps in groups

ence of the function of the concept that was being introduced so that students would not mismatch the concepts with their respective functions (Fig. 8.8).

With the scaffolding of the CLM materials, the CLM activities and games, as well as Wong Sir's elaboration on the science concepts, the students gradually became familiar with the concepts they had learned. In order to guide students to build up the interrelationship among the different concepts, Wong Sir assigned the students another group task – building a C + L map of "implantation." Some paper slips were distributed among the students on which there were printed vocabulary or sentences about description of the implantation process. To build the C + L map, students in different groups needed to make good use of their C + L cards, which represent the relevant concepts involved in the implantation process. With the C + L cards and paper slips, they needed to find out a way to piece up the different concepts (i.e., C + L cards) and the descriptive language (i.e., vocabulary and sentences on the paper slips) so that the C + L map describes the process of implantation correctly in terms of both the science knowledge and the academic language knowledge. For better demonstration of the C + L map, Wong Sir gave each group a large poster-sized piece of paper so that students were able to stick the C + L cards and paper slips on the paper and form their "map."

Excerpt 8.4: Building a C + L Map of"植入 Implantation"

	[*Students had built their C+L map and were invited to present their map in front of the class*]
T:	[*Ask Lily and Maggie*]Okay, who is going to present?
	[*Lily and Maggie both would like the other to present*]
T:	Who leads the presentation?
Sam:	Why not present together? You're partners.
T:	Which channel? A Putonghua channel or a Cantonese one?
Sam:	Cantonese.
Lucy:	TVB.
Jack:	ATV.
	[*Lily and Maggie decided to present together*]
T:	Okay. Give them a big hand.
Maggie:	[*Speaking to Shirley in **Cantonese***] You point to the words on the map and we present.
T:	Shh::!

(continued)

Lily & Maggie:	[*Started to present together in* ***Putonghua*** *with Shirley pointing to the words on the map*]After entering the female reproductive system, the sperm fuses with an egg produced by the ovary, a fertilized egg is formed. During this process, the fertilized egg experiences repeated cell division and develops into an embryo. When the embryo reaches the uterus, it is em .. .em ... (嵌 ... ... 嵌 ... ...)
Researcher:	Embedded (嵌入).
Lily & Maggie:	It is embedded in the uterine lining. This process is called implantation.
Ss:	[*Giving applause, showing thumb-ups and "like" cards*] Oh! Bravo!

After the groups had all made their C + L maps, Wong Sir posted them on the blackboard and invited the groups to present their maps one by one. In Lily's group, Shirley spoke only Cantonese, Lily spoke only Putonghua, and Maggie could speak both. Shirley wanted to play the role of guiding the audience to read their map by pointing at the concepts, but Lily and Maggie were both shy and wanted their groupmate to do the presentation. They finally decided to present the map together. Wong Sir realized the different L1s of the students and allowed them to select the language by asking a funny question "Which channel? A Putonghua channel or a Cantonese one?," which amused the audience as some mentioned the names of the local TV channels (i.e., TVB and ATV) as suggestions. Although the group members spoke different L1s, they cooperated very well in the C + L map-building and the presentation. Their clear and fluent presentation won the applause of both the teacher and the peers. It should be noted that, at the end of the C + L map-building activity, Wong Sir made a long summary about the importance of reflection on learning by connecting the concepts in prior knowledge with those in the new knowledge following the way of the C + L map.

T: Okay, students. We didn't play this game for differentiating the higher and lower achievers. I want you to know, I mean, in fact, in each lesson, for different subjects, teachers teach the same knowledge, but whether you learn it or not, whether you can use it, depends on what? It depends on whether you have actually used your mind to think about it. Or if you just sit there [*showing a stiff posture and a dazing expression*], with only your teacher talking, then the knowledge would just go out. But if it enters your mind and you catch it [*pointing to the C + L maps on the blackboard*], piecing up in your mind the images of the knowledge, then I think it will leave a deeper impression on you. Do you agree? So, I hope, through this activity, we don't just focus on who wins, but I hope you try to piece up what you've learned in the previous lessons. Through making your own C + L map, observing others make their maps; through presenting your own map and observing others present their maps, you reflect on what you've learned, you'll then get a deeper impression of the knowledge.

8.5 Discussion

After five science lessons in the collaboration unit, the students reported that the CLM activities were interesting and the provision of the CLM materials made learning easier for them. Wong Sir also commented that the CLM pedagogy helped to increase student participation and provided useful scaffolding to teaching and learning the science knowledge. These qualitative data collected from the post-intervention interviews with the teacher and the students were also evidenced by the quantitative data of the test results to some extent, which demonstrated that students had made progress in the development of both content knowledge and academic language. In this section, we shall discuss the above findings in this research.

8.5.1 Meaning-Making Is a Multimodal Process of Thematic-Patterns-Based Translanguaging, Trans-semiotizing, and Trans-registering

Observations of the science lessons in this research indicated that knowledge construction was developed based on the simultaneous meaning-making of different modes (Danielsson, 2016) – verbal mode (e.g., written texts in the CLM materials and textbook, and spoken communication between the teacher and students), visual mode (e.g., videos and diagrams in the CLM materials, textbook and student assignments), and actionable mode (e.g., the teacher's gestures and the actions of students in the CLM games). Each of these modes of presentation had its "meaning potential" (Van Leeuwen, 2005) or "affordance" for meaning-making (Danielsson, 2016). During the science lessons, the different modes were deployed simultaneously in a dynamic and dialogic manner; for example, when Wong Sir described the sperm searching for the egg in Excerpt 8.1, his verbal introduction served as the background information to the scenes in the video, which was the foreground that attracted students' attention. When he introduced the functions of the cilia in the inner wall of the oviduct, his gestures imitating the cilia passing down the egg became foregrounded whereas the video scenes became the background (Kress et al., 2001). The simultaneous orchestration of different modes as well as the alternating backgrounding and foregrounding between them did not just repeat the same meaning of the science knowledge (i.e., the thematic patterns) being discussed, but actually amplified the meaning (Kress et al., 2001; Lemke, 1998) by activating the unique "affordance" of different modes; namely, the visual modes (e.g., videos and diagrams) helped to visualize the position and interrelationship between components in the male or female reproduction system, as well as the vivid process of sperms trying to enter an egg; the actional modes simulated the processes or actions that may be unobservable (e.g., the movement of cilia passing down an egg), and slowed down or exaggerated the actions and movements that may be too subtle to be seen clearly (e.g., the fusing of the nucleus of an egg and a sperm); the verbal modes not only conveyed the abstract information that allowed repeated interpretation (e.g., written information in the CLM materials) but also allowed dialogic meaning negotiation (e.g., the teacher's spontaneous scaffolding and feedback as well as the peer discussions during the lessons). In this sense, knowledge construction was realized in a series of multimodal meaning-making activities in the science lessons (Danielsson, 2016).

In the multilingual multicultural science classroom, meaning-making was also a process of translanguaging (García & Li, 2014), trans-semiotizing (Lin, 2015; Lin & He, 2019), and trans-registering (Lin et al., 2020) practices that were dynamic, fluid, dialogic, and distributed within and beyond the science classroom. The meaning-making through different modes of representation was not limited by boundaries (Li & Lin, 2019) of time, space, language or culture, but interacted and developed into *flows* (Lemke, 2016) of information, energy, and emotion that were gradually entrained by the communicators and continually added to the expanding communicative repertoire (Lin, 2012) shared by the individuals (teacher and students) in different communities (pairs, groups, class, and family, etc.) (Lin et al., 2020). During the knowledge construction in the science lessons, the prior knowledge about "giving birth to life" was activated through brainstorming relevant concepts in a spider graphic organizer. This previous knowledge was brought back to real-time learning in the classroom and shared among the teacher and peers. The same topic was further discussed under the guidance of the teacher by watching videos of different registers (i.e., the animated cartoon for primary learners and the scientific video clip produced by professionals using high-tech video cameras). The prior knowledge about the science topic, previously expressed in everyday language (e.g., "有咗," a colloquial word of "pregnant" in Cantonese), was updated to a "science word" (e.g., "懷孕," a formal expression of "pregnant" in Chinese). Students captured the scientific nomenclature in their cartoon. The relevant process of giving birth to life was presented by the visual demonstration in the video scenes and the teacher's simultaneous verbal elaboration and gestural simulations. All these trans-registering and trans-semiotizing practices during the lessons had not only engaged the students but also motivated their imagination, which was embodied in their creative descriptions, questions, and drawings during their science learning.

It should be noted that, although in the science classroom multiple "named languages" co-existed (e.g., Cantonese, Putonghua, emerging Cantonese, and English), meaning-making was not limited by the static and rule-based "code-view" of language (Li, 2018); rather, it was a process of "*languaging*," which is characterized by "an assemblage of diverse material, biological, semiotic and cognitive properties and capacities which language agents orchestrate in real-time and across a diversity of timescales" (Thibault, 2017, p.82). In the three scenarios about Lily's science learning (Excerpts 8.2, 8.3 and 8.4), the different "name languages" including her L1 (Putonghua) constituted just one type of semiotic resources that had been incorporated into the "assemblage" of diverse semiotic, material, biological, and emotional artifacts (e.g., CLM materials, personalized C + L card, "stars," "like" cards and applause), acts (e.g., designing personalized C + L cards, labeling diagrams, matching functions, and building C + L maps), and events (e.g., games and knowledge competition) that were leveraged by the language agents – Lily, the teacher, and peers, during a process of seamless translanguaging and trans-semiotizing throughout the five science lessons. For example, in the matching functions activity (Excerpt 8.3), the different "named languages" (spoken Cantonese, spoken Putonghua, academic Cantonese, academic Putonghua, and written Traditional Chinese characters on the paper slips) and the visual information on the A3 paper – a diagram of the reproductive system, an "assemblage" of audio and visual, spoken and written, colloquial and academic, L1 and additional languages that were selected based on the underlying meaning relationship of concepts – the "thematic patterns" following the principle of "repetition with variation" during the meaning-making process (Lemke, 1990). It is thus essential to accentuate that the translanguaging, trans-semiotizing, and trans-registering practices were neither random nor arbitrary, but were guided by the thematic patterns that connected the basic thematic items (i.e., concepts) in the science theme – give birth to life. The thematic patterns as a "web of meaning relationships" (Lemke, 1990) provided the fundamental clue for selection of appropriate "affordance" from the shared communicative repertoire (Lin, 2012) for meaning-making. The C + L cards in the science lessons were the embodiment and carriers of such thematic patterns, which students frequently referred to during the CLM activities. Hence, the thematic patterns as information and artifact of the integration of content and language were unpacked and repacked in different CLM activities that facilitated knowledge construction in the multilingual, multicultural science classroom.

8.5.2 Academic Literacy Develops Through a Process of Thematic-Pattern-Based Trans-registering in Science Education

Just like content-based instruction (CBI) and CLIL, science education inevitably confronts the challenge of academic literacy development. This was a typical issue in the collaborative school in this study. As Wong Sir admitted, the students were facing both cognitive and linguistic difficulties, which resulted in their inadequate academic literacy in science learning. Just like every academic school discipline, there is a continuum between social language and academic language (Swinney & Velasco, 2011), along which there are different linguistic registers that construe meanings of different levels of complexity. Lemke (1990) emphasized the necessity to bridge colloquial and scientific languages in science education, and suggested that teachers may guide their students to understand the semantic and conceptual relationships in colloquial language first and then gradually help them to move toward academic language by replacing the colloquial language with the scientific and technical terms. It is also worth noting that in the teaching of academic language, teachers need to raise students' awareness of academic language by making the reasoning and form-meaning relationships explicit (Schleppegrell, 2009).

In Wong Sir's science lessons, academic language development was realized in a process of thematic- pattern-based trans-registering. This process was also facilitated by both the designed scaffolding (i.e., the CLM teaching materials) and the spontaneous scaffolding (i.e., the teacher's guidance and feedback during the CLM activities) (Gibbons, 2009; Lin, 2016). The CLM teaching materials allowed the students to learn science knowledge through different registers (e.g., C + L cards and sentence-making tables of academic register, and an animated cartoon, which presented a science topic in both every-day and academic registers). Wong Sir's spontaneous scaffolding was crucial for the cognitively and linguistically unprepared students. For example, he encouraged students to note down the "science words" in the animated cartoon, and kept reminding students to use the "專業字眼 professional vocabulary" to answer the questions. In his description of the scenes in the professional video clip (Excerpt 8.1), it can be seen that Wong Sir had tried to use metaphoric language (e.g., sperms as "brothers and sisters") and colloquial language (e.g., "砌低咗" colloquial Cantonese saying meaning "defeated/beat") to introduce the science concepts, so that students were able to understand the semantic and conceptual relationships first (Lemke, 1990). When presenting the sentence-making table, which was incorporated into a simple writing template of a comparison text-type (Fig. 8.4), the example Wong Sir provided (i.e., the choosing of a suitable husband) was typical of the everyday discourse in the local community, which helped students to grasp the technique of making comparison in academic registers. This technique was actually shown in Lily's presentation of testes (Excerpt 8.2) by making a proper comparison between the parallel concept ovary in the female reproductive system. Another useful CLM activity for academic literacy development was the C + L card design assignment. Instead of using just the same set of teacher-prepared C + L cards, which represented the academic register, Wong Sir allowed students to design their own C + L cards. As Fig. 8.5 shows, the students' personalized C + L cards represented the hybrid of a science register (i.e., scientific description of sex cells) and a cartoonist register (i.e., cartoonist expression of male and female in children's imagination). Such trans-registering was encouraging in science education as it engages students by exercising their creativity. With ongoing knowledge construction, students were constantly shaping their content and language knowledge and move continually toward academic language; for example, the student revised on her personalized C + L card the description of the size of sperm by more accurate use of language (i.e., from "smaller than an egg" to "much smaller than an egg").

8.6 Implications

Reflecting on this research, the data analysis has offered us evidence for some of our theoretical and pedagogical assumptions. For example, the traditional monolingual purism in MOI policies was not supported in the multilingual, multicultural science classroom in this study. We propose further empirical research on the theorization of the dynamic, dialogic, fluid, and distributed view of translanguaging, trans-semiotizing, and trans-registering (Lin et al., 2020) in meaning-making and knowledge construction of different school disciplines. We are impressed by the collaborative students' active responses and positive comments in the CLM activities as well as the teacher's summary on the usefulness of the CLM pedagogy (i.e., Excerpt 8.4, the building of C + L maps based on the thematic pattern-based C + L cards), which have corroborated the results in our previous studies (He & Lin, 2019; Lin & He, 2017). However, we also realized some limitations in this study; for example, owing to a tight school schedule, we only had one unit of science lessons for collaboration and only observed five lessons, which did not form a larger data source. Similarly, as the current study was based on the qualitative data of project research, although we collected sufficient evidence about the effects of the CLM pedagogy, we could not obtain the feedback of the teacher and students on the translanguaging practices. Therefore, future research may be aimed at the longitudinal observation of lessons and interviews that probe the perceptions of teachers and students on the translanguaging, trans-semiotizing, and trans-registering practices. Being both linguists and educators, we aim for a wider and more profound collaboration with teachers, researchers and specialists in science education and research to turn deficit-based curriculum genres to assets-based ones (Lin, 2020). Judging by the facilitating role of translanguaging, trans-semiotizing, and trans-registering practices in the content and language development of students from diverse backgrounds, future studies may explore the impact of translanguaging practices on classrooms of inclusive education, which is a crucial mission of world education today (UNESCO, 2020).

References

Bloome, D., Cater, S. P., Christian, B. M., Otto, S., & Shuart-Faris, N. (2005). *Discourse analysis and the study of classroom language and literacy events: A micro-ethnographic perspective*. Lawrence Erlbaum Associates.

Creswell, J. W. (2003). *Research design: Qualitative, quantitative, and mixed methods approaches* (2nd ed.). Sage.

Danielsson, K. (2016). Modes and meaning in the classroom – The role of different semiotic resources to convey meaning in science classrooms. *Linguistics and Education, 35*, 88–99.

García, O., & Li, W. (2014). *Translanguaging: Language, bilingualism and education*. Palgrave Macmillan.

Gibbons, P. (2009). *English learners, academic literacy, and thinking: Learning in the challenge zone*. Heinemann.

Halliday, M. A. K., & Hasan, R. (1976). *Cohesion in English*. Longman.

Halliday, M. A. K. (1993). Towards a language-based theory of learning. *Linguistics and Education, 5*(2), 93–116.

Halliday, M. A. K. (2013). Languages, and language, in Today's changing world. Research seminar given at the University of Hong Kong, October 23.

He, P., & Lin, A. M. Y. (2019). Co-developing science literacy and foreign language literacy through "concept + language mapping.". *Journal of Immersion and Content-Based Language Education, 7*(2), 261–288.

He, P., Lai, H., & Lin, A. M. Y. (2017). Translanguaging in a multimodal mathematics presentation. In C. M. Mazak & K. S. Carroll (Eds.), *Translanguaging practices in higher education: Beyond monolingual ideologies* (pp. 91–118). Multilingual Matters.

Hodgson-Drysdale, T. (2014). Concepts and language: Developing knowledge in science. *Linguistics and Education, 27*, 54–67.

Kress, G., Jewitt, C., Ogborn, J., & Tsatsarelis, C. (2001). *Multimodal teaching and learning: The rhetorics of the science classroom*. Continuum.

Kusters, A., Spotti, M., Swanwick, R., & Tapio, E. (2017). Beyond languages, beyond modalities: Transforming the study of semiotic repertoires. *International Journal of Multilingualism, 14*(3), 219–232.

Lemke, J. (1990). *Talking science: Language, learning and values*. Ablex.

Lemke, J. (1998). Multiplying meaning: Visual and verbal semiotics in scientific text. In J. R. Martin & R. Veel (Eds.), *Reading science: Critical and functional perspectives on discourses of science* (pp. 87–113). Routledge.

Lemke, J. (2016). *Translanguaging and flows*. Unpublished research manuscript.

Lemke, J. (2018). *Translanguaging and flows*. Webinar hosted at Simon Fraser University, Canada, 18 September 2018.

Li, W. (2018). Translanguaging as a practical theory of language. *Applied Linguistics, 39*(1), 9–30.

Li, W., & Lin, A. M. Y. (2019). Translanguaging classroom discourse: Pushing limits, breaking boundaries. *Classroom Discourse, 10*(5), 209–215.

Li, W., & Zhu, H. (2013). Translanguaging identities and ideologies: Creating transnational space through flexible multilingual practices amongst Chinese university students in the UK. *Applied Linguistics, 34*(5), 516–535.

Lin, A. M. Y. (2012). Multilingual and multimodal resources in L2 English content classrooms. In C. Leung & B. Street (Eds.), *English – A changing medium for education* (pp. 79–103). Multilingual Matters.

Lin, A. M. Y. (2015). Egalitarian bi/multilingualism and trans-semiotizing in a global world. In W. E. Wright, S. Boun, & O. García (Eds.), *The handbook of bilingual and multilingual education* (pp. 19–37). Wiley Blackwell.

Lin, A. M. Y. (2016). *Language across the curriculum & CLIL in English as an additional language (EAL) contexts*. Springer.

Lin, A. M. Y. (2019). Theories of trans/languaging and trans-semiotizing: Implications for content-based education classrooms. *International Journal of Bilingual Educationand Bilingualism, 22*(1), 5–16.

Lin, A. M. Y. (2020). From deficit-based teaching to asset-based teaching in higher education in BANA countries: Cutting through 'either-or' binaries with a heteroglossic plurilingual lens. *Language, Culture and Curriculum, 33*(2), 203–212.

Lin, A. M. Y., & He, P. (2017). *Arousing interest and increasing access: A Content + Language Mapping (CLM) approach to content and language integrated learning (CLIL)*. Paper presented at the International Conference of the International Society for Language Studies (ISLS), 15–17 June 2017, University of Hawaii at Manoa, USA.

Lin, A. M. Y., & He, P. (2019). *Integrating content and language learning in EMI education – exploring "Thematic patterns" as pedagogical strategies*. Presentation at Research and Development Project Dissemination Conference of SCOLAR: Advancing Excellence in Language Education: From Research and Development to Innovative Practice, at Hong Kong Science Park, Sha Tin, Hong Kong, on May 9th, 2019.

Lin A. M. Y., Wu Y. (Amy), & Lemke J. L. (2020). 'It takes a village to research a village': Conversations between Angel Lin and Jay Lemke on contemporary issues in translanguaging. In S. Lau & S. Van Viegen (Eds.), *Plurilingual pedagogies: Critical and creative endeavors for equitable language in education* (pp. 47–74). Springer.

Lincoln, Y. S., & Guba, E. G. (1985). *Naturalistic inquiry*. Sage Publications.

Maturana, H., & Varela, F. (1980). Biology of cognition. In Autopoiesis and cognition: The realization of the living. In R. S. Cohen & M. W. Wartofsky (Eds.), *Boston studies in the philosophy of science* (Vol. 42). Reidel.

Merriam, S. B. (1998). *Qualitative research and case study applications in education.* San Francisco: Jossey-Bass Publishers.

Miles, M. B., & Huberman, A. M. (1994). *Qualitative data analysis: An expanded sourcebook.* Sage.

Mortimer, E., & Scott, P. (2003). *Meaning making in secondary science classrooms.* Open University Press.

Osborne, J. (2014). Scientific practices and inquiry in the science classroom. In N. G. Lederman & S. K. Abell (Eds.), *Handbook of research on science education* (Vol. II, pp. 593–613). Routledge.

Pérez-Milans, M. (2016). Bilingual education in Hong Kong. In O. García, A. M. Y. Lin, & S. May (Eds.), *Bilingual and multilingual education: Encyclopedia of language and education* (Vol. 5, pp. 207–218). Springer.

Reeves, T. C. (2000). Enhancing the worth of instructional technology research through "design experiments" and other development research strategies. *International perspectives on instructional technology research for the 21st century, 27*, 1–15.

Schleppegrell, M. J. (2009, October). *Language in academic subject areas and classroom instruction: What is academic language and how can we teach it?* Paper prepared for the Workshop on the Role of Language in School Learning: Implications for Closing the Achievement Gap, Hewlett Foundation, Menlo Park, CA.

Swain, M., & Lapkin, S. (2013). A Vygotskian sociocultural perspective on immersion education: The L1/L2 debate. Journal of immersion and content-based language Education1 (1). 101–129.

Swinney, R., & Velasco, P. (2011). *Connecting content and academic language for English learners and struggling students grades 2–6.* SAGE Ltd.

Thibault, P. J. (2011). First-order languaging dynamics and second-order language: The distributed language view. *Ecological Psychology, 23*(3), 210–245.

Thibault, P. J. (2017). The reflexivity of human languaging and Nigel Love's two orders of language. *Language Sciences, 61*, 74–85.

Tytler, R., & Prain, V. (2010). A framework for re-thinking learning in science from recent cognitive science perspectives. *International Journal of Science Education, 32*(15), 2055–2078.

UNESCO. (2020). *Global education monitoring report 2020: Inclusion and education: All means all.* UNESCO.

Van Leeuwen, T. (2005). *Introducing social semiotics.* Routledge.

Wellington, J. J., & Osborne, J. (2001). *Language and literacy in science education.* Open University Press.

Working CLIL Digital (2018, May 29). Professor Do Coyle - What are the principles of CLIL? [Video].YouTube. https://www.youtube.com/watch?v=fS7VfRLOgnI&t=231s

Wu, Y., & Lin, A. M. Y. (2019). Translanguaging and trans-semiotizing in a CLIL biology class in HongKong: Whole-body sense-making in theflows of knowledge co-making, classroom Discourse, *10* (3–4), 252–273.

Chapter 9
Translanguaging in Middle School Science: Written Arguments About Issues of Biodiversity

Peter Licona and Gregory J. Kelly

Abstract The intersection of the cultural and linguistic diversification and science education reform in the United States provides researchers opportunity to investigate how emergent bilinguals engage in the epistemic practices of science. These practices include learning ways of formulating evidence in specific genres leading to the development of scientific argumentation. This study reports on how middle school (ages 12–13), English/Spanish emergent bilinguals constructed written arguments about issues of biodiversity, such as invasive and endangered species. Using a socioscientific issues pedagogical approach, students engaged with ill-defined and open-ended problems designed to elicit responses to societal issues related to science. Through translanguaging as a pedagogical strategy, the classroom teacher facilitated students in constructing and evaluating arguments using a scientific argumentation framework. Written arguments were analyzed to examine the ways that evidence was structured, the referents for decisions regarding socioscientific issues, and nature of the conclusions drawn. The analysis of the student work demonstrated ways that the communicative function of the academic task was accomplished, and often enhanced, by the uses of translanguaging among the student writers.

Keywords Translanguaging · Pedagogical strategy · Socioscientific issues · Scientific arguments

P. Licona (✉)
Elizabethtown College, Elizabethtown, PA, USA
e-mail: liconap@etown.edu

G. J. Kelly
Pennsylvania State University, State College, PA, USA
e-mail: gkelly@psu.edu

© The Author(s), under exclusive license to Springer Nature Switzerland AG 2021
A. Jakobsson et al. (eds.), *Translanguaging in Science Education*, Sociocultural Explorations of Science Education 27,
https://doi.org/10.1007/978-3-030-82973-5_9

9.1 Introduction

9.1.1 Diversification of American Schools

Schools in the United States, much like in other areas of the world, are continuing to experience cultural and linguistic diversification through increased migration and domestic increases of diverse populations. Nowhere else is this diversification more evident than in schools. In the United States, culturally and linguistically diverse students (CLDs) are the fastest growing K-12 demographic group. Culturally and linguistically diverse (CLD) is a blanket term which generally refers to students whose language and culture are different than the societally dominant language and culture. CLD, as a blanket term, encompasses a wide range of students, such as recent migrants, English Learners, and emergent bilinguals. As a result of the increase of CLDs, many schools are moving from monolingual to bilingual or multilingual instructional approaches. In the United States, the largest increase in CLDs has been in Latina/o students who speak Spanish as a home language. Recent statistics note that in 2017, Latina/o students represented 27% of public school enrollment and are projected to comprise 28% by 2029 (NCES, 2020).

9.1.2 Culturally and Linguistically Diverse Students in Science

The continuing diversification of our school-aged student population has historically presented our school systems with both challenges and opportunities in providing diverse students with accessible educational opportunities in all subjects, particularly science. Research indicates that CLDs often underperform on traditional science achievement outcomes (Lee & Luykx, 2006), raising questions about the opportunity gaps in the education system. The effect from failing to engage these students in science leads to long-term consequences such as a lack of representation in the science workforce (Landivar, 2013) and underrepresentation in STEM teaching (U.S. Department of Education, 2016). While issues of academic success and affiliation in the workplace are important, we must also realize the importance that scientific literacy plays in CLDs' ability to engage in meaningful ways with myriad socioscientific issues inherent in today's advanced scientific and technological society. Furthermore, CLDs have much to offer and can improve science and the associated practices through participation and innovation. Students' funds of knowledge (Moll et al., 1992) can be used to leverage learning but can also offer new perspectives on science. As science and science education have long been historically criticized for being less accessible for students from diverse cultural and linguistic backgrounds, it is important that educators investigate pedagogy and curriculum that render science education more accessible and responsive to culturally and linguistically diverse students.

9.1.3 English Learners and Emergent Bilinguals

Determining the appropriate terminology to refer to a particular group of students has always plagued the educational community. There has been significant progress in the United States in moving away from deficit-oriented terms such as LEP (Limited English Proficient) toward the use of English Language Learner (ELL) or English Learner (EL). Both ELL and EL refer to students who are in the process of attaining proficiency in English as a new, additional language (Wright, 2019). While the use of ELL and EL is an improvement from LEP, these terms do mask the important fact that many students are learning two, or more, languages simultaneously. Therefore, we choose to refer to the students presented in this study as emergent bilinguals (García et al., 2008). Emergent bilingual captures the fact that students are in the process of learning two or more language and frames students based on their assets and not on their deficits. Additionally, emergent bilingual is a more appropriate term given that this study takes place in a dual language learning environment.

9.2 Linguistic Diversity and Science Learning

9.2.1 Sociocultural Perspective on Learning

A sociocultural orientation to teaching and learning science recognizes the many ways that cultural norms and practices, discourse processes, and social interactions construct differential opportunities for student learning. This theoretical orientation situates science learning in contexts (political, institutional, cultural) to address the inequities faced by culturally and linguistically diverse students. In this chapter, we examine the interactions among students' spoken and written discourses in a dual language school. In this, and other educational settings, these interactions "shape and are shaped by discourses, develop situated definitions of what it means to be a scientist, reader, writer, a group member among other rules and relationships, and construct local knowledge that becomes common knowledge within the group or class" (Kelly & Green, 1998, p. 147). Learning and meaning-making can be thought of as occurring through interactions – interactions among individuals or as individuals interact with the cultural products of the classroom, such as books, practices, or content (Leach & Scott, 2003). Furthermore, this perspective acknowledges that society, the educational system, the school culture, home culture, and peer culture also influence learning environments both within and outside the classroom (González et al., 2005). Students and teachers are not only local members, in this case, of a science classroom but also members of other groups and may bring discourse practices, experiences, values, and practices that are similar to, or differ from, those used within the science classroom (Kelly & Green, 1998). This perspective takes on added significance as our science classrooms become more linguistically complex due to science education reform efforts and the continued cultural and linguistic diversification of our student demographics.

9.2.2 *Relevant and Responsive Pedagogies*

In response to failure of the educational system to provide equitable education, numerous pedagogical strategies, such as culturally relevant (Ladson-Billings, 1995) and culturally responsive (Gay, 2002) pedagogy, have been implemented to conceptualize and utilize a student's culture as a resource to leverage learning and success in school. While these pedagogies focusing on culture have proven to be very beneficial in facilitating success among diverse students, more recent pedagogies have focused on the importance of language in teaching and learning. Linguistically responsive teaching (Lucas & Villegas, 2013) draws attention to how teachers' understanding of the impact of language on teaching and learning can benefit students whose home language is different than that of classroom instruction. Lucas and Villegas (2013) frame linguistically responsive teaching as being comprised of both orientations and pedagogical knowledge and skills. The orientations are defined as sociolinguistic consciousness, recognition of the value for linguistic diversity, and the inclination to advocate for ELLs. The pedagogical knowledge and skills are defined as a repertoire of strategies for learning about linguistic and academic backgrounds of ELLs in English and native languages, an understanding of, and ability to apply, key principles of second language learning, the ability to identify the language demands of classroom tasks, and a repertoire of strategies for scaffolding instruction for ELLs. Taken collectively, the orientations, knowledge, and skills allow teachers to consider how language, situated in a cultural context, impacts how students learn.

Responsive pedagogies frame diversity as an asset and seek ways to draw on the knowledge and experiences of diverse students to achieve success in school. In a more critical approach to working with culturally and linguistically diverse students, Paris (2012) posits that relevant and responsive pedagogies do not go far enough in supporting these students. He suggests that pedagogies be more than responsive or relevant; they need to support students in sustaining and developing their home culture and language(s) while facilitating access to dominant cultural competence.

These pedagogies hold promise in leveraging student cultural and linguistic resources in the pursuit of science learning. Brown and Spang (2008) noted how a science teacher responding to student's out-of-school linguistic practices was able to create a hybrid method of language to leverage science learning. In an often-cited study, Warren et al. (2001) conducted research that resulted in findings that students' cultural and linguistic practices can be used as resources for learning. These resources fall along a continuum of everyday language and practices and the languages and practices used by scientists. Their research positions students' everyday language and cultural practices as serving as a basis for developing sense-making activities within the science classroom. At the core of the above research studies is that relevant and responsive approaches to science education allow students to use their cultural and linguistic resources in making meaning in science classrooms.

9.2.3 Translanguaging as Pedagogy

In recent years, translanguaging as pedagogy (García et al., 2016) has gained attention for its potential in providing emergent bilinguals with equitable learning opportunities. Translanguaging as a linguistic practice can be defined as the fluid and dynamic language practices bilinguals employ to make sense of their bilingual worlds (García, 2009). Teacher translanguaging embodies both linguistically responsive and culturally sustaining pedagogies. The mere act of a teacher employing translanguaging as a pedagogical strategy demonstrates the orientations and pedagogical knowledge and skills, mentioned above. As a culturally sustaining pedagogy, translanguaging supports students in sustaining and developing their out-of-school or home culture(s) and language(s) while facilitating access to dominant cultural discourses.

Translanguaging as an academic pedagogy holds much promise for engaging emergent bilinguals in the general academic practices of school. In this chapter, we examine specifically the potential of translanguaging in teaching and learning science. As many culturally and linguistically diverse students are not afforded consistent learning opportunities with school science as currently taught, translanguaging can be seen as a pedagogical approach to improve student engagement with and learning of science. Furthermore, given the language demands of science, translanguaging shows potential in providing emergent bilinguals with linguistic scaffolding into the language-intensive discourses of science.

Translanguaging as pedagogy refers to the dynamic, fluid, and responsive languaging practices of bilingual teachers in a concerted effort to allow all students access to the knowledge and practices being constructed in bilingual classrooms. Creese and Blackledge (2010) argue that translanguaging is a flexible bilingualism that is used by teachers as an instructional strategy to bridge classroom practices with those found in students' social, cultural, community, and linguistic worlds. García and colleagues (2016, p. 2) in relation to translanguaging as pedagogy describe the translanguaging classroom as "a space built collaboratively by the teacher and bilingual students as they use their different language practices to teach and learn in creative and critical ways." Their definition of translanguaging pedagogy includes four primary purposes: supporting students as they engage with and comprehend complex content and texts, providing opportunities for students to develop linguistic practices for academic contexts, making space for students' bilingualism and ways of knowing, and supporting students' bilingual identities and socioemotional development. Translanguaging as a science pedagogy takes on a more nuanced meaning given the sophisticated content and practices being promoted by recent science education reform efforts, such as *A Framework for K-12*

Science Education (NGSS Lead States, 2013) and the *Next Generation Science Standards* (National Research Council, 2012) in the United States.

The use of translanguaging, by teachers and students, in formal science learning environments is an emergent research area for both the science education and linguistic research communities. Nevertheless, while this potential has garnered interest, few studies exist that address translanguaging in formal K-12 science learning environments. Sayer (2013) investigated the use of a local vernacular drawing on English and Spanish, TexMex, in a second-grade science classroom in Texas. This study found that TexMex was used extensively by the students in both community and classroom meaning-making activities, including the formal learning of science content. Espinosa (2016) investigated the implementation of translanguaging in a middle school English/Spanish bilingual science classroom in New York City. In this work, the bilingual science teacher modeled the kinds of translanguaging practices learners could implement with texts and peers to access science-content knowledge, which included discussing and deepening their understanding of science vocabulary in both Spanish and English. While the above papers focused on learning of science content, the work of Esquinca et al. (2014) investigated how translanguaging assisted students in engaging in the practices of science (e.g., observing, questioning, hypothesizing, and explaining). Their study is important given recent science education reform efforts. Thus, translanguaging has been identified as a way to engage students in the discourse and practices of disciplinary knowledge.

9.2.4 Bilingual Education in the United States

In this section, we will discuss bilingual education, as this relates to the setting of the research study informing this chapter. We also mention bilingual education as bilingual learning spaces offer opportunities to investigate translanguaging in the classroom. Bilingual education has existed throughout the history of education in the United States since multiple languages have been mixing in formal and informal learning settings. Bilingual education is an approach to education difficult to define by one stereotypical or general program model and is simply a label for a complex phenomenon (Ovando & Combs, 2012). One form of bilingual education that is continuing to grow in the United States is termed dual language education or dual or two-way immersion. Dual language education consists of programs that enroll both English-dominant students and minority language-dominant students. Minority language refers to all languages not the societally, politically dominant language in a specific setting. In the case of the United States, the most common non-English language is Spanish, and the majority of dual language education environments are English/Spanish. One goal of dual language education is to frame both languages as equal and thus this approach conceptualizes linguistic diversity and bilingualism as normal and as resources to be used in teaching and learning. While, indeed, it is difficult to frame both languages as completely equal due to societal influences, dual language education is an attempt at allowing emergent bilinguals the opportunity to use different languages in making meaning, thus legitimizing multiple language use for academic purposes. In the pursuit of a loftier societal goal, dual language education attempts to position bilingualism as the norm and not the exception in the United States.

Ovando and Combs (2012) provided an excellent summary of dual-language education models. Two models are generally described in the literature: the 50–50 model, characterized by instruction conducted equally in both majority and minority languages, and the 90–10 model, characterized by 90% of the instruction being conducted in the minority language and 10% being conducted in the majority language. Both models generally operate under a language separation or language bracketing paradigm in which the use of each language is separated with little to no translation or repetition in the other language (Collier & Thomas, 2004). While difficult to ensure strict adherence to percentages of each language used, language bracketing is generally achieved by teaching different content subjects in different languages or using separate languages during different time periods. The rationale behind language bracketing is that it creates an environment which immerses students in a second language experience, without allowing them to fall back on the familiarity of their home language. The goals of dual language education generally include the promotion of bilingualism and biliteracy, promotion of more positive attitudes toward the opposite language, as well as reducing achievement gaps (Collier & Thomas, 2004; De Jong, 2002; Morales & Aldana, 2010).

9.2.5 Science Education Reform

Science education in the United States has a rich history of reform efforts, with more acute interest beginning in the late 1950s and early 1960s. While all reform efforts have been precipitated by a different event, such as the Soviet success with launching Sputnik, each has demanded change on the part of both teacher and student. The most recent science education reform movement has been the publication of *A Framework for K-12 Science Education* (National Research Council, 2012) and the subsequent *Next Generation Science Standards* (NGSS Lead States, 2013). Both the Framework and NGSS are redefining what it means to engage in science in order to align science education with the authentic scientific practices. These documents frame science and science education as an intertwining three-dimensional endeavor, which includes disciplinary core ideas, crosscutting concepts, and science and engineering practices. The inclusion of science practices is an attempt at moving away from strictly content-acquisition science education to engaging students in the authentic practices of science. For example, engaging in argument from evidence and constructing scientific explanations are two recommended practices for inclusion in K-12 science education. As can be expected, many of the science practices will require students to use language in new and highly sophisticated manners. These discourse intensive practices will place increased language demand on all students, but particularly on students such as English Learners (Lee et al., 2013) and emergent bilinguals. The intersection of science education reform and pedagogies for educating emergent bilinguals provides potential for investigating how translanguaging as pedagogy can scaffold emergent bilinguals into the discourse intensive science practices.

One scientific practice that is being promoted through the Framework and NGSS is scientific argumentation or engaging in argument from evidence. Scientific argumentation has a rich history in science education, and many researchers have championed its inclusion in formal science classrooms (Duschl & Osborne, 2002). Others have constructed frameworks for teaching students how to engage in argument from evidence (McNeill & Krajcik, 2012). Still other researchers have expanded argumentation to address socioscientific issues (Sadler & Donnelly, 2006). Across the different approaches to uses of argumentation is an understanding of the importance of developing abilities among students to construct and critique evidence. Yet, such argumentation practices are situated in the overall discourse ecology of complex classroom life, where discourse serves multiple roles in communication, structuring participation, and forming student identities (Kelly, 2014).

Additionally, the Framework also brings attention to the needs of CLDs and how science education should be made relevant to students' cultural realities. Considering that CLDs' out of school worlds are culturally and linguistically diverse, making science education relevant takes on new meaning that goes beyond conceptual relevance and should include cultural and linguistic relevance. In this way then, uses of discourse processes, such as argumentation, need to bridge the students' ways of being, knowing, and speaking with the normative goals of the educational program. This is particularly true for CLDs who often come from communities experiencing disproportionate health risks and environmental racism (Pellow & Vazin, 2019). In this way, while engaging in the examination and formulation of evidence invites students to use scientific discourse through purposeful activity, it also serves to develop scientific literacy relevant to assessing scientific information for citizenship.

9.2.6 Socioscientific Issues Pedagogical Approach

Closely related to larger science education reform efforts, specific pedagogical approaches have been forwarded to engage with authentic issues of science. One such approach, called Socioscientific Issues, has been defined to refer to the explicit use of socioscientific issues as pedagogy in science classes. Socioscientific issues (SSI) are controversial social issues with conceptual and/or procedural ties to science (Sadler, 2004). These issues involve aspects of personal and societal decision-making and the science that is tied into the issue. On one hand, these issues demand that individual or societal decisions be made that will have ramifications on the future activities of humans. On the other hand, socioscientific issues by nature are heavily laden with scientific concepts that demand a scientific perspective be taken in order to consider the consequences of such decision-making. The complex interaction of the social and scientific factors results in issues that are ill-structured, are open-ended, are contentious, and often have multiple solutions based on multiple viewpoints. While not a new phenomenon in the history of science, SSI have gained prominence in today's society as technological and scientific innovation, combined with an increasing human population bring about not only progress, but problems associated with these innovations. Zeidler (2014) argues for the inclusion of SSI in today's science classrooms in order to portray science as relevant and connected to everyday life. In addition, the inclusion of SSI into school science classrooms supports a vision of scientific literacy (Roberts & Bybee, 2014) which argues for citizen participation in the decision-making processes of a democratic society. As such, it is important that students – future decision-makers and participants in the democratic process – are aware of the connections between science and society that exist, demand attention, and necessitate informed decision-making.

9.3 Research Setting and Intervention in a Dual Language Middle School Science Classroom

9.3.1 Research Questions

The intersection of science education reform (e.g., inclusion of discourse-intensive scientific practices) and the continual cultural and linguistic diversification of our science classrooms offer unique opportunities to research how teacher translanguaging can scaffold student science learning. The goal of this research was to investigate how teacher translanguaging as pedagogy could scaffold student engagement in discourse-intensive scientific practices, such as scientific argumentation. The research questions guiding this investigation are:

- How does translanguaging as a pedagogical strategy facilitate teacher framing of scientific argumentation about socioscientific issues?
- How does translanguaging as a pedagogical strategy facilitate student uptake of scientific argumentation about socioscientific issues?

9.3.2 Research Setting

While dual language education is becoming more prominent in the education landscape of the United States, the full potential of this approach has not been fully recognized. To examine how dual language learning occurs in science, we sought previously established education professionals in a metropolitan area characterized by a large and growing Spanish-speaking, Latina/o population. In response to this demographic shift and the need to accommodate bilingual learners' needs, the AB

Dual Language School (a pseudonym) was founded in 2010 with the explicit focus of providing English/Spanish dual language instruction to elementary and middle school students in two smaller sized cities in a larger urban metropolitan area located in Northeastern United States. This larger metropolitan area, comprising three smaller urban centers, has experienced a rapid increase in the Latina/o population over the past several decades. The school operated under 50:50 dual language model, in which 50% of the instructional time was devoted to home language and 50% to target language. School demographic information at the time of the study was as follows: 81% of students were classified as Hispanic; 5% as Black, non-Hispanic; 3% as White, non-Hispanic; and 10% as two or more races. Over 80% of students qualified for free or reduced lunch (a proxy for lower economic status in the United States).

9.3.3 Study Participants

The participants in this study consisted of a seventh grade science teacher, Ms. Romero (a pseudonym) and two different sections of seventh grade science students (n = 46, ages 12–13). Ms. Romero, a first year science teacher with a teaching certification in elementary education, served as the school's sixth and seventh grade science teacher. Prior to this position as middle school science teacher, Ms. Romero was an English as a Second Language teaching aide in one of the school districts served by the AB Language School. Occasionally, this paper will refer to teachers, as Licona, a researcher, would engage in instruction and interaction with the students.

All seventh grade students in Ms. Romero's seventh grade classrooms were recruited from two local English-medium instruction school districts, and therefore the dual language nature of the school was likely the first time that Spanish was used with these students in a formal educational setting. Through an informal questionnaire administered by the teacher, students reported the following information about their home language use: 35% spoke only English in the home, 28% spoke only Spanish, and 27% used both English and Spanish in the home. According to the official school-based assessment of English proficiency, 26% of the seventh-grade students were classified as English learners. Given their range of English and Spanish language proficiencies and that 81% of students at AB Language identified as Hispanic, it is safe to assume that many of the seventh graders were heritage language speakers.

9.3.4 Curriculum Intervention

In order to investigate how reform-based science education is enacted in a dual language education setting, a curriculum was constructed to promote the use of "talk" in science sense making. Prior to the study, the implemented science curriculum represented a variety of scientific topics ranging from ecosystems concepts based on a FOSS science kit (FOSS, 2020), to time, distance, and velocity lessons based on materials found online. The school did not have an official seventh grade science curriculum, and the classroom teacher spent considerable time finding and planning fun and engaging science lessons for her students. Based on classroom observations of Ms. Romero's classroom before the study, little to no evidence of scientific argumentation was observed in the class prior to the curriculum intervention. This was not a surprise, nor is it a critique of the classroom teacher or school, as scientific argumentation is not a common practice in K-12 science classrooms (Duschl & Osborne, 2002).

The curriculum intervention was co-constructed by Licona (first author) and the classroom teacher. Licona conducted eight classroom visits prior to the implementation of the curriculum. Based on the observations and input from the classroom teacher, Ms. Romero, Licona constructed a new science curriculum for the seventh grade. Forming a co-expertise model, the curriculum design drew from Licona's research and expertise in science curriculum construction and pedagogy and from Ms. Romero's intimate knowledge of her students' academic, linguistic, and cultural characteristics.

Following recommendations from recent science education reform documents (e.g., Framework and NGSS) and grounded in literature supporting emergent bilinguals, Licona designed a curriculum intervention to engage students in science learning that promoted new discourse and meaning-making practices. The featured curriculum was premised on teaching sequence literature (Leach & Scott, 2002), where the term "teaching sequence" refers to a progression of short instructional activities that target curricular goals intended to build learner understanding and use of scientific concepts and practices (Leach et al., 2005).

The curriculum focused on contemporary issues of biodiversity through a socioscientific issues pedagogical approach in conjunction with a scientific argumentation framework. A socioscientific issues approach (Sadler, 2004) situates scientific concepts and practices within controversial and ill-defined issues that have relevance to both science and society. In this intervention, the class was considering issues related to biodiversity, such as invasive and endangered species. The intervention promoted the use of the claims-evidence-reasoning (CER) framework (McNeill & Krajcik, 2012) such that learners could engage in scientific argumentation, both verbally and in writing. This curriculum intervention represented a change in the established classroom practices and afforded learners greater opportunity to participate in the discourse practices of science, including argumentation around socioscientific issues. Most importantly, this approach allowed for learners to engage in discourse about science with the teacher and other learners as they attempted to make sense of the questions posed and how to respond to them using the specific argumentation framework.

In an effort to design relevance into the curriculum, three iconic animal species (the green turtle, the Puerto Rican coqui, and the timber rattlesnake) were chosen to be featured in the curriculum. The first species, the green turtle, is a culturally iconic species featured in the movie Finding Nemo, which was an enormously successful and popular movie among middle school students. The Puerto Rican coqui, a culturally iconic species in Puerto Rico, was chosen as many of the students were of Puerto Rican descent and either migrated from or had friends and family living in Puerto Rico. The coqui, as a collective species, is a very unique frog species often featured in Puerto Rican literature, art, and music. The timber rattlesnake is a culturally iconic species in the Mid-Atlantic states region of the United States (location of the research site), but is also a very recognizable species in the United States and throughout the world. The green turtle is an officially recognized globally endangered species and both the Puerto Rican coqui and timber rattlesnake are species of concern in the herpetological research community.

In order to engage students with the aforementioned species, information pamphlets (trifolds) were constructed that contained various data related to each species. Examples of species data included physical characteristics, adaptive features and behaviors, habitat, predators and prey, reproductive characteristics, and human pressures on the species. Students were also provided with in-class worksheets to guide their construction of scientific arguments based on data (evidence) from the trifold. The information trifold and in-class worksheets also served as resources for constructing whole class verbal scientific arguments.

Prior to the implementation of the curriculum, students were administered a pre-assessment that required students to respond to read two short passages and respond to a prompt about each passage. One scenario had students read about and consider a solution to the ecological issue of invasive species in the Florida Everglades. The second scenario had students read about and consider whether or not an endangered species should be able to be hunted. The pre- and post-assessments were exactly the same and were designed to measure written argumentation before and after the curriculum intervention, respectively. All curriculum materials, including pre- and post-assessments, were made available in both English and Spanish. Students were instructed by the classroom teacher that they were free to respond in English, Spanish, or both according to their preference or linguistic strengths. A summary of the curriculum implementation sequence is found in Fig. 9.1.

Lesson One: Pre-Assessment
Lesson Two: Endangered Species
Lesson Three: Impromptu Informal Classroom Debate
Lesson Four: Setting the Problem
Lesson Five: Ecosystems
Lesson Six: The Case of the Green Turtle
Lesson Seven: Scientific Explanations/Scientific Argumentation (CER Framework)
Lesson Eight: The Case of the Puerto Rican Coqui
Lesson Nine: Scientific Explanations/Scientific Argumentation (CER Framework Two)
Lesson Ten: The Case of the Timber Rattlesnake
Lesson Eleven: Post-Assessment

Fig. 9.1 Summary of the curriculum implementation

9.4 Research Approach and Data Analysis

9.4.1 Data Collection and Representation

9.4.1.1 Classroom Observations

The first phase of data collection consisted of eight full days of classroom observation, over the course of 5 months, prior to the implementation of the curricular intervention. The purpose of the classroom observations was manifold but focused on the following: to observe the implemented science curriculum and pedagogy, to understand the languaging practices of the teacher and students, and to understand the established and developing classroom culture. These classroom observations served to inform the construction of the curriculum intervention. Based on the linguistic patterns (e.g., translanguaging) inherent in the classroom, the language demands of reform-based science education, and the use of socioscientific issues approach it was decided that translanguaging as pedagogy would be of value for teaching the students.

Following the pre-curriculum observations, the researcher (Licona) continued to engage in classroom observations, but in a different manner. Throughout the curriculum observation phase, and during all the time in the classroom, the researcher's role fluctuated between observer and participant; in each instance, the researcher was a participant observer with greater or lesser active participation based on the situated contexts of activity and strategic decisions about when and how to engage (Spradley, 1980/2016). Throughout this phase of the research, the researcher observed the classroom events and recorded field notes and was often drawn into or led the class discussions or activities, as he was framed as an expert in science and science education by the classroom teacher.

9.4.1.2 Audiovisual Recording of Classroom Events During Curriculum Implementation

Audiovisual recordings of classroom events were collected during an extended period (19 consecutive days) of time during which the curriculum was implemented. Two cameras were employed: one camera was placed in the back corner of the classroom and captured the teacher, the whiteboard, and the students; the second camera placed in the front of the classroom captured the activities and discourse of paired and small groups of students. Recordings from both cameras were used to document and analyze classroom events and discourse.

9.4.1.3 Construction of Event Maps

Event maps (Licona, 2018) were constructed, during the viewing of video records, and documented the events and discourse of the classroom. Event maps are constructed from an emic perspective, identifying how the participants construct and names their activities. See Fig. 9.2 for an example of a part of an event map. This example demonstrates how video records were analyzed and documented. The example shows the time, major events, interactional space, language, resources, and researcher notes. Event (Topic) refers to the main activity as defined by the participants' talk and actions and catalogued with a time stamp. Interactional Space refers

Time	Event (Topic)	Interactional Space	Language (E, S, or B)	Resources	Researcher Notes
800	IRE with individual student CER in Spanish. Some difficulty with this.	Teacher – Student	T – Spanish		Does CER translate well into Spanish? Look at paired video camera view. Translangauging here – T switches in an out of Spanish to see if students understand the meaning of the word claim.

Fig. 9.2 Sample taken from an event map

to how the activity was enacted; in this case, the teacher was working with an individual student. Language refers to the language being used and the speaker, in this case Spanish was used by the teacher for most of the interaction. Resources refer to physical materials being used during the event. Researcher notes refer to ideas or questions that arose during video analysis. Through a series of passes through the corpus of video data, the event maps were refined and notes of relevance recorded. Through this process, an analytic focus emerged around teacher translanguaging during whole class construction of verbal arguments, as this became a salient issue for understanding the communicative ecology of the classroom. The purpose of the final pass was to identify instances and patterns of the use of English and/or Spanish by both the teacher and students concerning the substantive concepts regarding the socioscientific issues. The final analysis was then conducted to generate full transcripts (see transcript one) of teacher translanguaging and the associated student discourse.

9.4.1.4 Collection of Student Artifacts

Throughout the curriculum implementation, students were afforded multiple opportunities to construct both verbal and written arguments. The analysis for this chapter focused primarily on the student written work but considered ways this work was framed and supported by teacher discourse. Written assignments differed from the verbal exchanges – for the written assignments, the students had time to make reference to resource material, think through the implications of the socioscientific issues, and have a record to which they could return at a later time. Student written argument assignments were implemented through a think-pair-share pedagogical approach in which students first engaged with the materials, then discussed and critiqued the work of another student, and finally shared their work in a whole class setting discussion. Students completed guided worksheets to scaffold their work and employ the CER framework. Students engaged in five written argument assignments over the course of the curriculum. Students were first given a pre-assessment that had them respond to two biodiversity issues, one related to invasive species and the other related to an endangered species. The pre-assessment was to gauge students' initial performance of constructing a scientific argument in response to a socioscientific issue. Students were then presented with a lesson on constructing scientific arguments using the claim-evidence-reasoning (CER) framework (McNeill & Krajcik, 2012). In addition to a lesson on the CER framework, students engaged in an assignment that had them analyze the quality of arguments based on the CER components. This assignment required students to critique the claim, evidence, and reasoning of various arguments of differing quality. As a permanent scaffold, posters, in English and Spanish, demonstrating the CER framework were hanged on the classroom walls. Students then constructed arguments based on responding to three other biodiversity issues – the green turtle, the Puerto Rican coqui, and the timber rattlesnake. Finally, students completed a post-assessment of the same two prompts as the pre-assessment. Each of the students' five argument assignments were collected and later coded according to a coding scheme based on the CER framework.

9.4.2 *Data Analysis*

9.4.2.1 Coding of Written Arguments

Based on the curriculum intervention, most students constructed five written arguments to complete the assignments – pre-assessment invasive species and endangered species prompts, three cases of endangered or threatened species (green turtle, Puerto Rican coqui, timber rattlesnake), and the post-assessment invasive species and endangered species prompts. All written arguments were entered into a spreadsheet for analysis and coding. Arguments were analyzed and coded considering the following four dimensions: structural components, coherence, content, and appropriateness.

Structural components refer to the presence of the individual components – claim, evidence, and reasoning – of the argumentation framework (McNeill & Krajcik, 2012). This framework was chosen for both teaching and analytical purposes because of its accessibility for student learning. Following this framework, a claim was defined as a statement or conclusion that answers the original question/problem. There were claims that did not answer the original posed task question, but were coded as claims, nonetheless. Evidence was any data viewed by the student writer as supporting the claim. From an analytic point of view, we considered evidence as any data that supported the claim, regardless of whether or not the evidence was of a scientific nature or if the claim did or did not answer the original question. Reasoning was a justification that connected the evidence to the claim and in principle drew from and applied scientific principles. Much like the claims, and evidence, for analytic purposes, we took reasoning from the point of view of the student – the analysis was based on the internal logic of the student writers. Reasoning was coded as any statement that connected the evidence to the claim in the student writing, regardless of whether or not the principles were of a scientific nature from a normative point of view.

Argument coherence referred to how each of the three components – claim, evidence, and reasoning – were tied together in mutually supportive ways in the argument (Kelly & Bazerman, 2003). In other words, as analysts, we posed the questions: Did the evidence support the claim and did the reasoning connect the evidence to claim? One level of coherence concerned the relationship between the claim and evidence; a second more fully coherent argument demonstrated lexical and conceptual ties across the claims, evidence, and reasoning. The full coherence across CER would indicate a stronger argument.

Content of the argument referred to the particular perspective or vantage point from which the student was arguing. As socioscientific issues can be viewed through multiple lenses – scientific, economic, political, personal, or ethical – students were free to argue from different perspectives and thus arguments would emanate from different loci. Arguments were coded based on the content of the evidence as well as the content of the reasoning. We coded the content of student arguments as scientific, drawing on scientific evidence and reasoning, or other, drawing on non-scientific evidence and reasoning. There was not a valence on the types of perspective brought into the position taken by the students. For example, the socioscientific issues purposefully situated the science in a large problem space where economic, political, ethical, or even personal frames for understanding were relevant. The curriculum was design in this way to invite student participation in discourse involving science and larger societal issues.

Appropriateness of the student response to the prompt referred to whether or not the argument presented addressed the original question. The CER framework (McNeill & Krajcik, 2012) defines a claim as a statement that responds to the original question. Rather than discount the students' responses that were off subject, we chose to examine the appropriateness of the response in addressing the question, but also understanding the student point of view. For example, there were examples of student arguments that demonstrated internal coherence, but did not fall within the domain of the original prompt question.

9.5 Emergent Themes of Translanguaging in Classroom Discourse

The results presented in this chapter emerged from a larger ethnographic study of translanguaging that included analysis of both verbal and written discourse. As is typically the case in studies of classroom discourse, the framing of the norms and expectations of the interactions, and the thematic development of writing practices, were constructed over time through the discursive work of the teachers and the students as a collective classroom culture. In this way, the student written work was framed, supported, guided, and interpreted through the overall classroom discourse practices. To frame the socioscientific issues, and the uses of argumentation, the classroom teacher engaged in translanguaging throughout the course of the curriculum intervention. This translanguaging included both verbal and written modes. Verbal modes included languaging in small group and whole class settings. Written translanguaging was conducted as the classroom teacher used the classroom whiteboard to record student responses during verbal construction of arguments. In this way, the classroom teacher made visible the value of translangauging for the communicative purposes of the task at hand and signaled to the students that this was a valued practice in the classroom. Thus, the teacher's actions set the focus on accomplishing the important work of understanding and communicating the scientific, ethical, and social considerations of the topics embedded in the socioscientific issues. Due to the prominence of the teacher's stance, and valuing of communicating understanding, we focus on themes identified through her discursive work.

Three major functions or themes of teacher translanguaging emerged during the coding process: constructing classroom culture, facilitating the academic task, and framing epistemic practices. Constructing the classroom culture refers to teacher translanguaging in order to construct, develop, and maintain the norms and expectations of the classroom culture (e.g., redirecting student behavior, responding to non-academic classroom situations, sustaining bilingual practices of the classroom). Facilitating the academic task refers to teacher translanguaging in order to assist students engaging in and completing general classroom academic tasks (e.g., assigning homework, explaining directions, introducing academic tasks). Framing epistemic practices refers to instances of teacher translanguaging to support members of the community learning to develop ways of knowing related to scientific argumentation about socioscientific issues (Kelly & Licona, 2018). Each of these functions supported the bilingual learning environment; and each of these functions, constructing classroom culture, facilitating the academic task, and framing epistemic practices, was mutually constitutive of the activities in the classroom. For this chapter, we present the ways that the epistemic practices, particularly around producing and critiquing evidence regarding the socioscientific issues, were framed using translanguaging as science pedagogy.

Epistemic practices are defined as the ways members of a group propose, justify, assess, and legitimize knowledge claims (Kelly, 2008; Kelly & Licona, 2018). In this study, epistemic practices center on three different ways the students came to make sense of uses of evidence. The first uses of epistemic practices focused on scientific argumentation and included the generation and evaluation of claims, evidence, and/or reasoning. These practices invited the students into ways of understanding how to construct, communicate, and critique evidence. The second focus of epistemic practices examined the appropriateness of a scientific explanation. In this case, students determined if and how the overall argument and its individual components responded to the issue in an appropriate manner. A third set of epistemic practices concerned a consideration on the students' part of the purpose of the scientific argumentation/explanation framework. These practices differ than the previous, as the focus was on interpreting the communicative purpose of the audience, including understanding the intended audience and making inferences about the ways that evidence might be persuasive. In this chapter, we present examples of the classroom verbal discourse and student written work to illustrate ways that translanguaging facilitated the development of epistemic practices among the students.

9.5.1 Translanguaging in Classroom Spoken Discourse

Transcript 9.1 shows an instance of the teacher translanguaging while engaging in whole class construction of a scientific argument. This episode occurred as the teacher and whole class constructed an argument linking the destruction of green turtle eggs to the potential extinction of the species. Specifically, this episode demonstrates the teacher developing a line of reasoning with a smaller group of students.

Transcript 9.1 Teacher translanguaging during whole class argument construction

Turn/ line #	Original transcript	Spanish To English translation	Functioning of translanguaging
1.01 1.02 1.03 1.04 1.05 1.06 1.07 1.08 1.09 1.10 1.11 1.12 1.13 1.14 1.15 1.16	Ms Romero: Do you see the connection of how my evidence of where the eggs were being destroyed to how they're going to become endangered? Alright. Te lo dice en español. Escuche. Escuchan. La afirmación tuya era que el anima la tortuga estaban en peligro de extinción. ¿Verdad? La evidencia que pusieron era que los huevos están siendo destruidos. ¿Cómo los huevos se están destruyendo está causando la tortuga estar en peligro?	*I will tell you in Spanish. Listen. Listen. Your claim was that the animal, the green turtle was in danger of extinction. Right? The evidence that you put was that the eggs are being destroyed. How is it that the eggs that are being destroyed is causing the turtle to be endangered?*	The teacher begins in English and then translanguages after realizing some students may not have understood. The teacher translanguages in order to reaffirm the claim of "yes, the green turtle is an endangered species." She then reiterated the evidence that students used to support the claim. She finally asked how turtle eggs being destroyed puts them in risk or in an endangered status.
2.01	Emelio: No crecen.	*They won't grow.*	This student responds that turtles won't grow.
3.01 3.02 3.03 3.04 3.05 3.06 3.07	Ms Romero: Well. No crecen, pero nunca. ¿Si están destruyendo puede salir e huevo?	They don't grow, but never (or won't grow at all). If they (the eggs) are being destroyed can they (the turtles) emerge from the egg?	The teacher then questions the student's reasoning and asks a follow-up question.
4.01	Lydia: No.		Another student answers no in response to the teacher's question.
5.01 5.02 5.03	Ms. Romero: ¿Que. le pasan? ¿Y qué pasa a la población de la tortuga?	*What happens to it? And what happens to the turtle population?*	The teacher continues in Spanish to continue developing reasoning and asks what will happen if eggs are destroyed.
6.01	Emelio:Se mueren.	*They die.*	The original student then responds to the teacher's question.
7.01 7.02	Ms. Romero: ¿Y qué pasa si siguen muriendo?	*And what happens if they continue dying?*	Again, the teacher continues in Spanish to develop the line of reasoning.
8.01 8.02	Daria: There's not gonna be any more.		A third student responds appropriately in English indicating she is correctly following the line of reasoning.

(continued)

Transcript 9.1 (continued)

Turn/ line #	Original transcript	Spanish To English translation	Functioning of translanguaging
9.01	Emelio: Se van a extinguir.	*They will go extinct.*	The original student links eggs being destroyed to an eventual extinction of the species.
10.01 10.02 10.03 10.04 10.05 10.06	Ms. Romero: Se van a extinguir. ¿Tú viste la conexión de la evidencia al peligro para extinción? ¿Entendiste? Eso que es razonamiento. Razonamiento. ¿Entienden?	*They will go extinct. Did you see the connection from the evidence to danger of extinction? Did you understand? That is reasoning. Reasoning. Understand?*	The teacher then asks if the students are able to see the connection from the claim and evidence as being linked by reasoning. She then defines reasoning as connecting the evidence to the claim.

The above transcript demonstrates how the classroom teacher employed translanguaging to construct a line of reasoning with her students – she wanted to develop the inferences from the destruction of the eggs (1.03–1.04), to the students' response about turtles not growing (2.01) dying (6.01), and to the idea of extinction (10.01–10.05). The teacher was constructing this line of reasoning in English with the whole class and then switched to Spanish (1.06) upon realizing that some students may not have understood the discourse in English. The teacher, working with three students in the whole class conversation, built a line of reasoning that connected the green turtle's eggs being destroyed to the eventual extinction of the species due to lack of reproduction. This reasoning was used to connect the evidence of egg destruction to the claim of the green turtle being classified as an endangered species. While most of this interaction is conducted in Spanish, one participant responded appropriately in English indicated that she was following the discourse in Spanish. The example shows how the teacher was able to adjust her discourse to attend to her perception of the understanding among the students. The switch to Spanish facilitated her communicative goal of collectively building an argument about the importance of the turtle's eggs to the survival of the species.

9.5.2 Language Use in Student Written Arguments

Student written argumentation improved across the course of the curriculum intervention, starting with the pre-assessment and ending with the post-assessment. The illustrative example presented here focuses on prompt one (invasive pythons in the Florida Everglades) from pre- and post-assessment written arguments. For this task, the students were prompted to make a decision on whether or not invasive python species in the Florida Everglades should be hunted and killed by humans, given the ethical and ecological considerations for removing the destruction to native fauna by the invasive species. Students were introduced to the problem through a brief presentation and then were asked to respond to a reading prompt. The pre-assessment occurred before any formal introduction to scientific argumentation or frameworks of argumentation. The post-assessment occurred after the introduction of and engagement with the CER argumentation framework. Additionally, students were able to employ the CER framework to three other species before engaging with the post-assessment python issue.

There were modest changes in the ways that the students progressed in the construction of written arguments from the pre- to the post-assessment. For the invasive python prompt, all students included a claim in both the pre and post assessment, and the majority included uses of evidence on both the pre- (85%) and post-assessment (91%) arguments. The most significant change over the course of the science lessons, and manifested in the assessments, was the increase in students providing reasoning. For analytic purposes, reasoning was defined as scientific principles linking evidence to the claim. Only 9% of students provided reasoning on the pre-assessment whereas 41% provided reasoning on the post-assessment. Additionally, argument coherence improved from the pre to the post assessment, with an improvement from 4% on the pre-assessment to 30% on the post assessment. Nearly all students constructed arguments through appropriate use of the subject-matter domain, with values of 98% and 100% for the pre- and post-assessments, respectively.

Interestingly, as related to language, the majority of students' responses on the pre- and post-assessment prompts were written entirely in English. Responses composed entirely in Spanish were comprised 11% and 14% of pre- and post-assessments, respectively. For these students, many of whom were still learning English, the choice was made to write in English, despite the uses of translanguaging in the classroom discourse, and the value of such translanguaging for communicating about science and constructing arguments. This shows that students were free to respond to the issue in their preferred language and were not bound by monolingual policies. Nevertheless, despite the uses of Spanish in many homes, the students predominantly wrote in English. This may have been due to the fact that for many students, this was their first year in a dual language program. Thus, they may have brought with them the language ideology of their previous schooling experience.

The referents in the students' arguments varied across domains but generally made use of science or ethics as a basis for their reasoning. The content of arguments for the pre-assessment was 68% scientific and 2% ethical; the content for post-assessment was 80% scientific and 5% ethical. These totals do not add to 100%, as some arguments failed to provide sufficient reasoning to be analyzed. We highlight scientific and ethical responses in order to demonstrate that these responses accounted for the majority of argument content. The shift to the use of more scientific referents in the argument demonstrates a further understanding among the students of the purposes of the task and the ways that science can be used to be persuasive. Recall, the task for the student writing was open ended and did not specify which perspective (scientific, ethical, personal, etc.) students were required to draw on to construct their arguments.

In comparing the pre- and post-assessment for one student, Juan, (see Fig. 9.3), some differences are evident. In the pre-assessment the student has included two claims [C] which complement each other and respond appropriately to the original question. Additionally, the student has provided two pieces of evidence to support the claims. This argument would be considered incomplete, as only claims and evidence were provided. This claim and evidence do demonstrate coherence yet are not explicitly linked with reasoning. In the post-assessment, the student included two very similar claims which respond appropriately to the original prompt. The student provided five pieces of evidence, which supported the claim. Most importantly, the student provided reasoning which links the evidence to the claim. In this case the reasoning suggests that invasive species have the potential to cause other endemic species to become extinct. This argument would be considered complete in that all C, E, R components were provided. Furthermore, this argument shows coherence, as all of the components support each other. Both arguments are from a scientific perspective.

In comparing the pre- and post-assessment for this example, there are also improvements in the student reasoning (see Fig. 9.4). For example, in the pre-assessment, the student, María, has provided two claims – one claim of "no" and another claim that the snakes should be placed somewhere else. The student did provide reasoning to back her original claim of "no." Interestingly, the student's pre-assessment argument drew from a religious or ethical perspective, in that snakes deserve to live and not to be harmed in a way similar to humans. Only one piece of evidence has been provided. The student did not provide reasoning. In the post-assessment, the student had changed her stance and produced a claim of "yes." The student provided multiple pieces of evidence backing the claim. Additionally, the student furnished her reasoning, noting that that an invasive species without predators has the potential to reproduce very rapidly, thus taking over and ecosystem to the peril of other organisms. While the reasoning is not able to connect all the evidence to the claim, in fact this is a more sophisticated argument. Furthermore, this argument is of a scientific perspective, which is a change from the pre-assessment religious or ethical perspective. Like the first example of student work in English,

Pre-Assessment [C] This is a good idea because [E] this animal is overpopulating and [E] endangered other animal to the point of extinction. [C] So I think they should send hunters to kill some of them and make a law stating that people cannot let go of their snake into the wild.
Post-Assessment Claim: [E] the Florida Everglades is a fragile ecosystem. [C] It is a good idea to send hunters into the Florida everglades ecosystem [E] people keep letting go snakes into the wild. [E] The snakes don't belong there so [E] it is causing other animals' populations to decrease and [E] other animals can't reproduce as fast as them. [R] if this keeps happening other animals will go extinct. (C) Hunters should get paid to go in and kill and remove these snakes from the everglades so everything will be normal.

Fig. 9.3 Student pre- and post-assessment argument comparison

Pre-Assessment [C] Yo creo que no porque [E] dios hizo los animales y como las personas para que disfrutarán, no para que lamentaran, [C] - aun las pongan en un lugar y mantenerlas en habito como darle poca comida y que allí pueda cambiar. Translation: [C] I believe no because [E] God made the animals, and like humans, to be enjoyed and not to be hurt, [C] even put them in a place and maintain them in a habitat and give them a little food and there they can change.
Post-Assessment Afirmación [C] si, deben matar a la serpiente. Evidencia - porque [E] la serpiente no tiene depredadores y [E] la serpiente esta seguir comiendo todos animales y [E] están creciendo mucho. Razonamiento - [E] estas serpientes no tienen enemigo ni depredador y [R] como no tiene depredador con capaz de reproducirse mas rápido. Translation: Claim [C] yes, they should kill the snake. Evidence – because [E] the snake does not have predators and [E] the snake is continuing to eat all the animals and [E] it is growing a lot. Reasoning – [E] these snakes don't have any enemies nor predators and [R] since it does not have predators it has the capacity to reproduce very rapidly.

Fig. 9.4 Student pre- and post-assessment argument comparison

the arguments by María were monolingual. The students were given the choice of how to articulate their arguments, and in majority of the cases across the class, students made the choice to write in only one language, with occasion use of a word or two from the second language. The practice of translanguaging was not evident in this, the more formal genre of science communication, even if valued to support the development of the genre through verbal exchanges.

9.6 Discussion

Scientific argumentation is a discourse-intensive scientific practice which places increased language demands on all students, particularly those students who may are in the process of learning English, such as English Learners and emergent bilinguals (Lee et al., 2013). While students bring linguistic strengths within and across their language usage, the field of science education needs to be attentive to fostering scientific practices in ways that builds from the assets contributed from the students. Formulating evidence in various genres takes time and practice, and regardless of language background and experience, students need scaffolds to engage in scientific argumentation. This is of paramount importance for our linguistically diverse students whose home language may not be English. There are four points of discussion that relate translanguaging to the development of epistemic practices supporting argumentation: building linguistically responsive pedagogy; teacher discursive work and translanguaging; dual language environments to support science learning; and revisiting scientific argumentation.

9.6.1 *Building Linguistically Responsive Pedagogy*

Responsive pedagogies (Gay, 2002) have been developed and implemented to scaffold culturally and linguistically diverse students. As suggested by Lucas and Villegas (2013), properly implemented translanguaging can serve as linguistically responsive pedagogy. In this chapter, we focused specifically on the pedagogical knowledge and skills directly related to scientific argumentation around socioscientific issues. The pedagogical knowledge needed for teaching in this setting were multi-fold. First, the teacher needed a repertoire of strategies for learning about linguistic and academic backgrounds of the emergent bilingual students including their abilities in English and native languages. This knowledge of students provides ways for teachers to make decisions about choices of pedagogy that build on the strengths students bring to the classroom and provide scaffolds across languages to support meaning-making. The classroom teacher had extensive knowledge of her student linguistic backgrounds, as evidenced by her verbal interactions with her students and her knowledge of student language use in the home. The strategies employed by the teacher to learn about her students' linguistic backgrounds included language proficiency assessment data (i.e., official designation as an English Language Learner), informal conversations with students, and an informal home language survey administered to the students. Being aware of her students' linguistic backgrounds, both in and out of the classroom, allowed the teacher to translanguage in response to this knowledge.

Second, the teacher had an understanding of and ability to apply key principles of second language learning. In this case, we saw the teacher employing providing scaffolding in both English and Spanish. As the students were a mix of Spanish as home language and English as home language, translanguaging served as a scaffold for language learning. Using a student's home language, whether English or Spanish, to develop fluency in an additional language is supported in many programs, such as transitional, developmental, or dual language education, addressing language learners (Wright, 2010). Through translanguaging as pedagogy provided students with a linguistically relevant supportive space for students to engage with language learning.

Third, it was important for the teacher to identify the language demands of classroom tasks and develop a repertoire of strategies for scaffolding instruction for ELLs. This form of responsive pedagogy is complicated by the curricular goals and assessment strategies of the classroom. In the examples from this chapter, the problem space for talking science was purposely expanded to include socioscientific issues that might be of interest to the students. By building in this complexity and recognizing the need to foster development of epistemic practices supporting argumentation, the language demands of participation were augmented. While we view these decisions as potentially beneficial for the students, their language development, and science knowledge, the teachers' understanding of the language demands were crucial for developing linguistically responsive pedagogy.

9.6.2 Teacher Discursive Work and Translanguaging

While the knowledge of students and their abilities was requisite for providing scaffolds for science learning, the linguistically responsive pedagogy was accomplished through the discursive work of the teachers (Romero and Licona). As the students demonstrated differing linguistic abilities across English and Spanish, Ms. Romero used knowledge of this to cue her translanguaging in a responsive manner in many interactions with students in her dual language classroom. Importantly, by understanding the language demands of complex scientific practices, such as scientific argumentation, Ms. Romero employed translanguaging as a scaffold to promote learning across English and Spanish. Learning how to frame arguments and understand an audience for evidence is challenging for all speakers and writers. In this case, the classroom discourse supported the development of epistemic practices as members of the classroom came to propose, justify, assess, and legitimize knowledge claims (Kelly & Licona, 2018). To engage in these practice, the teachers and students worked to define local meanings for what was meant by evidence, argument, and explanation. To build common understandings of these key terms, the teacher needed to engage in discursive work, which included translanguaging to support the development of the relevant epistemic practices.

9.6.3 Dual Language Environments to Support Science Learning

As home, or first, language is a key component to learning a new, or second, language, the teacher used students' home language to scaffold learning in the new language. This is an essential component in dual language learning environments and demonstrated an understanding and ability to apply key principles of second language learning. While the school's stated language paradigm was of equal language use 50:50 (Spanish/English), with an implied language bracketing format, the teacher's classroom would best be described as a translanguaging classroom (García et al., 2016). The teacher freely and fluidly translanguaged basing on her responsive choices to her students' linguistic needs. The teacher used both English and Spanish in her oral discourse with students, as well as in writing during her use of the classroom whiteboard. The mixing of languages was very common in the classroom and mirrored the linguistic characteristics of the geographical area in which the students resided. As all students in the class were learning both English and Spanish, the teacher's use of translanguaging afforded all students, regardless of proficiency in English or Spanish, opportunities to engage in language learning experiences. In this classroom, translanguaging served to accomplish the various communicative functions of the classroom. The pedagogical goals drove decisions about how to employ language to build common understandings of key constructs (evidence), paradigms (CER), and practices (building coherence in arguments). The free and fluid mixing of language has been recommended by scholars in order to frame languages as equal (García et al., 2016). Additionally, scholars posited that a translanguaging pedagogy can push back against the ideology of English as the only legitimate language of science and promotes science as a bilingual or multilingual pursuit (Infante & Licona, 2018).

9.6.4 Revisiting Scientific Argumentation

Scientific argumentation has been noted for its linguistic complexity (Halliday & Martin, 1993; Lemke, 1990) and that, as a scientific practice, it poses particular demands on students formulating evidence in a non-home language, such as emergent bilinguals or English Language Learners (Lee et al., 2013). The language demands associated with scientific argumentation go beyond traditional content-acquisition science pedagogies. In the case of this curriculum, students engaged in both written and oral argumentation, thus drawing on reading, writing, speaking, and listening domains of language. The classroom teacher was able, through translanguaging, to improve students' argumentation skills, most notably with their use of reasoning in constructing more coherent arguments around socioscientific issues. The use of translanguaging scaffolded student argumentation in both English and Spanish. While we identified ways that assessing and critiquing evidence was supported, we also noticed that efforts to improve argumentation need to consider the discourse ecology of the classroom. As argumentation is one of many discourses of science, it needs to be understood as a component of a rich classroom culture fostering different discourse genres.

While this chapter focused on translanguaging in facilitating student engagement in epistemic practices of scientific argumentation, we posit that translanguaging as pedagogy has potential in facilitating emergent bilinguals gaining access to the varied science content and practices being foregrounded by recent science education reform. Furthermore, as dual language education continues to gain acceptance in the United States, translanguaging as pedagogy holds promise as an alternative to language bracketing paradigms that dominate dual language learning environments.

9.7 Concluding Remarks

The confluence of discourse-intensive science education reform and the increasing population of emergent bilinguals necessitate the exploration of pedagogical approaches that facilitate access to science learning. As a linguistically responsive science pedagogy, translanguaging has been shown to facilitate student access to and engagement in scientific argumentation. This study demonstrates that through teacher translanguaging emergent bilinguals were able to engage in and improve their performance of scientific argumentation. While scientific argumentation is one discourse intensive scientific practice, translanguaging as science pedagogy warrants further exploration to determine its full potential in providing emergent bilingual access to the full extent of recent science education reform. The pedagogical potential of translanguaging is particularly relevant given the poor educational opportunities currently afforded the fastest-growing student demographic group in the United States. The applicability of scientific literacy goes beyond classroom exercises and allows individuals to engage with the myriad socioscientific issues that have personal, environmental, and societal implications. Furthermore, translanguaging as science pedagogy builds on students' assets and opens up the potential for developing emergent bilinguals' identity with science and continued interest in learning science.

References

Brown, B. A., & Spang, E. (2008). Double talk: Synthesizing everyday and science language in the classroom. *Science Education, 92*(4), 708–732.

Collier, V. P., & Thomas, W. P. (2004). The astounding effectiveness of dual language education for all. *NABE Journal of Research and Practice, 2*(1), 1–20.

Creese, A., & Blackledge, A. (2010). Translanguaging in the bilingual classroom: A pedagogy for learning and teaching? *The Modern Language Journal, 94*(1), 103–115.

De Jong, E. J. (2002). Effective bilingual education: From theory to academic achievement in a two-way bilingual program. *Bilingual Research Journal, 26*(1), 65–84.

Duschl, R. A., & Osborne, J. (2002). Supporting and promoting argumentation discourse in science education. *Studies in Science Education, 38*, 39–72.

Espinosa, C. M. (2016). Reclaiming bilingualism: Translanguaging in a science class. In O. García & T. Kleyn (Eds.), *Translanguaging with multilingual students: Learning from classroom moments* (pp. 174–192). Routledge.

Esquinca, A., Araujo, B., & de la Piedra, M. T. (2014). Meaning making and translanguaging in a two-way dual-language program on the US-Mexico border. *Bilingual Research Journal, 37*(2), 164–181.

FOSS. (2020). *Full option science system*. https://www.fossweb.com/foss-modules. Accessed 4 June 2020.

García, O. (2009). Emergent bilinguals and TESOL: What's in a name? *TESOL Quarterly, 43*(2), 322–326.

García, O., Kleifgen, J. A., & Falchi, L. (2008). *From English language learners to emergent bilinguals*. Equity matters. Research review no. 1. Campaign for Educational Equity, Teachers College, Columbia University.

García, O., Ibarra Johnson, S., & Seltzer, K. (2016). *The translanguaging classroom: Leveraging student bilingualism for learning*. Caslon.

Gay, G. (2002). Preparing for culturally responsive teaching. *Journal of Teacher Education, 53*(2), 106–116.

González, N., Moll, L. C., & Amanti, C. (Eds.). (2005). *Funds of knowledge: Theorizing practices in households, communities, and classrooms*. Lawrence Erlbaum Associates.

Halliday, M. A. K., & Martin, J. R. (1993). *Writing science: Literacy and discursive power*. University of Pittsburgh Press.

Infante, P., & Licona, P. R. (2018). Translanguaging as pedagogy: Developing learner scientific discursive practices in a bilingual middle school science classroom. *International Journal of Bilingual Education and Bilingualism*. https://doi.org/10.1080/13670050.2018.1526885

Kelly, G. J. (2008). Inquiry, activity, and epistemic practice. In R. Duschl & R. Grandy (Eds.), *Teaching scientific inquiry: Recommendations for research and implementation* (pp. 99–117). Sense Publishers.

Kelly, G. J. (2014). Analysing classroom activities: Theoretical and methodological considerations. In C. Bruguière, A. Tiberghien, & P. Clément (Eds.), *Topics and trends in current science education: 9th ESERA conference selected contributions* (pp. 353–368). Springer.

Kelly, G. J., & Bazerman, C. (2003). How students argue scientific claims: A rhetorical-semantic analysis. *Applied Linguistics, 24*(1), 28–55.

Kelly, G. J., & Green, J. (1998). The social nature of knowing: Toward a sociocultural perspective on conceptual change and knowledge construction. In B. Guzzetti & C. Hynd (Eds.), *Perspectives on conceptual change: Multiple ways to understand knowing and learning in a complex world* (pp. 145–182). Erlbaum.

Kelly, G. J., & Licona, P. (2018). Epistemic practices and science education. In M. Matthews (Ed.), *History, philosophy and science teaching: New research perspectives* (pp. 139–165). Springer.

Ladson-Billings, G. (1995). Toward a theory of culturally relevant pedagogy. *American Educational Research Journal, 32*(3), 465–491.

Landivar, L. C. (2013). Disparities in STEM employment by sex, race, and Hispanic origin. *Education Review, 29*(6), 911–922.

Leach, J & Scott, P. (2002) Designing and evaluating science teaching sequences: An approach drawing upon the concept of learning demand and a social constructivist perspective on learning. *Studies in Science Education, 38*, 115–142.

Leach, J., & Scott, P. (2003). Individual and sociocultural views of learning in science education. *Science & Education, 12*(1), 91–113.

Leach, J., Ametller, J., Hind, A., Lewis, J., & Scott, P. (2005). Designing and evaluating short science teaching sequences: Improving student learning. In K. Boersma, M. Goedhart, O. de Jong, & H. Eijkelhof (Eds.), *Research and the quality of science education* (pp. 209–220). Springer.

Lee, O., & Luykx, A. (2006). *Science education and student diversity: Synthesis and research agenda*. Cambridge University Press.

Lee, O., Quinn, H., & Valdés, G. (2013). Science and language for English language learners in relation to next generation science standards and with implications for common core standards for English language arts and mathematics. *Educational Researcher, 20*(10), 1–11.

Lemke, J. (1990). *Talking science: Language, learning, and values*. Ablex Publishing Corporation.

Licona, P. R. (2018). Translanguaging about socioscientific issues in middle school science. In *Theory and methods for sociocultural research in science and engineering education* (p. 73).

Lucas, T., & Villegas, A. M. (2013). Preparing linguistically responsive teachers: Laying the foundation in preservice teacher education. *Theory Into Practice, 52*(2), 98–109.

McNeill, K. L., & Krajcik, J. S. (2012). *Supporting grade 5–8 students in constructing explanations in science: The claim, evidence, and reasoning framework for talk and writing*. Pearson Allyn & Bacon.

Moll, L. C., Amanti, C., Neff, D., & Gonzalez, N. (1992). Funds of knowledge for teaching: Using a qualitative approach to connect homes and classrooms. *Theory Into Practice, 31*(2), 132–141.

Morales, P. Z., & Aldana, U. (2010). Learning in two languages: Programs with political promise. In *Forbidden language: English learners and restrictive language policies* (pp. 159–174).

National Center For Education Statistics. (2020). *Racial/Ethnic enrollment in public schools*. https://nces.ed.gov/programs/coe/indicator_cge.asp. Accessed 1 Aug 2020.

National Research Council. (2012). *A framework for K–12 science education: Practices, crosscutting concepts and core ideas*. National Academies Press.

NGSS Lead States. (2013). *Next generation science standards: For states, by states*. The National Academies Press.

Ovando, C. J., & Combs, M. C. (2012). *Bilingual and ESL classrooms: Teaching in multicultural contexts*. McGraw-Hill.

Paris, D. (2012). Culturally sustaining pedagogy: A needed change in stance, terminology, and practice. *Educational Researcher, 41*(3), 93–97.

Pellow, D., & Vazin, J. (2019). The intersection of race, immigration status, and environmental justice. *Sustainability, 11*(14), 3942.

Roberts, D. A., & Bybee, R. W. (2014). Scientific literacy, science literacy, and science education. In N. G. Lederman & S. K. Abell (Eds.), *Handbook of research on science education, volume II* (pp. 559–572). Routledge.

Sadler, T. D. (2004). Informal reasoning regarding socioscientific issues: A critical review of research. *Journal of Research in Science Teaching, 41*(5), 513–536.

Sadler, T. D., & Donnelly, L. A. (2006). Socioscientific argumentation: The effects of content knowledge and morality. *International Journal of Science Education, 28*(12), 1463–1488.

Sayer, P. (2013). Translanguaging, TexMex, and bilingual pedagogy: Emergent bilinguals learning through the vernacular. *TESOL Quarterly, 47*(1), 63–88.

Spradley, J. P. (1980). *Participant observation*. Harcourt Brace Jovanovich College Publishers.

U.S. Department of Education, Office of Planning, Evaluation and Policy Development, Policy and Program Studies Service. (2016). *The state of racial diversity in the educator workforce*. Washington, DC.

Warren, B., Ballenger, C., Ogonowski, M., Rosebery, A. S., & Hudicourt-Barnes, J. (2001). Rethinking diversity in learning science: The logic of everyday sense-making. *Journal of Research in Science Teaching, 38*(5), 529–552.

Wright, W. E. (2010). *Foundations for teaching English language learners: Research, theory, policy, and practice*. Caslon Publishing.

Wright, W. E. (2019). *Foundations for teaching English language learners: Research, theory, policy, and practice*. Caslon Publishing.

Zeidler, D. L. (2014). Socioscientific issues as a curriculum emphasis. Theory, research, and practice. In N. G. Lederman & S. K. Abell (Eds.), *Handbook of research on science education, volume II* (pp. 697–726). Routledge.

Chapter 10
Contradictions Confronting Hybrid Spaces for Translanguaging in the Lebanese Context: A CHAT Perspective

Sara Salloum

Abstract The Lebanese educational system is multilingual with science taught and assessed in a foreign language (English or French). Deploying a foreign language in science instruction have raised concerns due to Lebanon's low performance on international assessments, with research showing that fluency in the foreign language and other socioeconomic factors contribute significantly to students' TIMSS science performance. A content literacy model and translanguaging, or the deployment and leveraging of students' entire linguistic repertoires, supports diverse students' deep science learning; however, in the Lebanese context, enacting translanguaging practices runs against monoglossic norms that favor the foreign languages. Such norms have raised tensions around equity and access to quality science instruction and the value of Arabic as a symbol of national identity. This paper utilizes cultural historical activity theory (CHAT) to conceptually and empirically identify tensions as "contradictions" within the Lebanese context that confront enactment of a content literacy model and translanguaging. CHAT is a transformative approach to understanding human practice as a goal-oriented collective activity, in which contradictions are inevitable and become an impetus for change and development. Exploring contradiction can inform contextually responsive interventions for enhancing the quality of science education in Lebanon.

Keywords Translanguaging · Multilingual science education · CHAT · Activity theory

S. Salloum (✉)
University of Balamand, Balamand al-Kurah, Lebanon
e-mail: sara.salloum@balamand.edu.lb

© The Author(s), under exclusive license to Springer Nature Switzerland AG 2021

A. Jakobsson et al. (eds.), *Translanguaging in Science Education*, Sociocultural Explorations of Science Education 27,
https://doi.org/10.1007/978-3-030-82973-5_10

10.1 Introduction

The significant role of language in development and meaning-making is a legacy of Vygotsky's sociocultural approach, with language acting as a major mediating artifact and cultural tool in science and the science classrooms. Language and science learning are interdependent: learning science involves not only conceptual understandings but also developing the discourse of science or science's discursive skills (Lemke, 2001; Nygård Larsson & Jakobsson, 2020). Such view of science learning has been termed a content literacy model in which teachers and students navigate hybrid spaces that integrate content knowledge (represented through multiple modalities), literacy skills, and discursive skills from students' different social spaces (e.g., their homes, schools, peer groups, or communities) (Martínez-Álvarez, 2019; Nygård Larsson, 2018). In science classrooms, a "hybrid space" is where students make meaning of science concepts and gradually develop the discourse of science by moving back and forth between informal everyday experiences and languages on the one hand and formal scientific terminology and discourse on the other (Lee et al., 2019; Nygård Larsson & Jakobsson, 2020). Multilingual settings, even as they entail challenges for both learners and teachers, can also entail hybrid spaces for richer learning and meaning-making through multiple modalities and translanguaging or the deployment of the full range of students' linguistic repertoires (e.g., Jakobson & Axelsson, 2017; Martínez-Álvarez, 2019).

The Lebanese educational system is multilingual[1]: French or English are introduced with Arabic in first grade, and the foreign language (English or French) is designated as the language of instruction for science and mathematics. Due to a multitude of factors (specifically SES), students in a science classroom may exhibit various foreign language proficiency levels. A content literacy model and translanguaging can hold promise in Lebanese multilingual science classrooms as these open up hybrid spaces in which students' entire communicative resources (e.g., their home language and cultural backgrounds) are deployed for constructing deeper meanings in science. Yet, translanguaging, especially in terms of deploying the home language, can go against strongly held monoglossic notions by Lebanese teachers and administrators, which would create resistance and tensions against its enactment. Moreover, at the policy level, designating a foreign language as medium of science instruction entails additional tensions between a foreign languages' importance in providing access to scientific and technological innovations *and* Arabic's value as a symbol of national and regional identities (Amin, 2009).

I contend that such tensions find their way into the science classroom and translate into disparities in performance of students with different foreign language proficiency levels (Jakobson & Axelsson, 2017). Deploying a foreign language in teaching and assessing mathematics and science have raised concerns due to Lebanon's low performance on international comparative assessments such as the

[1] Mainly Arabic, French, and English are designated as languages of learning and teaching (along with Armenian in schools that serve the Armenian community).

TIMSS, where it has been found that how often learners spoke the foreign language at home and other socioeconomic factors[2] contributed significantly to students' performance in science. Indeed, research indicates that students with diverse foreign language proficiency levels receive little support in the Lebanese science classrooms (Salloum & BouJaoude, 2020a; Shuyab, 2016). To better understand such tensions, this paper uses cultural historical activity theory (CHAT) to uncover and characterize tensions as manifestations of "contradictions" within the Lebanese educational setting with regard to adopting and enacting translanguaging and a content literacy model. CHAT is a transformative approach to understanding human practice as a goal-oriented purposeful activity, in which contradictions are inevitable and can become driving forces for change (Sannino et al., 2016). In the Lebanese context, exploring contradictions can help us pursue systematic and contextually responsive appraoches that address implications of linguistic diversity for quality science education.

10.2 Theoretical Frameworks and Guiding Principles: Translanguaging and CHAT

10.2.1 Translanguaging

10.2.1.1 Translanguaging as Ideology and Pedagogy

The term "translanguaging" has first emerged to characterize pedagogical practices in the Welsh setting, where a symbiotic form of bilingualism was sought. It was later formulated as an ideology and practical theory that examines fluid and complex social interactions within multilingual settings from the perspective of the speaker (e.g., Creese & Blackledge, 2015; García, 2009; Wei, 2018). Translanguaging conceptualizes the complex ways in which plurilingual individuals deploy and blend different languages and modalities across social contexts to construct meaning and negotiate social identities; in essence translanguaging challenges traditional monoglossic orientation or "one language at a time" ideologies (Martínez-Roldán, 2015; García & Wei, 2014). It is defined as the "the deployment of a speaker's full linguistic repertoire without regard for watchful adherence to the socially and politically defined boundaries of named (and usually national and state) languages" (Otheguy et al., 2015, p. 283, italics in original). With translanguaging, a social (and pedagogical) space is created where plurilingual individuals' linguistic performances, personal history and values, experience and environment, and cognitive capacity are coordinated into meaningful lived experience (Wei, 2011). This translanguaging space is a source for identity positioning and for expressing criticality

[2] For example, parent's education level and number of household books and devices.

and creativity through generating novel representations and knowledge (García & Leiva, 2014).

More recently translanguaging is being further developed as a pedagogy that supports plurilingual learners' construction of meaning by encouraging and empowering them to use all available linguistic resources and repertoires (García, 2009; Karlsson et al., 2019). Translanguaging as a pedagogy purposefully builds on language overlap and connections among "social, cultural, community, and linguistic domains" of participants lives (Creese & Blackledge, 2010, p. 112). Translanguaging empowers learners by focusing on meaning-making and building on and enhancing experiences and so comprises an important mediator of identities and cognitive complex activities (Creese & Blackledge, 2015; García, 2009; Lin & He, 2017; Wei, 2018). Moreover, translanguaging promotes an inclusive pedagogy that recognizes diverse students' "language and meaning-making practices as a resource to learn and to show what they know, as well as to extend these" (García & Wei, 2014, p. 227). There is a growing body of research on translanguaging in multilingual classrooms (e.g., Martínez-Álvarez & Ghiso, 2017) and multilingual science classrooms (e.g., Karlsson et al., 2020; Lin & He, 2017; Probyn, 2019; Ryu, 2019). The notion that translanguaging opens up spaces in the classroom for recognizing and leveraging plurilingual learners' full range of communicative and linguistic resources along with dimensions of their experiences and identities is especially important for meaningful and equitable science education as I discuss below.

10.2.1.2 Translanguaging in Science Education

García and Lin (2016) and Blackledge and Creese (2014) characterized translanguaging using Bakhtin's theoretical and practical notion of "heteroglossia" or the coexistence of different languages, language forms, and dialects. Bakhtin (1981) developed the notion of "heteroglossia" to describe and theorize relationships and social tensions among language forms and the social, historical, and political aspects influencing them. Essentially, translanguaging as heteroglossic challenges prevailing colonial monoglossic orientations (Lin & He, 2017; Probyn, 2019). Bakhtin (1981) also introduced dialogicity or multi-voicedness, where language, meanings, and thoughts are both socially negotiated and dialogically based: utterances are part of a chain of communication, they respond to previous utterances and anticipate others. Such utterances are suffused with heteroglossia, and it is their interactions that produces deeper meanings. For science education, "heteroglossia" and dialogicity are relevant as they involve understanding the intersections and connections between simple and complex speech genres (Salloum & BouJaoude, 2019; Mortimer & Scott, 2003): Simple primary speech genres are our everyday experiences *and* speech (e.g., in the home and international language) and complex secondary speech genres are scientific terminology and discourse that emerge in complex and developed communication (Bakhtin, 1987). Dialogic interactions and discursive mobility among students' everyday ways of thinking and talking and the formal scientific content and discourse is needed to bridge the two genres for productive *meaning-making* of science concepts and the development of science discourse (Mortimer & Scott, 2003; Nygård Larsson, 2018).

Translanguaging enables dynamic but *purposeful* navigation and bridging of different speech genres (everyday language and experiences and the specialized science content and discourse) and thus promotes dialogic interactions that cultivate productive *meaning-making* in science (Nygård Larsson & Jakobsson, 2020; Probyn, 2019). Accordingly, learners' linguistic capital is leveraged for realizing greater access, equity, social justice, and an asset-oriented perspective in science education (Poza, 2018; Probyn, 2019). Indeed research in science education is demonstrating the importance of translanguaging in creating spaces for dynamic and joint negotiations that help students relate and contextualize science content within prior experiences (e.g., Karlsson et al., 2020). Moreover, quality science learning requires students to make meaning across various modalities, where understandings are developed through active engagement with multiple modes and media such as different forms of text, hands-on experiments, physical models, natural world exploration, and/or digital materials (Kress et al., 2014; Lemke, 2005). Accordingly, Probyn (2015) extended translanguaging with the notion of flexible "bridging discourses" (Gibbons, 2006) to illustrate types of scaffolding necessary for plurilingual students as they navigate everyday and scientific understandings and language (both written and oral) and hands-on experiences with theoretical understandings and scientific explanations.

Yet, celebrating and appropriating the power of learning through the flexible deployment of students' full linguistic resources can run against strong monoglossic norms and thus emerge as a controversial pedagogy in schools and among teachers (Otheguy et al., 2015; Probyn, 2015). Moreover, especially in contexts like Lebanon and the MENA region, how languages are deployed in social and educational settings has political and power implications (discussed below). Such issues would influence the realization of hybrid spaces for translanguaging and the enactment of a content literacy model in the science classroom. Cultural historical activity theory (CHAT) can help us characterize tensions as manifestations of 'contradictions' within the Lebanese educational settings towards spaces for translanguaging and inform potentialities for change.

10.2.2 CHAT: Cultural Historical Activity Theory

The inseparability between the individual and society is a key to Vygotsky's legacy. Vygotsky proposed the concept of social mediation, where higher mental processes develop as individuals interact with others through mediating tools, instruments, and signs. Moreover, given the cultural-historical nature of mediating tools, documenting individual learning without considering the circumstances that mediate learning would fail to capture important insights about learning and development (Martínez-Roldán, 2015). Accordingly, the socially organized activity surrounding the learner is proposed as the major unit of analysis within cultural historical activity theory or CHAT (Leont'ev, 1981). An activity is distinct from single "actions" (Engeström & Sannino, 2010) and encompasses various actions by different individuals within a community toward a goal or an *object*. Consequently, the construct of "activity system" was developed as a system of collaborative and purposeful human practice within the broad historical sociocultural context (Engeström, 1999).

An activity system involves a *subject* or a *group of subjects* who seek to direct the activity towards an *object* or targeted problem space with the help of *instruments* (tools and mediating artifacts). An activity system also outlines the *Community* of individuals and groups sharing the same *object*, the *Division of labor* or distribution of tasks and power among them, and the implicit and explicit *rules* that constrain their actions (Engeström & Sannino, 2010). A sample classroom activity system is depicted in Fig. 10.1 and appendix A (Salloum & BouJaoude, 2020b). The *object* of any activity is internally contradictory and characterized by ambiguity, interpretation, and the potential for productive change (Engeström, 2001). *Subjects* with their distinctive histories and perspectives interact across the elements of their activity systems (e.g., *instruments, rules, community*, and *division of labor*) and encounter contradictions among them. Such contradictions require individuals or groups to transform social realities and themselves through an ongoing materially and socially mediated process known as expansive learning, whereby the system reformulates new *objects* and *instruments* that develop and transform the activity (Engeström, 1999).

In the interest of multi-voicedness and to address diversity and dialogue between different traditions or perspectives, Engeström (2001) elaborated activity systems to include networks of interacting systems (Fig. 10.2). A network may include activity systems with *subjects* having different perspectives [e.g., the perspectives of students and science teachers and/or across schools (Martínez-Álvarez, 2019)]. And so "contradictions" become not only tensions within an activity system, but also among

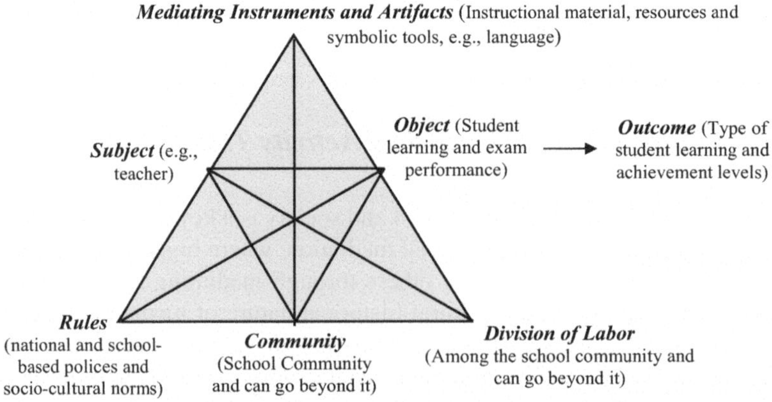

Fig. 10.1 The classroom as an activity system

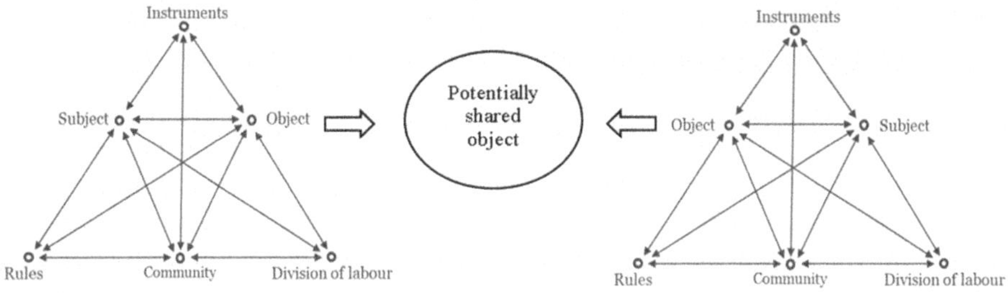

Fig. 10.2 Network of activity systems

interacting activity systems in a network, and such contradictions can also become the basis for expansive learning and development across systems.

CHAT's sociocultural origins, its focus on discursive manifestations of historically accumulated contradictions, and its transformative approach have led education researchers to utilize it as a cross-disciplinary framework for examining students and teachers' discursive practices and their transformation (actual or potential) through culturally and socially mediated processes (e.g., Goodnough, 2018; Martínez-Álvarez, 2019). In multilingual science education, where science and foreign/dominant language learning intersect with identity, social justice, and language-in-education policies, CHAT affords close examination of discursive practices and historically accumulated contradictions within them, revealing potentialities for transformation and change (e.g., Salloum & BouJaoude, 2020b; Martínez-Álvarez, 2019). For example, translanguaging along with multimodal resources and activities have been explored, through CHAT, as mediation tools for student and teacher learning. Martínez-Roldán (2015) and Martínez-Álvarez and Ghiso's (2017) found that students in a bilingual after-school programs benefited from translanguaging and fluid use of language as *instruments* that challenge deficit language ideologies. Teachers' particular translanguaging practices, though, can raise contradictions that reinforce the hegemony of English when students' performance on tests becomes a sought-after *object* (as opposed to deep science understandings). Similarly, Martínez-Álvarez (2019) found that expansive learning of minoritized students in a science afterschool program was facilitated through multimedia, culturally relevant artifacts, students' active participation, *and* spaces that accept bilingual learners' expansive ways of understanding science.

The abovementioned research have thus far explored translanguaging as a mediating *instrument*; in this paper I alternatively consider "Translanguaging Spaces for Multimodal Content Literacy Models" as a broad *aspired-to Object* for the activity systems of interest. Actually the peculiar reversal of *object* and *instrument* is a particularity of schooling according to Engeström et al. (2002), and so by setting such an *object* a priori and exploring conceptually and empirically contradictions that can initiate expansive learning and development, I aspire that translanguaging practices will later develop into *novel mediating instruments* in the Lebanese science classrooms.

10.3 A CHAT-Inspired Conceptual Review: Contradictions in the Lebanese Context

Contradictions are historic and current systemic phenomena that cannot be observed directly; they need to be identified through their manifestations as a paradox, tension, inconsistency, conflict, or dilemma (Engeström & Sannino, 2011). Therefore, identifying contradictions in the Lebanese context requires both a historical and current overview through conceptual and empirical analyses. For a conceptual analysis, I present first a brief historical overview of the Lebanese context. I then outline a CHAT-inspired review of relevant literature from Lebanon to discern manifestations of "contradictions" around the influence of the language-in-education policy. I selected papers that examine the Lebanese multilingual educational context and its interconnectedness with other aspects of cultural life (Diab, 2006, 2009; Shaaban & Ghaith, 1999, 2003; Orr & Annous, 2018) and the influence of language-in-education policy in general and on science education and/or pedagogical practices (Amin, 2009; BouJaoude & El-Hage, 2016; Dagher & BouJaoude, 2011; Shuyab, 2016; Bahous et al., 2011). Areas that manifest contradictions were conceptually discerned from these papers, and four analytical and interrelated areas emerged (explained below):

I. Power distribution: Identity and cultural tensions
II. Utility of language: Tensions and conflicts among languages
III. Equity and language of instruction in science (and mathematics)
IV. Quality of instruction: Authoritarian vs. interactive/dialogic teaching

10.3.1 A Historical Overview of Language-In-Education Policies in Lebanon

Lebanon's language-in-education policy has been influenced by the country's strategic geo-political position as a small Mediterranean country of several religious and cultural minorities[3] (Shaaban & Ghaith, 1999). Major influences include competing foreign missionaries between the seventeenth and twentieth Century, the French mandate over Lebanon (1920–1943), and the independence of modern Lebanon in 1946 with an economy reliant on commerce, tourism and service. Under the Ottoman rule, French and British missionary schools, that reflected their countries' socio-political interests in the area, exposed the Lebanese communities to Western languages and cultures, with the intention of cultivating religious ties between these communities and the West (Shaaban & Ghaith, 2003). During the French mandate (1920–1943), both French and Arabic were declared as national languages. After Lebanon's independence in 1946, Arabic was declared as the

[3] Lebanon is home to 18 recognized religious sects: four Muslim, 12 Christian, Druze, and Judaism.

country's only national language as an expression of national pride and decrees strengthened the Arabic language in schools (Shaaban & Ghaith, 1999): Arabic was established as the language of instruction in public schools for grades 1 through 6 in addition to either English and French as a compulsory foreign language. In secondary school, science and mathematics are taught and assessed in the foreign language. The 1994 educational reform that followed the country's 15-year civil war maintained the educational system's trilingual character, with science taught in a foreign language. Currently distribution of students between French or English-medium schools is around 52% for French-medium schools and 46% in English-medium schools (NCERD, 2019).

10.3.2 Contradiction Area I: Power Distribution and Identity Tensions

Utilizing a foreign language as the medium of instruction, especially that of the ex-colonizer, typically involves power and national identity tensions grounded in ideological and socio-political agendas (Dagher & BouJaoude, 2011). Various languages have historically competed and gained status in Lebanon, which has raised concerns about Arabic and the Lebanese "national identity" being marginalized (Dagher & BouJaoude, 2011; Diab, 2006, 2009; Orr & Annous, 2018; Shaaban & Ghaith, 1999, 2003). Diab (2006) and Shaaban and Ghaith (1999) suggested that decrees following Lebanon's independence reflected a *hasty expression* of national pride, rather than careful planning and so it remained that French and English endured strongly in the Lebanese educational system (Shaaban & Ghaith, 2003). Research in higher education (specifically in the private sector[4]) suggests that religious background is a determinant of linguistic identity construction, with university students from Christian communities seeming more likely to construct an identity that is ethnically and culturally distinct form the rest of the Arab world. Such views may be remnants of the French missionaries' influence in Lebanon, which gave certain Christian groups the opportunity to develop their foreign languages, thus opening access to employment privileges denied to most Muslims and other Christian sects and giving rise to power *tensions* among groups (Shaaban & Ghaith, 2003). Such tensions were manifested in a *contradiction* in the last educational reform (1994–97), where decisions on the language of instruction became a contentious and much-debated issue (Shuyab, 2016): one group stressed the Arabic language and identity of Lebanon and the other advocated a multicultural and multilinguistic Lebanon. This contradiction resulted in a compromise, where it was agreed that elementary (grades 1–6) mathematics and science textbooks be written in the three languages

[4]This research is also in English medium universities, this observation may emerge even stronger in French-medium private universities established by missionaries, or differently in the public Lebanese University.

(Arabic, English, and French), after that textbooks are only offered in in one of the foreign languages; a decision that would have equity implications as indicated below.

10.3.3 Contradiction Area II: The Utility of Language

Conflicting and diverging views about the utility of the various languages in education, administration, society, and media are also present in Lebanon (Shaaban & Ghaith, 2003). Research shows that even as private university students use colloquial Arabic for everyday communication, there exists a general acceptance across all groups of the importance and status of learning a foreign language over Arabic (Orr & Annous, 2018; Diab, 2009; Shaaban & Ghaith, 2003): English is valued as an international language for science, trade, and technology, and French as a language of education and culture (Shaaban & Ghaith, 2003). Orr and Annous (2018) discussed such acceptance of the language-in-education policy as "linguistics imperialism," where students and the community accept that intellectual progress is highly dependent on proficiency in a foreign language, especially English. Support of the policy springs from perceptions around unsuitability of Arabic as a twenty-first-century language and the dire need for a foreign language for immigration and future careers (Diab, 2006; Orr & Annous, 2018; Shaaban & Ghaith, 2003).

In light of the above historical and cultural backdrop and our contemporary context, science education in Lebanon involves a significant *tension*, where discussions around a foreign language as the medium of instruction invariably entail a struggle between the importance of the foreign language for gaining access to modern scientific and technological innovations (e.g., STEM majors) *and* deploying Arabic in science as "a basis for, and symbol of, national and regional identities" (Amin, 2009, p. 62).

10.3.4 Contradiction Area III: Educational Equity

Shuyab (2016) stated that the educational reforms in Lebanon saw poor nationalism and religious intolerance as the main threats to social cohesion, while ignoring social justice, power distribution, and equity issues as instigators of instability. And so, these reforms do not recognize the structural barriers and marginalization resulting from designating a foreign language to teach science and mathematics on disadvantaged at-risk groups. She added that even as learning a foreign language is considered a main objective of the Lebanese curriculum to reflect Lebanon's multicultural and multilingual image, it is a key factor leading to school dropout amongst disadvantaged groups in Lebanon. School dropout rates (especially for boys) increase in grades 6 and 7 *when* science and mathematics *have* to be taught in a foreign language. In lower socioeconomic status schools, teachers cite students' limited language proficiency and exposure to the foreign language outside school as "a main problem" (Bahous et al., 2011). Indeed, designating a foreign language for science instruction is considered as one of the key challenges inherent in our school system to providing quality science education (BouJaoude & El-Hage, 2016; Dagher & BouJaoude, 2011). Accordingly, a *paradox* or manifestation of a contradiction emerges whereby two of the goals of the Lebanese curricula oppose each other: access and equity vs. a multicultural and multilingual image.

10.3.5 Contradiction Area IV: Quality of Instructional Practices

An important aim of the last reform was moving teaching from teacher-centered to student-centered (NCERD, 1997). The general goals of science education include promoting the values, knowledge, and inquiry skills involved in modern scientific knowledge and literacy. Yet research in Lebanon has illuminated a number of issues and challenges toward realizing these goals: Authoritarian content and test-centered approaches persist in science classrooms, with scientific inquiry practices not sufficiently reflected in classroom practices (BouJaoude & El-Hage, 2016). Authoritarian rather than dialogic interactions dominate in many science classrooms, and these mostly target factual and algorithmic knowledge needed for the middle school high-stakes tests (administered in the foreign language) (Salloum & BouJaoude, 2019). Similar trends exist in foreign language learning with *a conflict* between the teaching approaches advocated in the curriculum and the actual teaching and learning practices (Bahous et al., 2011). Such *conflict* poses challenges to quality science education: Foreign language instruction that promotes a skills-based rote approach rather than language as a meaningful communicative tool will not support deep content learning and literacy.

10.4 Methodology

Activity system representations serve as schematic and structural models that describe and organize series of practices (Engeström, 2001). To illuminate series of practices and perspectives, data from previous research and ongoing research projects conducted in public and lower SES private schools were used and reinterpreted. This data included:

- Collective case studies from the classrooms of five science teachers in five schools (two public, two lower SES private, and an elite private school): Classroom observations (~16 science sessions per teacher), teachers, and school leaders interviews, and student focus groups
- Documented formative intervention meetings from an ongoing project (science and language teachers)

- Interview data with eight science teachers on role of textbooks in supporting students with diverse foreign language proficiency levels

Particularly for this paper, I conducted nine critical dialogues with three science education faculty, three language and educational leadership specialists, and three leaders from the two government institutions that oversee "teacher learning and development" and "teacher supervision." Critical dialogues were used as they clarify position and create new understandings around political and moral questions that entail reflexivity and paradigmatic shifts (Karlsson, 2001; Schwandt, 2001). Centering questions guided the critical dialogues and these lasted between 1 h to a little less than 2 h (Appendix B).

Through a recursive and iterative process and peer debriefing with a colleague, the different data layers and the contradiction areas identified through the conceptual analysis of the literature were used to construct two *representative* networks of interacting activity systems: one for lower SES private schools and one for public schools. In Lebanon, only around 31% of the students attend free public schools, with the rest attending private schools. Private schools though vary vastly in their tuition fees, with some being almost free through government or religious group subsidies Schools also vary in their level of secularism (many started as missionary schools) and human and physical resources available to them. Lebanese students from lower socioeconomic strata attend free public or lower or free tuition private schools. Essentially, these two groups of schools (public and lower SES) serve learners whose exposure to the foreign language outside school is generally less than their affluent counterparts; and so would need additional support in multilingual science classrooms to overcome the structural barriers associated with teaching science in a foreign language.

Data analysis was multilayered as it needed to inform the different elements in the activity system (Fig. 10.1). First, patterns in practices around language deployment and conceptual learning in the classrooms were discerned from observation data and also from recounts in critical dialogues. Second, interview data (teachers, students and principals) was revisited to identify patterns across the participants' views on the influence of teaching and learning science in a foreign language and their perceptions on the role of different scaffolds, including home language and literacy strategies. Finally, the nine critical dialogues were analyzed thematically to present the different perspectives and tensions that participants raise about quality science education, role of language in science education, and potentialities for change and development. Each CHAT-inspired network includes three activity systems with different *subjects:* students, science teacher, and larger school system (Figs. 10.3 and 10.4). The two representative networks here do not represent one specific school nor serve as an "exact" representation of the whole group of schools[5]; rather they holistically present and organize general trends in practices and perspectives discerned from the data and literature on the role and influence of deploying a

[5] Meaning an individual public school may diverge from this CHAT representation, but most would exhibit several aspects depicted in it.

foreign language for science teaching and learning, specifically for learners with a diverse range of foreign language proficiency. In doing so I hope to explore contradictions within and across activity systems, and discuss ones that can lead to potential expansive learning.

10.5 Findings

10.5.1 Condensations of Findings from Previous and Ongoing School-Based Research

I present first summaries from previously collected classroom observation data in Table 10.1 (Salloum & BouJaoude, 2020a; Salloum & BouJaoude, 2020b). Moreover, teachers, students, and principals' perspectives from interviews are presented narratively below.

In terms of perspectives, students across the contexts expressed a preference to communicate and ask questions in their home language to better understand concepts, even as they acknowledge the importance of English language proficiency for future studies and careers. As for teachers, teachers in all the studies (mentioned above) expressed concerns over students' language proficiency due to its impact on students' understandings, their ability to express understandings, and their performance on national examinations (for example students not understanding test items). Most private schools emphasize and consistently enforce the rule that teachers only deploy the foreign language in science classroom. Most private school teachers endorsed this rule and some mentioned that deploying two languages can confuse students, even as they acknowledged that students find it "easier" to respond in the home language. Moreover, all teachers mentioned that due to the diglossic nature of Arabic, scientific terms are easier to introduce in English as students (and teachers) may not know their meaning in standard Arabic.

Table 10.1 Summary of observation analysis (Middle and lower middle SES schools)

Lower SES Private Schools	Public Schools
Pedagogical practices: Mostly teacher-directed IRF (initiation, response, follow-up) with interactions being mostly factual and algorithmic, rather than conceptual; little active teaching methods used **Language practices:** Teachers adamantly deployed English to explain science concepts and respond to students, even when students exhibited difficulties. Home language deployed rarely and only as a last resort due to student frustration. Students deployed home language (Arabic) to respond and ask for clarifications, and English mostly to give shorter factual responses	**Pedagogical practices:** Mostly teacher-directed chains of IRF with interactions being both factual and conceptual with frequent multimodal teaching using simple props **Language practices:** Teachers and students moved back and forth between international and home language for concept development, with teachers summarizing concepts in English to conclude

The “English-only rule” is encouraged by public school principals but not fully enforced, and so the participant public school teachers argued for deploying the home language frequently and flexibly for “students to understand.” They acknowledged though that deploying home language may have its limitations, as students may translate their everyday or simplified language to English, rather than learning how to write and speak “scientifically.” Finally, the public school teachers said that they “get mixed messages” from the different overseeing bodies with regard to using the home language and that they did not receive systematic or specific teacher preparation for multilingual science education. The public schools teachers intentionally and flexibly deployed the home language, and such flexibility served as a mediating tool for higher dialogic and conceptual interactions (Amin & Badreddine, 2020; Salloum & BouJaoude, 2020a). The participant public school teachers’ intentional and creative use of the home language to scaffold conceptual learning did not constitute translanguaging though, as translanguaging necessitates both an ideological stance on the heteroglossic nature of multilingualism and the *planning* of dynamic integrated learning that leverages students’ entire language resources (García & Lin, 2016). Notwithstading, the teachers did raise insightful concerns on how language flexibility supports students’ development of science discourse, i.e., writing and speaking “scientifically,” thus indicating their need for more strategic and engaged language flexibility.

10.5.2 Themes from Critical Dialogues

The science education (SE) faculty work in private universities. Dr. Jad has 30 plus years with highly recognized work. Dr. Ramzi has 15 plus years of experience and Dr. Alex a little less than 10 years. All are well-established researchers. Dr. Ramzi has done work on science education in multilingual settings. Dr. Lina has a degree of applied linguistics and over 25 years of experience as a university faculty. Dr. Camille is in sociolinguist and has done work on supporting language learning for refugees and low SES students. Dr. Nada is a curriculum and leadership specialist and is a leader on equity and equality in education research. As for the government-based teacher educators and supervisors, Dr. Amal is a retired foreign language educator and Dr. Enja and Ms. Stella are science educators.

Thematic overviews from the critical dialogues, outlined below, highlight the salient and intricate particularities expressed by the participants on the role and influence of language-in-education policy and language deployment. These dialogues informed the construction of the CHAT networks. Due to the aspired “dialogic” aspect of our conversations, I also thread some of my own personal reflections in the themes below.

10.5.2.1 "Becoming" Aware of Issues around Language in Science Education

The three science education faculty and science educators described experiences that made them "become aware" of the large role played by language in science education. Interestingly for the SE faculty such awareness did not initially emerge academically and/or theoretically, rather it materialized 'indirectly' from either working on the ground (e.g., with teaching candidates and research) (Ramzi), NGO-supported PD work (Jad and Ramzi), or personal reflections on nature of science (Alex). The participants recounted that even as we all studied science in a multilingual school system, language and science were always seen in "isolation of each other" (Alex), both in secondary school and college. For Alex, whose home language is an endangered minority language, explicit discussions about their language's profound connection to their history, identity, and rights were always present throughout his education, but never in the science and mathematics classes. Similarly, the science educators, Enja and Stella, mentioned that in school and as science majors in college, they were not "sensitized" to the role of language in science education, neither in science classes nor in language classes. Enja added, "the way we learn languages does not illuminate the function of language;" it was through teaching that she understood the difference between typical foreign language proficiency (e.g., fluency and knowing vocabulary) and students' ability to comprehend, analyze, and extract relevant information from multi-modal science texts.

10.5.2.2 Authoritarian Orientations and Language Issues in Curricula and Teaching

All participants cite the highly prevalent authoritarian and algorithmic approaches to teaching science as disguising the role of language and the discursive aspects of science teaching and learning. They also acknowledge that such teaching is across the board for other subjects as well, including the foreign and national language (standard Arabic), where Ramzi mentioned that, students are not "enculturated in a way to use language to think with." Similarly, Alex added that during his own education as a science student, language was used as a technical tool rather than a meaning-making tool. *So what contributes to the prevalence such an authoritarian approach*? All participants illuminated several interrelated aspects at the cultural, system, curricular, and classroom levels. At the cultural and system level, Jad mentioned a strongly held "romantic" perception of Lebanon's high educational performance, especially in science and mathematics, and that such belief is partially hindering us from engaging in deep-rooted reforms across subjects. Such romanticized view may have stemmed from bright spots in the history of our educational system, as Lebanon has one of highest literacy rates in the region and after its independence (1946), the state invested vastly in public education and teacher preparation leading to a period of notable progress until the 1975 civil war (Shyaub, 2016).

At the curricular level, Ramzi added that science curricula, particularly in Lebanon, are not designed with a sense of progression that carefully targets how concepts progress from one sophistication level to another. At the classroom level, Amal and Lina underscored the importance of preparing language and science teachers especially in primary education to use interactive and integrated teaching approaches. Amal though, as a veteran teacher and government-based leader, noted that low teacher accountability, especially in primary public education, is greatly hurting students in high-needs schools.

10.5.2.3 Language Policy and Equity

All participants touched upon language policy and equity, with Nada and Alex critiquing a notable absence of social justice orientations in our curricula and school system. Jad and Alex pointed out the large gaps in access to human and physical resources (e.g., more competent teachers and effective instructional materials) among schools based on sector (public and private) and area (e.g., rural). Such gaps have important implications for students whose exposure to the foreign language is minimal outside school and who need to develop their foreign language proficiency at school. Ramzi emphasized that a policy that designates a foreign language for teaching science, but does not attend to how schools can bridge gaps in resources, becomes "discriminatory." The most vocal on inherent inequities in the curriculum was Nada as this is her area of research for Lebanese and refugee students. She mentioned that the equity implications of teaching science (and mathematics) in a foreign language are ignored and "discussions about effect of language of instruction on widening inequalities in Lebanon are undermined." Her work with students in high needs schools revealed that students even by grade 7 and 8 have not received enough support to develop the foreign language skills needed to stay in school. She added that our policies reflect confused and tempestuous efforts to promote a "middle class mentality" that regards *all* Lebanese as plurilingual, having a natural aptitude for languages and who aspire to *study* and *work abroad* (Orr & Annous, 2018). She and Jad added that such policies manifest a disconnect with realities on the ground and maintain "ivory towers" mentality biases. She mentioned, for example, that foreign texts selected by university faculty as high school instructional material were too difficult for the language teachers themselves. Jad further raised issues around "hegemony" of the foreign languages in schools and higher education, where similarly, Nada and Ramzi noted that designating a foreign language for *science* and *mathematics* associates this language with an enhanced modern status at the expense of the national language; Alex referred to this as a "hidden curriculum." Ultimately, Nada added that such policies promote a deficit perspective that "blames" students and their families and gives little regard to community learning.

10.5.2.4 Perspectives on Translanguaging

SE faulty and the science educators underscored the importance of the home language as a powerful meaning-making resource, for example, Alex, who was not very familiar with *translanguaging*, related this aspect to constructivism, "students cannot manipulate language and make use of language the way they do in their home language." Jad noted that many private schools maintain an "identity" as anglophone or francophone (either as sociocultural identity or a religious affiliation) and that this may counter the flexible deployment of language and proclaim a superior status for the foreign language. Even as both Jad and Ramzi see the importance in language flexibility, they pointed that translanguaging as a pedagogy in science education requires additional conceptual and practical clarifications. They mentioned that when discussing translanguaging as a pedagogy, particularly in science education, discussants diverge greatly in terms of how intentional is it, the level of strategizing involved, and its potential targets in a science classroom. An important aspect of our dialogues together was to collectively understand what translanguaging means to us in the Lebanese context and what it could entail, I return to this in the last section. Jad and Ramzi's point connects to a precaution put forth by Poza (2018) about oversimplifying translanguaging pedagogies to a linguistic free-for-all without sufficiently addressing the need to gradually master desired target forms, such as definition of concepts with precise science terminology and appropriation of disciplinary discourse. Interestingly the two linguistics specialists had very diverging views on translanguaging with Camille being an earnest advocate and Lina being very cautious about mixing languages, suggesting that mixing languages is impacting negatively student's language learning, "students cannot even speak a language without shifting to other languages."

10.5.3 Networks of Interacting Systems

Translanguaging spaces for content literacy were set a priori as an aspired-to *object*, and the conceptual analysis of literature and the empirical data were used to identify contradictions around the realization of the aspired *object.* These contradictions are summarized below for each school group. Figs. 10.3 and 10.4 visually represent the network for low SES private schools and public schools respectively with contradictions represented as numbered "lightning" arrows (e.g., C1, C2, etc.). The figures also include contradiction boxes that summarize each numbered contradiction.

10.5.3.1 Contradiction in Lower SES Private Schools

Contradictions Related to Instruments Several contradictions toward "translanguaging for content literacy" emerged related to mediating *instruments* at the teacher and the larger school system levels. With the teachers as *subject*, a major contradiction was the dominance of highly teacher-directed authoritative interactions as *instruments*, in addition to both teachers and school leaders perceiving language flexibility as a deficit practice rather than as a meaning-making practice (**C1** in Fig. 10.3). At the school system level, a major contradiction was the official curriculum's limitation in reflecting an explicit intention for meaningful progression towards deep content literacy and providing interdisciplinary spaces for science and language teachers to collaborate on transversal competencies (e.g., literacy-rich science practices) (**C2**).

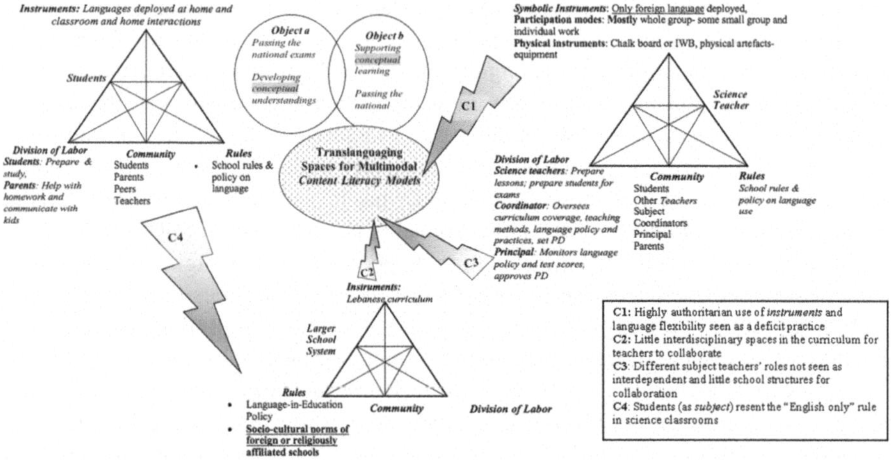

Fig. 10.3 Lower SES private schools' activity system network

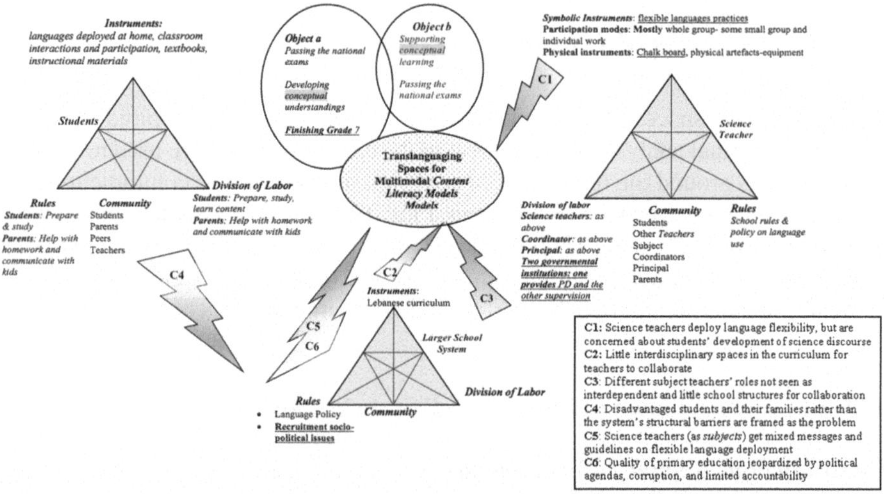

Fig. 10.4 Public schools' activity system network

Contradictions Related to Community and Division of Labor Even as language difficulties are acknowledged by language and science teacher and school leaders; the different subject teachers do not see their roles and responsibilities as interdependent (teacher level) (**C3**). Moreover, there are little systematic purposeful structures at the school level for collaboration among science and language departments for deep content literacy (**C3**).

Contradictions Related to Rules A major contradiction according to many of the critical dialogues participants is limited careful and well thought-out planning around language deployment and language policies at all levels: Classroom, school, and larger education sector. This contradiction was manifested by students (as *subject*) resenting the rule set by many private school of *only* deploying the foreign language in science classes (**C4**). Alternatively, students' parents are resistant to using Standard Arabic for science and mathematics as they maintain that schools need to build students' foreign language to give them a competitive "edge" for future careers and travel (Shaaban & Ghaith, 2003).

10.5.3.2 Contradictions in Lower SES Public Schools

Contradictions Related to Instruments As with the low SES private schools, contradictions in the public schools emerged at the teacher and the larger public school system levels. With the teachers as *subject*, there was also a contradiction manifested in the dominance of highly teacher-directed interactions, but here home language was deployed flexibly for meaning-making. Teachers, though, worried that flexible language practices are not helping students develop science discourse; so it remained that such *instrument* is perceived as deficit rather than asset-oriented (**C1** in Fig. 10.4). At the school level, the contradiction around the official curriculum's limitation was even more evident due to the uneven distribution of resources and lower teacher accountability in the public sector (**C2**).

Contradictions Related to Community and Division of Labor Again here contradictions around who is responsible for supporting students' language development and helping them overcome the structural barriers resulting from the language-in-education were evident. At the teacher level, there are little systematic professional development and structures for collaboration among science and language departments set by schools (**C3**). Rather many school leaders and teachers "frame" students as the problem rather than the policy: e.g., "their parents are illiterate," "they are not exposed to the FL outside school" (Nada). As for many male students in remote and very economically depressed areas (as *subjects*), their aspiration is to leave school at grade 7, so that they join the army at the age of 18 (**C4**).

Contradictions Related to Rules The contradiction of limited careful planning around language deployment and language policies was manifested by the teachers (as *subjects*) getting mixed messages from school administrators and official supervisors and inspectors on deploying the home language in science instruction (**C5**). At the school system level, the official curriculum does not tackle how language-in-education policies widen gaps among factions and result in lack of equity. Moreover, attention to quality of teaching, especially at the primary level, where students need to develop adequate foreign language proficiency to remain in school, is jeopardized by political agendas, corruption, and lack of accountability in teachers recruitment and performance (**C6**).

10.5.4 Potentialities for Change

The critical dialogues explicitly explored potentialities for change, and the participants, especially Camille, the sociolinguistics specialist, and the government-based teacher educators recounted several success stories from language and science classrooms. Stella, from the teacher supervision body, mentioned notable progress as she and her public school teachers engaged in prolonged coaching on supporting diverse students. She recounted, for example, collaborative work with a grade 6 public school science teacher on creating graphic organizers to help students discuss scientific concepts, understand terminology, and develop systematic thinking (e.g., classification). Particularly, Camille worked with language teachers and content teachers in high-needs and refugee schools on teaching with an "awakening to languages" approach (AtL) supported by translanguaging as a meaning-making tool. She described successes she and her teachers experienced with students who were the least motivated and considered by school personnel as "trouble-makers" or "cognitively challenged." Their interactive approach included: bringing students' experiences and life knowledge into the classroom through drawings, mini-videos, and audio recordings; using texts in different languages; and purposeful scaffolds for reporting and expression. These positive experiences represent what Engeström (1999) refers to as small innovative cycles that can potentially lead to expansive learning. There is a need though for more planned and systematic transformations to resolve the contradictions outlined above, with new *objects* reformulated to promote multimodal content literacy and to develop innovative instruments for strategic translanguaging (discussed below).

10.6 Discussion and Concluding Thoughts

A large and maybe "idealistic" transformation at the national level for equitable education would involve equitable allocation of resources and addressing limited accountability and other system-level corruption aspects, especially in public schools. Many government-based leaders are highly competent and dedicated, but their impact remains limited in scope due to such issues. Another maybe more "realistic" transformation is curriculum reform. Fore and foremost and to address contradictions around the official curriculum as a useful *instrument*, reform needs to pay close attention to social justice and equitable access and reformulate these as sought-after *objects* for the national curriculum. As most critical dialogue participants asserted, we need to better understand the impact of language-in-education policies on diverse learners, for example Nada, the leadership and curriculum specialist, emphasized the need for longitudinal research that uncovers the profile of students most affected by the language policies and structural barriers emerging from them. A major consideration in such reform would be to minimize such barriers and provide *all* learners with the needed supports. Especially in primary education, new structures for *division of labor* and teacher collaboration need to be established for active and integrated science and language education. For instance, such curriculum would emphasize dialogic and open-ended classroom structures with spaces for diverse leaners to participate agentically by drawing on their entire resources (Siry et al., 2022).

As for the science curriculum, there is a need for more careful consideration of meaningful progression, both for concepts and the discourse of science as mentioned by SE faculty and Enja. Enja, for example, stressed that multidisciplinary transversal competencies be used as entry points in science curricula (e.g., drawing conclusion and constructing explanations based on evidence), which according to her would motivate science and language teachers to break the boundaries of their subjects and collaborate. Her point is akin to the importance of framing science curricula and learning through science practices[6] and considering the language demands of such practices (Lee et al., 2019). A curriculum that attempts to address such demands would reformulate deep content literacy as an *object*, whereby interrelations between thinking and language demands of science are underscored, along with the need for new *instruments* to address them, e.g., purposeful and strategic translanguaging (Nygård Larsson, 2018).

Nevertheless, we saw from the critical dialogues and literature (e.g., Probyn, 2019; Poza, 2018), that contradictions or historical and sociocultural tensions are also enmeshed in translanguaging itself as a pedagogy in science education. Contradictions around translanguaging from classroom and interview data involved a perception of translanguaging as a deficit practice (**C1** in Figs. 10.3 and 10.4), and even the public school teachers, who adopted language flexibility, raised concerns around such flexibility's potential for promoting students' development of science discourse. Therefore, an important implication of this research is that opening up hybrid spaces for translanguaging in the science classroom *needs* to address and tackle head-on "necessary" tensions among students' everyday language and experiences (primary genres) and the specialized content and discourse of canonical science (e.g., Aguiar et al., 2010). Hybrid spaces for bridging these genres need to be made meaningfully explicit for students with purposeful progression towards the specialized science discourse, lest we further disadvantage certain groups of students rather than empower them (Serder & Jakobsson, 2016). For example,

[6] https://ngss.nsta.org/PracticesFull.aspx

translanguaging would need to *strategically* support students in gradually developing and using higher levels of precision that go beyond the use of specialized vocabulary. Precision means that students can communicate sophisticated ideas, for example, engaging in arguments based on evidence (Lee et al., 2019). As students develop sophisticated ideas through translanguaging and other multimodal experiences, the science teachers gradually help them communicate through the specialized registers of science.

As science education faculty, educators, and community, we are engaging more in discussions and research on the potentiality of translanguaging as a transformative pedagogy, and this will hopefully further our understandings on its effectiveness as a strategic mediating and empowering tool in multilingual science classrooms. To summarize, *strategic translanguaging* in science education, with careful planning, would address the abovementioned tensions and open: (a) a hybrid space where students' experiences and their communicative repertoires are leveraged and where students collaboratively take linguistic risks without fear of humiliation and marginalization (García, 2009; Poza, 2018); (b) a space for dialogic pedagogy, inquiry, and discourse, where deeper levels of science knowledge and discourse are constructed through active engagement and flexible, situated, and interactional utilization of language resources and multimodal tools and media (Siry et al., 2022; Probyn, 2019; Martínez-Álvarez, 2019); (c) a space that affords content coherence, flow, and continuity among languages and modalities (Lin & He, 2017; Karlsson et al., 2020; Probyn, 2019); and finally, (d) a space with ample opportunities for students to develop precision through meaningful interactions with target language practices (Nygård Larsson, 2018; Poza, 2018). In the Lebanese context, for example, even as the diglossic nature of Arabic poses certain challenges, there needs to be more discussions and research on how we can deploy and leverage our national language in both its forms. Only then would we a have a "truly multilingual system" that develops learners' content literacy and competencies in all language, which as Ramzi asserts "is not unrealistic after 12 year of schooling, *it needs* planning and intention."

Another important contradiction to address is related to teacher learning, empowerment and governance (mentioned by all participants; specifically Alex). From a CHAT perspective, transformations across activity systems occur with the agentive actions of the *subjects* toward new reformulated *objects* that necessitates the development and implementation of new *instruments* like translanguaging (Salloum & BouJaoude, 2020a, b; Martínez-Álvarez, 2019). Science and language teachers need to be empowered by engaging in practical problem-solving and curricular decisions (Goodnough, 2018). As we saw above, as science education faculty and educators, we were not exposed enough during our teacher education to the role of language. Both pre-service and in-service teacher education need to include planned and systematic translanguaging pedagogies (Probyn, 2019); teachers need to consider how to effectively use diverse and flexible participation modes, interactive and multimodal content delivery, and students' inquiry (e.g. Lee et al., 2019). As importantly, spaces need to be created for teachers and leaders to reflect collaboratively and critically and question discriminatory policies at the classroom, school and system level that are affecting their practices and students' science learning, achievement, and later life trajectory (Vieira & Kelly, 2014).

An intricate history with socioeconomic, cultural, and political considerations intersects in science teaching and learning in Lebanon. These complex intersections created contradictions that need close attention for authentic transformation and expansive learning. Contradictions included monoglossic orientations to multilingualism in education that originated from a colonial era, a middle-class blind spot, and deficit views. These are resistant to change, but as more voices are requesting such change, hopefully it will start.

Appendices

Appendix A

Rural Private School Activity System with Contradiction Areas (physical science)

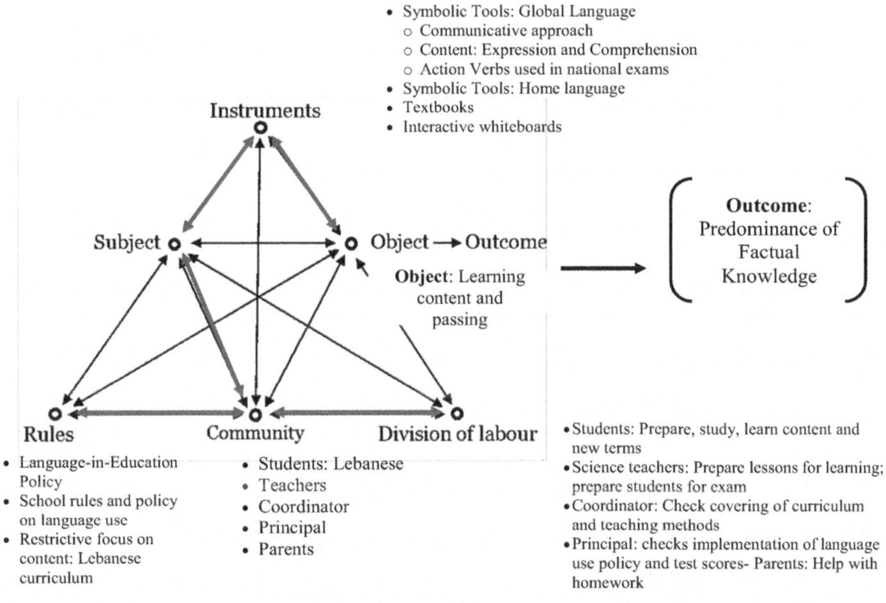

Contradiction areas	Aspects of contradictions
Subject – Community	Teachers almost exclusive use of English, students as part of the community respond in home language (spoken Arabic), resisting almost exclusive use of English
Rules – Community	Students as part of the community resent the school's policy of "only English" in the classroom, even when they acknowledge the utility of the language-in-education policy
Instruments – Object	Use of instruments leading to surface rather than deep content learning: • Interactive features of the interactive white board were not used • Home language not used to scaffold learning • Textbooks are not used as a resource for students
Community – Division of labor	No coordination among language and science teachers

Appendix B

General Perceptions on Role of Language

Science Educators:

- What do you believe affects students' performance in science? (For example, Lebanese students' TIMSS and PISA results are below the international average.)
- What solutions do you see and why?
- In your work, how have you tackled these issues?

For All:

- In a recent issue in Annahar,[7] it was highlighted that Lebanese students' performance in the languages was less than optimal based on international assessments:
  - How do you explain these results?
  - How do you think this affects students' performance in school in general?
  - How may this affect their performance in the content areas (e.g., science and mathematics)?
  - What solutions do you see for this issue and why?
- In Lebanon, science and mathematics are taught in a second language, what are your ideas about this policy?
  - What advantages and/or disadvantages do you see in teaching science in a foreign language?
  - What impact (positive or negative) do you think this policy has on students?
  - How have you encountered this matter in your research?
- Should teachers be allowed to use Arabic when teaching science in a foreign language? If so why, if not why not.

[7] A renowned local newspaper.

- If Arabic is used, what forms of Arabic would be used (formal or home language)?
- As a scholar, what are your ideas about the notion of "content literacy"?
  - How may it different than 'literacy' in general?
  - Do you see a value for this notion in the Lebanese context? Why or why not?
- It has been suggested that allowing students to use different languages (spoken Arabic, English, formal Arabic, Armenian, etc.) can promote learning science (researches call this process translanguaging). What is your view about this process?
- What do you think will work in the Lebanese context to promote quality science learning?

References

Aguiar, O. G., Mortimer, E. F., & Scott, P. (2010). Learning from and responding to students' questions: The authoritative and dialogic tension. *Journal of Research in Science Teaching, 47*(2), 174–193.

Amin, T. G. (2009). Language of instruction and science education in the Arab world: Toward a situated research agenda. In S. BouJaoude & Z. Dagher (Eds.), *The world of science education: Arab states* (pp. 61–82). Sense.

Amin, T., & Badreddine, D. (2020). Teaching science in Arabic: Diglossia and discourse patterns in the elementary classroom. *International Journal of Science Education, 42*(14), 2290–2330.

Bahous, R. N., Bacha, N. N., & Nabhani, M. B. (2011). Multilingual educational trends and practices in Lebanon: A case study. *International Review of Education, 57*(5), 737–749.

Bakhtin, M. M. (1981). Discourse in the novel. In M. M. Bakhtin (Ed.), *The dialogic imagination: Four essays by M. M. Bakhtin* (pp. 259–422). University of Texas Press.

Bakhtin, M. M. (1987). *Speech genres and other late essays* (edited by Michael Holquist). University of Texas Press.

Blackledge, A., & Creese, A. (2014). Heteroglossia as practice and pedagogy. In A. Blackledge & A. Creese (Eds.), *Heteroglossia as practice and pedagogy* (pp. 1–20). Springer.

BouJaoude, & El-Hage. (2016). Science education research and practice in Lebanon: Current status, challenges, and future prospects. In M. H. Chiu (Ed.), *Science education research and practice in Asia* (pp. 41–54). Springer.

Creese, A., & Blackledge, A. (2010). Translanguaging in the bilingual classroom: A pedagogy for learning and teaching. *The Modern Language Journal, 94*(1), 103–115.

Creese, A., & Blackledge, A. (2015). Translanguaging and identity in educational settings. *Annual Review of Applied Linguistics, 35*, 20–35.

Dagher, Z. R., & BouJaoude, S. (2011). Science education in Arab states: Bright future or status quo? *Studies in Science Education, 47*(1), 73–101.

Diab, R. R. (2006). University students' beliefs about learning English and French in Lebanon. *System, 34*(1), 80–96.

Diab, R. R. (2009). Lebanese university student's perceptions of ethnic, national, and linguistic identity and their preferences for foreign language learning in Lebanon. *Linguistics Journal, 4*, 101–120.

Engeström, Y. (1999). Activity theory and individual and social transformation. In Y. Engeström, R. Miettinen, & R.-L. Punamaki (Eds.), *Perspectives on activity theory* (pp. 19–38). Cambridge University Press.

Engeström, Y. (2001). Expansive learning at work: Toward an activity theoretical reconceptualization. *Journal of Education and Work, 14*(1), 133–156.

Engeström, Y., Engeström, R., & Suntio, A. (2002). Can a school community learn to master its own future? An activity theoretical study of expansive learning among middle school teachers. In G. Wells & G. Claxton (Eds.), *Learning for life in the 21st century: Sociocultural perspectives on the future of education* (pp. 211–224). Blackwell.
Engeström, Y., & Sannino, A. (2010). Studies of expansive learning: Foundations, findings and future challenges. *Educational Research Review, 5*(1), 1–24.
Engeström, Y., & Sannino, A. (2011). Discursive manifestations of contradictions in organizational change efforts. *Journal of Organizational Change Management, 24*, 368–387.
García, O. (2009). *Bilingual education in the 21st century*. Wiley-Blackwell.
García, O., & Leiva, C. (2014). Theorizing and enacting translanguaging for social justice. In A. Blackledge & A. Creese (Eds.), *Heteroglossia as practice and pedagogy* (pp. 1–20). Springer.
García, O., & Lin, A. M. Y. (2016). Translanguaging in bilingual education. In O. García, A. M. Y. Lin, & S. May (Eds.), *Bilingual and multilingual education: Encyclopedia of language and education* (Vol. 5, pp. 117–131). Springer.
García, O., & Wei, L. (2014). *Translanguaging: Language, bilingualism and education*. Palgrave Macmillan.
Gibbons, P. (2006). *Bridging discourses in the ESL classroom: Students, teachers and researchers*. Continuum.
Goodnough, K. (2018). Addressing contradictions in teachers' practice through professional learning: An activity theory perspective. *International Journal of Science Education, 40*(17), 2181–2204.
Jakobson, B., & Axelsson, M. (2017). Building a web in science instruction: Using multiple resources in a Swedish multilingual middle school class. *Language and Education, 31*(6), 479–494.
Karlsson, O. (2001). Critical dialogue: Its value and meaning. *Evaluation, 7*(2), 211–227.
Karlsson, A., Nygård Larsson, P., & Jakobsson, A. (2019). Multilingual students' use of translanguaging in science classrooms. *International Journal of Science Education, 41*(15), 2049–2069.
Karlsson, A., Nygård Larsson, P., & Jakobsson, A. (2020). The continuity of learning in a translanguaging science classroom. *Cultural Studies of Science Education, 15*(1), 1–25.
Kress, G., Jewitt, C., Ogborn, J., & Tsatsarelis, C. (2014). *Multimodal teaching and learning: The rhetorics of the science classroom*. Continuum.
Lee, O., Llosa, L., Grapin, S., Haas, A., & Goggins, M. (2019). Science and language integration with English learners: A conceptual framework guiding instructional materials development. *Science Education, 103*(2), 317–337.
Lemke, J. L. (2001). Articulating communities: Sociocultural perspectives on science education. *Journal of Research in Science Teaching, 38*(3), 296–316.
Lemke, J. L. (2005). *Talking science: Language, learning, and values*. Ablex.
Leont'ev, A. N. (1981). The problem of activity in psychology. In J. V. Wertsch (Ed.), *The concept of activity in Soviet Psychology* (pp. 37–71). Sharpe
Lin, A. M. Y., & He, P. (2017). Translanguaging as dynamic activity flows in CLIL classrooms. *Journal of Language, Identity and Education, 16*(4), 228–244.
Martínez-Álvarez, P. (2019). What counts as science? Expansive learning actions for teaching and learning science with bilingual children. *Cultural Studies of Science Education, 14*(4), 799–837.
Martínez-Álvarez, P., & Ghiso, M. P. (2017). On languaging and communities: Latino/a emergent bilinguals' expansive learning and critical inquiries into global childhoods. *International Journal of Bilingual Education and Bilingualism, 20*(6), 667–687.
Martínez-Roldán, C. M. (2015). Translanguaging practices as mobilization of linguistic resources in a Spanish/English bilingual after-school program: An analysis of contradictions. *International Multilingual Research Journal, 9*(1), 43–58.
Mortimer, E. F., & Scott, P. H. (2003). *Meaning making in secondary science classrooms*. Open University Press.
National Center for Educational Research and Development. (1997). *Manahej Al-Ta'alim Al-A'am Wa Ahdafaha* [Public educational curricula and goals]. Author.

National Center for Educational Research and Development (NCERD). (2019). *Al-nashra al-Ihsai'ya* [Statistics bulletin]. Author.
Nygård Larsson, P. (2018). "We're talking about mobility:" Discourse strategies for promoting disciplinary knowledge and language in educational contexts. *Linguistics and Education, 48*, 61–75.
Nygård Larsson, P., & Jakobsson, A. (2020). Meaning-making in science from the perspective of students' hybrid language use. *International Journal of Science and Mathematics Education, 18*(5), 811–830.
Orr, M., & Annous, S. (2018). There is no alternative! Student perceptions of learning in a second language in Lebanon. *Journal of Language and Education, 4*, 79–91.
Otheguy, R., García, O., & Reid, W. (2015). Clarifying translanguaging and deconstructing named languages: A perspective from linguistics. *Applied Linguistics Review, 6*(3), 281–307.
Poza, L. E. (2018). The language of ciencia: Translanguaging and learning in a bilingual science classroom. *International Journal of Bilingual Education and Bilingualism, 21*(1), 1–19.
Probyn, M. (2015). Pedagogical translanguaging: Bridging discourses in South African science classrooms. *Language and Education, 29*(3), 218–234.
Probyn, M. (2019). Pedagogical translanguaging and the construction of science knowledge in a multilingual South African classroom: Challenging monoglossic/post-colonial orthodoxies. *Classroom Discourse, 10*(3–4), 216–236.
Ryu, M. (2019). Mixing languages for science learning and participation: An examination of Korean-English bilingual learners in an after-school science-learning programme. *International Journal of Science Education, 41*(10), 1303–1323.
Salloum, S., & BouJaoude, S. (2019). The use of triadic dialogue in the science classroom: A teacher negotiating conceptual learning with teaching to the test. *Research in Science Education, 49*(3), 829–857.
Salloum, S., & BouJaoude, S. (2020a). Language in teaching and learning science in diverse Lebanese multilingual classrooms: Interactions and perspectives. *International Journal of Science Education, 42*(14), 2331–2363.
Salloum, S., & BouJaoude, S. (2020b). Understanding interactions in multilingual science classrooms through Cultural-Historical Activity Theory (CHAT): What do contradictions tell us? *International Journal of Science and Mathematics Education, 19*, 1333–1355. https://doi.org/10.1007/s10763-020-10114-5
Sannino, A., Engeström, Y., & Lemos, M. (2016). Formative interventions for expansive learning and transformative agency. *Journal of the Learning Sciences, 25*(4), 599–633.
Serder, M., & Jakobsson, A. (2016). Language games and meaning as used in student encounters with Scientific Literacy test items. *Science Education, 100*(2), 321–343.
Schwandt, T. A. (2001). Understanding dialogue as practice. *Evaluation, 7*(2), 228–237.
Shaaban, K., & Ghaith, G. M. (1999). Lebanon's language-in-education policies: From Bilinigualism to Trilingualism. *Language Problems and Language Planning, 23*(1), 1–16.
Shaaban, K., & Ghaith, G. M. (2003). Effect of religion, first foreign language, and gender on the perception of the utility of language. *Journal of Language, Identity, and Education, 2*(1), 53–77.
Shuyab, M. (2016). Education for social cohesion attempts in Lebanon: Reflections on the 1994 and 2010 education reforms. *Education as Change, 20*(3), 225–242.
Siry, C., Wilmes, S., te Heesen, K., Sportelli, D., & Heinericy, S. (2022). Young children's trans-modal participation in science investigations: Drawing on a diversity of resources for meaning making. In A. Jakobsson, P. Nygård Larsson, & A. Karlsson (Eds.), *Translanguaging in science education*. Springer.
Vieira, R. D., & Kelly, G. J. (2014). Multi-level discourse analysis in a physics teaching methods course from the psychological perspective of activity theory. *International Journal of Science Education, 36*(16), 2694–2718.
Wei, L. (2011). Moment analysis and translanguaging space: Discursive construction of identities by multilingual Chinese youth in Britain. *Journal of Pragmatics, 43*(5), 1222–1235.
Wei, L. (2018). Translanguaging as a practical theory of language. *Applied Linguistics, 39*(1), 9–30.

Chapter 11
Translanguaging in Science Education in South African Classrooms: Challenging Constraining Ideologies for Science Teacher Education

Annemarie Hattingh, Carolyn McKinney, Audrey Msimanga, Margie Probyn, and Robyn Tyler

Abstract South Africa is a multilingual country and the Language in Education Policy (1997) allows for any of the 11 official languages to be used as the language of learning and teaching. However, language ideologies originating in the colonial era, coupled with the global dominance of English, have meant that the majority of schools have chosen English from grade 4 or even earlier – despite the fact that few children by this stage achieve the necessary proficiency in English to fully access the curriculum.

This chapter seeks to challenge the dominant monolingual and anglonormative language ideologies by exploring some alternative translanguaging practices of teachers and learners in multilingual classrooms and how these support opportunities to learn science. Illustrative vignettes are presented from four research studies: the grade 4 transition year from home language isiXhosa instruction to English and how this constrains learners' participation and meaning-making; learners' languaging practices in small-group work and how they mobilize their linguistic resources for effective discussion of science concepts; translanguaging practices in teacher-led whole-class talk and how this engages learners in meaning-making and scaffolds learning across the mode continuum; and how written translation activities enable learners to deepen their conceptual understandings. Collectively, these

---

A. Hattingh (✉) · C. McKinney
University of Cape Town, Cape Town, South Africa
e-mail: annemarie.hattingh@uct.ac.za; carolyn.mckinney@uct.ac.za

A. Msimanga (Deceased)
Sol Plaatjie University, Kimberly, South Africa

M. Probyn · R. Tyler
University of the Western Cape, Cape Town, South Africa
e-mail: mprobyn@uwc.ac.za

© The Author(s), under exclusive license to Springer Nature Switzerland AG 2021

A. Jakobsson et al. (eds.), *Translanguaging in Science Education*, Sociocultural Explorations of Science Education 27,
https://doi.org/10.1007/978-3-030-82973-5_11

vignettes offer insights for teacher education and development in multilingual classrooms.

Keywords Translanguaging · Science education · Multilingual education · Language ideologies · Teacher education · South African education

11.1 Introduction and Context

South Africa is a multilingual country with 11 official languages that include the two former colonial languages –- English and Afrikaans – and nine indigenous African languages: Sepedi, Sesotho, Setswana, siSwati, Tshivenda, Xitsonga, isiNdebele, isiXhosa, and isiZulu.

The Language in Education Policy (Department of Education, 1997) allows for any of the 11 official languages to be used as the medium of instruction or *language of learning and teaching* (LoLT). Although English is the home language of less than 10% of learners, language ideologies originating in the colonial era, coupled with the present-day dominance of English nationally and globally, have meant that the majority of schools have chosen English as the LoLT, from no later than the fourth grade. In so doing, they silence what McKinney (2017, p. xv) has described as "the most valuable resource a child brings to formal schooling" namely their home language/s.[1]

It is a source of great concern that South African learners' performance in mathematics and science falls far below the levels of their counterparts in other parts of the world – as is evident in successive Trends in International Mathematics and Science Studies (TIMSS) (Mullis et al., 2020). One of the contributing factors identified in the South African TIMSS reports is that of the language medium when this is not the home language of the learners (Howie, 2001; Reddy, 2006). This is not surprising given that research has demonstrated that at least 6 years of academic learning through one's home language is necessary in order to cope with academic concepts in a new language, even in well-resourced contexts (Bamgbose, 2000; Thomas & Collier, 1997). Research into science education specifically also shows that science learners for whom English is a second language struggle with the challenges of building registers for the language of learning and teaching. They have difficulty engaging meaningfully in sense-making discussions and to then articulate

[1] We have opted to use the term 'home language/s' to refer to the language or languages that a learner or teacher most often uses at home and in their community and which will be the language or languages most familiar to them. We are aware that the term is imprecise and does not capture the fluid multilingual languaging practices in many urban townships. In addition, the variety of named language/s used by a learner at home or in their community may differ from the official variety and many learners may be most familiar with an urban vernacular that encompasses elements of several named languages. By contrast, outside of the multilingual urban environments, the official African languages have strong geographic bases and so teachers and learners frequently share a common home language.

their understanding of science concepts (Rojas-Drummond & Zapata, 2004). South African research suggests that some learners who seemed to be struggling with understanding science do much better if provided with the opportunity to engage in their own languages (see for example Probyn, 2015; Rollnick & Rutherford, 1996). As Reddy (2006) has noted, it is difficult to disentangle the language medium from other co-occurring factors such as teacher subject knowledge and the socio-economic status of learners. Nevertheless, it is only English-speaking learners and a small minority of Afrikaans speakers who learn through the medium of their home language and so for the majority of learners who are African-language speakers, language in the classroom combines with poverty to doubly disadvantage.

In South African classrooms there is a striking mismatch between the language resources of teachers and students, the vast majority of whom are bi- and multilingual, and an English-only approach to the implementation of the language policy. This mismatch is enabled by a national curriculum that expects all learners to learn in their so-called *home language* for the first 3 years of school and then to switch suddenly to monolingual English instruction from year 4 onwards.[2] In the first 3 years, children are expected to learn English as a subject for between 2–4 h a week whereas all other curriculum material is delivered through the home language. From year 4 onwards, the home language is learned as a subject and the rest of the curriculum is delivered in English. There are no textbooks, classroom resources or assessments available in African languages from year 4 onwards.[3] As stated earlier, according to official Language in Education policy (Department of Education, 1997), schools can choose their language of instruction, which could be any of the 11 official languages (including bilingual options), but in reality, the national and provincial government education departments do not support languages of instruction other than English and Afrikaans. In addition, there is little teacher education to date that equips teachers to engage with learners' most familiar languages in ways that might support epistemic access and social equity. Given the difficulties teachers face in teaching through an unfamiliar language, the use of code-switching or translanguaging in oral discourse is common. However, such practices are not sanctioned by school management and government education departments, who ask teachers to refrain from bilingual strategies such as translation and code-switching in classrooms (see for example Western Cape Education Department [WCED] minute, 2014a; WCED, 2014b). Therefore, South African science classrooms can currently be described as *adaptive translanguaging spaces* (Garcia & Li, 2014). In the main, there is no official, supported space for translingual practice (Canagarajah, 2018), but this practice continues and has learning benefits, as we will discuss.

---

[2] The exception to this is Afrikaans home language learners who are able to continue learning through Afrikaans as the official language of instruction throughout schooling. This is a small minority of learners.

[3] An encouraging development reported in the media is that in 2020, grade 12 learners in one South African province were for the first time able to choose whether to write some of their school leaving examinations in English or isiXhosa (https://www.dispatchlive.co.za/news/2020-09-07-eastern-cape-mother-tongue-exam-choice-now-a-reality/)

The various translingual practices or languaging-for-learning (Guzula et al., 2016) to be found in South African science classrooms will be theorised in this chapter under the umbrella term, *translanguaging*. Accounting for the kinds of translanguaging that occur under constrained conditions in our context requires an understanding of the notion from critical applied linguistics of *language ideologies*, which we unpack in the following section. In order to provide empirical examples of the varied and contextually bound nature of our understanding of translanguaging in our context, we offer vignettes from four South African classrooms. Each study provides a different perspective on translanguaging for science learning. A variety of learning activities (groupwork and whole-class discussion); modes and registers of expression (oral, written, gesture, academic registers, everyday registers); and classroom linguistic make-up (homogenously bilingual and highly multilingual) across the vignettes allows us to present a broad understanding of translanguaging. The four vignettes provide insights into the constraints of *anglonormative ideologies* on classrooms and the possibilities of translanguaging in multilingual science classrooms in South Africa. Furthermore, we provide some indicators of possible teacher education for multilingual teaching and learning towards creating *established translanguaging spaces* (Garcia & Li, 2014).

We now turn to the common theoretical tools that underpin the analysis in our chapter.

11.2 Language Ideologies and Translanguaging

We have shown in the introduction that there is official support and enforcement of an English-only approach to teaching and learning science in South Africa, despite the linguistic diversity of children in schools. Language ideologies, or our beliefs about language, we would argue, play a central role in enabling a situation where the most valuable resource a child brings to formal schooling, language, can be consistently ignored and recast as a problem. Language ideologies can be defined as:

> … the sets of beliefs, values and cultural frames that continually circulate in society, informing the ways in which language is conceptualised and represented as well as how it is used. Such ideologies are constructed through discourse, that is, systems of power/knowledge (Foucault, 1980). (Makoe & McKinney, 2014, p. 659).

Dominant language ideologies support a number of *monolingual myths*, all of which work to deeply constrain the potential to enable languaging-for-learning (Guzula et al., 2016). These myths include the ideas that:

- Languages are pure, autonomous and clearly bounded entities (Blommaert, 2006).
- Monolingualism, or a high level of proficiency in a single named language, is the norm.
- Nations are made up of speakers of one language: one language, one nation, one geographical territory (Ricento, 2000).

- Linguistic purity is inherently superior, or good language use keeps named languages separate from each other whereas deficient language use is mixed.
- Bi/multilingualism is understood as *multiple monolingualisms*, or as equivalent proficiency in two or more named languages, so-called *balanced bilingualism* (Grosjean, 1982).
- Bi/multilingualism is considered undesirable/a problem (Ruiz, 1984). In South African science classrooms this is expressed as seeing the use of African languages in the classroom as detrimental to learner development of the English-language skills that they need to take the high stakes examinations (Msimanga et al., 2017).
- Proficiency in English is regarded as normative and valued above proficiency in other languages – what McKinney has referred to as *anglonormativity* (2017, p. 80). This to a large extent reflects South Africa's colonial history and the post-colonial global dominance of English.
- Finally, as Ag and Jørgensen (2013) have pointed out, a "consequence of the monolingualism ideology is the belief that every person must have a particularly close relationship to one 'language,' almost invariably the 'mother tongue' of the person" (p. 527). However, in many multilingual communities in South Africa an individual's linguistic experiences and affiliations may shift in time and space and the notion of a single fixed mother tongue or home language does not fit with far more fluid languaging realities (Makalela, 2016).

It is these monolingual myths that in part underpin the language policy choices made by schools, and entrench practices that limit access to the curriculum and skew educational advantage to the minority of learners who happen to be proficient in the former colonial languages that dominate the linguistic hierarchy, namely English and Afrikaans.

A translanguaging perspective shifts our analytical gaze from named languages and whether one or another language should be chosen as the language of learning and teaching, to a recognition of linguistic repertoires and how translanguaging in the classroom may open up access to learning. Translanguaging as a concept has expanded from its beginnings in bilingual education in Wales where it was used to describe the variation in language of input and output in learning (Williams, 1996). It has been put to work by sociolinguists to describe the various uses of linguistic and other resources by multilingual people in order to communicate and make meaning in the world (Kusters et al., 2017; Makalela, 2016). Translanguaging has also been expanded as a term within applied linguistics and education studies, some would argue to its detriment in that it now refers to everything and therefore nothing (Bhatt & Bolonyai, 2019; Jaspers, 2018).

Our use of the term translanguaging in this chapter relates specifically to teaching and learning activities in which there is movement between linguistic (and sometimes other semiotic) resources, either between individuals, with one individual using one language and another using a second language; or between modes, with one individual reading in one language and writing in another; or within utterances by the same individual. The connecting thread between these different activities that we term translanguaging is that different linguistic resources are leveraged in order to deepen conceptual understanding in science. Many of the translanguaging practices reported on in education contexts – what Probyn (2015) describes as *pedagogical translanguaging* – take the form of spontaneous, usually oral practices in which multilingual teachers and/or learners engage their full semiotic repertoires in order to understand and express content. García and Li (2014), drawing on Williams (2012), describe this as *natural translanguaging* and distinguish it from *official translanguaging* (ibid.), which describes planned activities involving language alternation.

In our analysis, we use the terms *spontaneous* and *planned* translanguaging as those activities that explicitly require learners to translanguage are not officially sanctioned in the South African context. Vignette 4 below showcases an example of planned translanguaging in the form of a translation exercise. Baynham and Tong King Lee (2019) argue that translation is an applied practice that should be viewed through the theoretical lens of translanguaging. Although the distinction between natural and official translanguaging is helpful, our data indicate that spontaneous or natural classroom translanguaging practices can also be used strategically for learning. The case of Mr. Mafunda (Vignette 3) pertains as he drew on his multilingual repertoire spontaneously, but was very clear in interviews that he did this deliberately to enable access to concepts. Therefore, we view natural or spontaneous translanguaging and official or planned translanguaging as occurring along a continuum.

Scholars have also argued that translanguaging should be conceived of as having a critical literacy dimension (Janks, 2010) offering transformative potential and a social justice agenda (Flores, 2014; Garcia & Li, 2014; Otheguy et al., 2015). We agree that although translanguaging can be transformative, this is not inevitable; the meaning potentials of every instance of language use are always embedded within the context of use. Without challenging dominant anglonormative ideologies, the transformative potential of translanguaging is limited (Jaspers, 2018). The data we present are intended to show a variety of translanguaging practices to demonstrate that a broad conceptualisation of translanguaging is important if we are to maximise its transformative potential.

11.3 Snapshots of Practice

The four vignettes presented in this chapter offer snapshots of classroom practice that provide insights into the constraints and possibilities of languaging practices in multilingual science classrooms in South Africa and some indicators for teacher education for multilingual teaching and learning. The vignettes have been drawn from different research projects conducted by the authors (see McKinney, 2017; Msimanga & Lelliot, 2014; Probyn, 2017; Tyler, 2018) and collectively they offer complementary perspectives on teaching and learning science that illustrate the argument being made for an acceptance of translanguaging practices to counter the educational disadvantage resulting from anglonormative ideologies and monoglossic practices.

The first vignette illustrates how monolingual language ideologies constrain possibilities for meaning-making and conceptual understanding in a year 4 science lesson in a school in the Western Cape Province, where the learners share a common home language, isiXhosa, and have recently switched to English as a language of learning and teaching. Vignettes 2 and 3 both provide examples of spontaneous translanguaging practices in classrooms: the second vignette is situated in an urban township school in Gauteng Province, where there is a high degree of multilingualism in the community and the school and it describes the translanguaging practices of learners working on a science problem in small-group discussion. The third vignette is situated in a rural school in the Eastern Cape Province, where the community is linguistically homogenous, the dominant language being isiXhosa, and learners only experience English at school. The research describes the translanguaging practices of the teacher and learners in whole-class interactions. The fourth vignette is also situated in the Western Cape Province. The research describes a planned translanguaging intervention where the researcher created opportunities for learners to work with translingual science texts.

11.3.1 Vignette 1: Constraining Effects of Anglonormative Ideologies on the Teaching and Learning of Science

This vignette illustrates the constraints of anglonormativity on classroom talk and science learning. The excerpt of classroom talk below is from a science lesson with a class of year 4 learners (grade 4, 9- 10-year-olds) in a relatively poorly resourced township school, where the teacher and children shared a home language in isiXhosa (McKinney, 2017). Typical of the majority of primary schools in South Africa (Heugh, 2013; Probyn, 2015), these children would have experienced a sudden transition to English as the language of instruction at the beginning of grade 4; thus, all their textbooks, classroom resources and assessments were available in English only. The lesson took place mid-way through the academic year and was described by the science teacher as a revision lesson, i.e. she was revising content previously taught. The excerpt is from the beginning of the lesson.

(In the excerpts, T = Teacher, L = Learner and Ls = Learners. Languages other than English are written in italics and glossed in the third column; actions are bracketed in italics)

Excerpt 11.1 Year 4 Science Lesson 'Matter Is Anything That Occupies Space'

Turn	Speaker	Speech	Gloss
1.	T:	All right! Who can tell me … what is matter? Hey? What is matter? *(nominates learner)*	
2.	L:	Matter is anything that occupies space.	
3.	T:	Matter is anything that occupies space. All of you …	
4.	Ls:	*(in chorus)* matter is anything that occupies space.	
5.	Ls:	*(teacher writes the sentence on the board while learners repeat the sentence)* matter is anything that occupies space.	
6.	T:	*(nominates a learner)* are you a matter?	
7.	L:	Yes.	
8.	L:	OK. Are you a matter? Why?	
9.	L:	Yes.	
10.	T:	Why do you say you are a matter? Stand up.	
11.	L:	*(silent)*	
12.	T:	Come … who can tell me … you all know that matter is anything that occupies space. Why do you say that? *(nominates another learner)*	
13.	L:	Because I occupy space …	
14.	T:	Because he occupies space …	
15.	T:	He says he is matter because he occupies space … if I take all these chairs and I take you outside … will there be space here?	
16.	L:	Yes.	
17.	T:	Huh?	
18.	L:	Yes.	
19.	T:	But now, we took all the desks … (that) are here, we have occupied⇧ … what?	
20.	L:	Space.	
21.	T:	We have occupied the space. Very good!	
22.	T:	Uh, do you know that air is also matter?	
23.	Ls:	Yes, Miss.	
24.	T:	Do you know that?	
25.	Ls:	Yes, Miss.	
26.	T:	Why do you say that air is also matter? We say matter is anything that occupies⇧ … space. Why do we say that air is also matter?	
27.	L:	Because it occupies space.	
28.	T:	Where? Here … Ja, it also occupies ⇧ … what? … occupies space. Air occupies space. Is there any air inside this room now?	
29.	Ls:	Yes!	
30.	T:	*(announcement interrupts lesson)* … meaning that the air is inside this room … *(teacher writes the two sentences on the board: 'Matter occupies space. Air occupies space')* OK, we say: Matter occupies space. Matter is also around us. *Andithi?* OK?	didn't I say?

In this excerpt, the teacher and learners speak only in English, apart from a single question tag in isiXhosa in turn 30. We see the children demonstrating their ability to produce the verbal definition of *matter* in response to the teacher's question, "What is matter?" (turn 1). The learner's response in the full sentence, "Matter is anything that occupies space" and the teacher's injunction to the class to repeat this exact wording suggests that she is eliciting the definition that she had previously taught the class. In turn 3 the teacher affirms the individual learner's answer through repetition and by inviting the rest of the class to repeat the answer in choral response with the initiation "All of you". That the learners are accustomed to choral response is evident in their continued repetition of the definition "Matter is anything that occupies space" as the teacher writes the sentence onto the chalkboard with her back towards the class. The teacher then progresses to extend the learners' ability to answer a different question using the learned definition of "matter" as "anything that occupies space" by asking "Are you a matter?" (turn 6). The learner's inappropriate response "Yes" to the follow up question "Why?" in turn 8 and silence in turn 11, shows that the learners are unable to spontaneously answer the question in English as to why they themselves might constitute matter. In turn 12, after repeating her question as to why the learners constitute matter, the teacher says "You all know that matter is anything that occupies space" and the child in turn 13 is then able to apply this reasoning/wording in answering the question of why he constitutes matter: "Because I occupy space". In turns 19–20, with the teacher's rising intonation cueing the learners to complete the sentence "We have occupied?", learners give the appropriate answer, "space" (turn 20).

Cued choral response through elicitation following questions and rising intonation is characteristic of what Chick (1996) has described as *safe-talk* and has been identified in post-colonial schooling contexts further afield, where there is a mismatch between the language of instruction and the linguistic resources children bring with them to schooling (e.g. Hornberger & Chick, 2001). Chimbutane (2011, p. 28) outlines the key pattern in safe-talk as.

> … that of teacher prompt and pupils' choral responses (…), that is, teachers routinely provide cues to which pupils respond in chorus. The prompts or cues used by teachers to trigger such pupils' chorusing responses include yes/no questions and oral gap-filling exercises.

It is evident from the excerpt that the learners have been successfully socialised into cued elicitation and choral responses. It could also be argued that they are learning to produce a scientific register in phrases like "Matter is anything that occupies space" and in being able to answer questions such as why does x constitute matter, or that they are learning to talk like scientists (in English). The children may be able to achieve full marks in answer to a test question, "What is matter?" by providing the written definition "Matter is anything that occupies space." But there is no evidence that the children understand this definition, i.e. have developed conceptual understanding of the scientific concept *matter*, as they do in Vignette 2 when they explore scientific ideas using their full linguistic repertoires.

In turn 22, the science teacher makes an effective pedagogical decision that has the potential to deepen conceptualisation of the concept *matter* when she states, "Do you know that air is also matter?" However, the power of this pedagogical decision to tackle the challenging idea of air as matter (and as occupying space) is limited by the restriction to monolingual English. Anglonormativity thus limits epistemological entry into the abstract and often counterintuitive world of science where what is studied is frequently invisible and untouchable. Without the opportunity to engage with the concepts using their full linguistic repertoires, these learners are limited to parroting answers and rote learning.

In contrast, the following three vignettes present examples of translanguaging practices that offer ways of constructing meaning in science classrooms that open up rather than close down epistemic access and engagement with learning.

11.3.2 Vignette 2: Spontaneous Translanguaging by Learners in Group Work in a Multilingual Township School

In contrast to Vignette 1, the second vignette shows how learners mobilize their linguistic resources for engagement with science knowledge when afforded the opportunity. Excerpt 11.2 is the transcript of a discussion by a group of learners in a multilingual township high school chemistry classroom in an urban area (taken from Msimanga & Lelliott, 2014 and Mudadigwa & Msimanga, 2019). There were four learners in the group: Thandi, an isiZulu-speaking girl; Senzo and Muzi, isiZulu-speaking boys; and Tapelo, a Setswana-speaking boy. This part of their discussion focused on understanding the task and agreeing on a strategy to solve the problem: an acid–base reaction. English speech is indicated in plain text, isiZulu speech in italics and Setswana speech in bold.

Excerpt 11.2 Solving an Acid–Base Reaction Problem in a Group

Turn	Speaker	Speech	Gloss
1.	Thandi:	*(pointing to an equation in her book) Nanku bhekani bhekani Ukuthi nje kufanele sibhalanse i-equation ngendlela esijayele ukubhalansa ngayo i-equation* and then *ke sesizo fomuleyitha i-gama laleyo salt* and because of *ama results esowathola* out of *i-equation niya andastenda*?	Here here look this is it it's just that we have to balance the equation in the way we always do and then we can formulate the name of the salt from the results we get out of the equation. Do you understand me?
2.	Muzi:	*Phinda futhi.*	Say that again.
3.	Thandi:	*Sibhalansa i-equation masiqeda ukubhalansa i-equation i-end result esizoyithola sizo neyima i salt leyo* because out of *i-equation* let's say we get sodium.	We balance the equation and then after balancing the equation the end result that we get we name the salt based on that equation let's say we get sodium
4.	Tapelo:	**Lamoutlwisisa?**	Do you understand her?

(continued)

Turn	Speaker	Speech	Gloss
5.	Senzo:	*Ushukuthi njengoba iso soqala sibhalanse le-equation?*	Are you saying that … since it is like this, we first need to balance the equation?
6.	Thandi:	*Hhmm masiqeda lentole le answer le siyitholile sineyima isalt*	Uh huh and then from the answer we can name the salt
7.	Muzi:	*Ssibhalansa yona kuqala?*	So we balance this one first?
8.	Thandi:	*kuzoba yigama* let's say *kuzoba yi nitric* …	It's going to be the name of let's say say nitric …
9.	Senzo:	*Oh sibhalansa le kuqala?*	Oh so we balance this one first?
10.	Thandi:	Ja	Yes
11.	Tapelo:	**Kanti a reetsi reaction pele? Keona reyi balanse?**	But don't we need to do the reaction first? And then balance it?
12.	Thandi:	*Maubhalansile sesibhalansile sesizoneyima …*	After balancing it we will then name …
13.	Tapelo:	Ja **e ya-*naming*** I do understand **mara** I'm saying … you know what I'm saying? **Mara ha riyibhalansa retswantsi rebhalanse reaction this side pele?**	Yes I do understand that about naming but what I'm saying is do we not need to balance the reaction this side first?

Altogether, the group made 517 utterances in this discussion, split almost equally between the four group members. Thus, 342 isiZulu utterances (by three speakers) and 102 Setswana utterances by the one Setswana-speaking learner, as well as 73 English utterances between them. The excerpt illustrates how Thandi externalizes her understanding of the task as the others probe for explanations, all in their home languages. Senzo (turns 5, 9), Muzi (turns 2, 7) and Tapelo (turns 4, 11, 13) seek clarity, questioning Thandi and asking her to repeat or explain her thinking. Thandi effectively articulates and makes explicit to others her thinking and understanding of the task. The differences in each learner's home language, isiZulu or Setswana, do not pose a problem for the group. In turn 11 Tapelo asks Thandi questions in Setswana and Thandi replies in isiZulu. Tapelo has no problem understanding her as is evident in turn 13 and the engagement continues. Speakers also translanguage within utterances (turn 13) and this does not hinder communication. In the process the three boys are making connections and constructing their own understanding of both the task and the content. Not only do they check with Thandi to understand her explanations, but they reconsider some of their initial understandings and are (sometimes) persuaded by her arguments. Tapelo's query in turn 13 changes the direction of the engagement as he challenges Thandi on what should come first, naming the salts or balancing the equation. The discussion continued in this manner until the four agreed on a strategy to use the names of the reactants to help them to construct the equation, balance the equation and then name the products. In the end, they articulated the chemical equation correctly and named the salt formed as required.

There is considerable research evidence on the nature and role of group discussion in science. As Alexopoulou and Driver (1996) note, "these joint actions and communication with others" are critical not only for internalisation of scientific practices and discourse but also for sense-making and understanding of the science itself (p. 1099). In multilingual settings, learners mobilise their linguistic resources for effective engagement with science content in the small groups (Karlsson et al., 2019; Msimanga et al., 2017; Tyler, 2018). From the nature of engagement in the learner discussion illustrated here it can be argued that learners stay on task and that they do engage with the nuances of the task at hand effectively using their home language/s. The four managed to get to the science concepts under discussion and solved the problem posed to them using their full linguistic repertoires. Our argument is that translanguaging does have the potential to facilitate engagement with difficult concepts and that it is a legitimate resource to create opportunities for conceptual understanding in science lessons.

11.3.3 Vignette 3: Pedagogical Translanguaging During Teacher-Directed Whole-Class Talk in a Rural Bilingual School

The vignette in this section is drawn from a case study that was part of a larger multiple case study of eight grade 8 science teachers who were teaching in rural and township schools where the language of learning and teaching (LoLT) was English but the home language of all the learners and teachers was isiXhosa (Probyn, 2017). This is the typical linguistic scenario in the eastern half of the Eastern Cape Province where the case studies were located. Typically, learners have very little exposure to English outside the classroom and therefore low proficiency in English and should be considered emergent bilinguals. As a result, many teachers resort to using the learners' home language to bridge the communication gap.

The purpose of the research was to investigate the spontaneous/unplanned languaging practices of teachers and learners and how these appeared to construct or constrain opportunities to learn science.

In the study, the practice of one particular teacher, Mr. Mafunda (a pseudonym), stood out from the rest as offering greater opportunities to learn science; thus, data from his classroom practice are presented here. One factor that immediately distinguished Mr. Mafunda's practice from those of the other teachers was that he made far greater use of translanguaging in the observed lessons than did the other teachers: 53% of his classroom talk was in isiXhosa, whereas the other teachers only used isiXhosa for between 0 and 13% of classroom talk. However, a fine-grained analysis of lesson transcripts and video recordings of lessons indicated that although the translanguaging in Mr. Mafunda's lessons contributed positively to the opportunity to learn science, what also distinguished his lessons was the clear coherence of the science content where key facts were linked to generalisable concepts and concepts exemplified by rich factual detail (Donovan & Bransford, 2005); and his engagement of learners in knowledge construction through dialogic discourse (Alexander, 2001; Gibbons, 2006; Mortimer & Scott, 2003). It was the intermeshing of these aspects of languaging and science content that supported opportunities to learn science.

The post-lesson interviews with teachers revealed a marked difference in the underlying language ideologies of the teachers in the sample and thus a possible reason for the notable difference in translanguaging practices between Mr. Mafunda and the other teachers in the sample: two of the eight teachers did not believe that the learners' home language had a positive role to play in the classroom and so they stuck to using English throughout their lessons. Five teachers admitted that they had to resort to using isiXhosa when they felt that learners were "lost", but that they felt conflicted about this and most often did so covertly. By contrast Mr. Mafunda regarded the learners' home language as a resource to be exploited and explained that he would introduce new ideas in the learners' home language and once they understood the concepts would scaffold the transfer of meaning to English.

The school that was the site of this case study was situated in a poor rural community consisting of mud and thatched huts and some small breeze block houses dotted over rolling grassy hillsides. Some families had vegetable gardens and kept a few cattle and goats, but most subsisted on social grants and many parents worked in cities and left children in the care of grandparents. In this context, children have little or no exposure to spoken or written English outside the classroom.

The science lesson to be discussed was one of a series of five consecutive lessons on mixtures, compounds and separating mixtures. Lessons followed a typical science *curriculum macrogenre* (Mortimer & Scott, 2003), an explanatory arc with a review of key ideas, introduction of new idea/s, a practical activity, presentation of observations, development of explanation of observations, generalisation, application to real life and then a written activity.

The particular lesson to be discussed was on the topic of separating a mixture of salt and water. Translanguaging and discourse patterns shifted across the lesson with presentational talk for the review or consolidation of ideas conducted mainly in English following patterns of teacher instructional monologue and brisk Initiation–Response–Evaluation (IRE) interaction pattern questioning sequences (Mehan, 1979), whereas the exploratory talk for developing new ideas (Barnes, 1992) was conducted mainly in isiXhosa and dominated by dialogic interaction patterns (Alexander, 2001; Gibbons, 2006), as the teacher engaged learners in meaning-making: linking ideas into lines of argument that modelled scientific thinking. In addition, the key ideas from the lesson were clearly linked to the generalising principle, establishing conceptual coherence within and across lessons (Donovan & Bransford, 2005).

The shifts in translanguaging and discourse structures are illustrated with excerpts below (the original isiXhosa text is written in italics and the English gloss given alongside).

During the first stage of the lesson, the teacher led the learners in a brisk review of the key ideas from the previous lesson when they had separated a mixture of soil and water through the process of filtration. Known facts were elicited with a typical IRE discourse structure, and the science content was elicited by Mr. Mafunda mainly in English, with regulative talk conducted in isiXhosa. This is illustrated in Excerpt 11.3.

Excerpt 11.3 Review of Known Facts

Turn	Speaker	Speech	Gloss
1.	T:	How did we separate *i*-muddy water? ... Which process did we use to separate *i*-muddy water? *Masiphakamise izandla Laprocess besiyisebenzisa izolo* Hands up, hands up, hands up. Yes? *(nominates a learner)*	How did we separate muddy water? ... Which process did we use to separate muddy water? Let's raise our hands. That process we used yesterday. Hands up, hands up, hands up. Yes?
2.	L:	Filter paper.	Filter paper.
3.	T:	We used a filter paper then. What did we call the process ... process of separating ehh that muddy water using a filter paper? *Sithe yintoni?* What did we call it? Process of separating ehh muddy water *okanye* that mixture *yamanzi ne* soil, muddy water using a filter paper. *Sithe yintoni?*	We used a filter paper then. What did we call the process ... Process of separating ehh that muddy water using a filter paper? What did we say it was? What did we call it? Process of separating ehh muddy water or that mixture of water and soil, muddy water using a filter paper. What did we say it was?
4.	L:	Filtration.	Filtration.
5.	T:	Filtration. *Masitsho sonke.*	Filtration. Let's all say it.
6.	Ls	Filtration.	Filtration.
7.	T:	So the separation of a mixture using a filter paper *... masitsho*	So the separation of a mixture using a filter paper ... let's say
8.	Ls:	Separation of a mixture using a filter paper.	Separation of a mixture using a filter paper.
9.	T:	*Heke!* Is called *intoni*? I-filtration.	Good! Is called what? Filtration.

During the practical stage of the lesson Mr. Mafunda demonstrated how to make a salt solution and how to separate it by heating the solution in an evaporating dish so that the water evaporated and the salt remained behind. While the learners gathered round him and observed, Mr. Mafunda engaged them in exploratory dialogic talk, mainly in their home language isiXhosa. The key ideas were unpacked by a series of questions that led learners through observing and naming the scientific process as he was demonstrating. Learners' responses provided the springboard for the next questioning triad and ideas were elicited first in isiXhosa and then transferred to English. Excerpt 11.4 below illustrates these points. Mr. Mafunda first elicited a prediction of how to separate the salt from a salt solution (turn 3) in isiXhosa, based in part on learners' existing knowledge of the stages of matter and how water behaved when heated. The key terms *heat* and *evaporation* were first named in isiXhosa "*ukutshisa*" (turns 4–11) and "*azakutsha*" (turns 13–19) and then transferred to English. Mr. Mafunda insisted on precise meanings: *heat* not *boil* (turn 11), a feature of scientific discourse. In turns 20–21, Mr. Mafunda explained the precise meaning of the word *prediction* in isiXhosa.

Excerpt 11.4 Exploratory Talk While Predicting How to Separate a Mixture of Water and Salt or Salt Solution

Turn	Speaker	Speech	Gloss
1.	T:	*Heke!* We're going to separate now *into esiyifunayo* we want to have *le* salt back, *iyavakala*?	Good! We're going to separate now into what we want, we want to have the salt back, is it clear?
2.	Ls:	Yees.	Yees.
3.	T:	We want to have *le* salt back. *Ngubani onosajesta intoba singayenza kanjani lonto leyo?* How can we get this salt back? Mhmm?	We want to have the salt back. Who can suggest as to how we can do that? How can we get this salt back? Mhmm?
4.	L:	*Utshisa amanzi … Izakushekela phaya ezantsi.*	You heat water … It will remain in the bottom.
5.	T:	*Utshisa amanzi izakushekela phaya ezantsi.* So you say *ukutshisa … Kukuthini ukutshisa*? *Xa utshisa amanzi uyawathini*? *Uwenza abe njani*?	You heat water and it will remain in the bottom. ... So you say to heat … What is to heat? When you heat water what are you doing to it? How are you making it?
6.	Ls:	*Abeshushu.*	Hot.
7.	T:	*Abeshushu*, so *uzakuthi kukwenza ntoni*?	Hot. So that is to what?
8.	Ls:	Boil.	Boil.
9.	T:	Heh? *Andiyiva.*	Heh? I can't hear.
10.	Ls:	Boil water.	Boil water.
11.	T:	*Heke!* You heat, *masingathi* boil, *masithi* you heat … you heat water *ne*?	Good! You heat, let us not say boil, you heat … you heat water, okay?
12.	Ls:	Yees.	Yees.
13.	T:	*Heke*! *Kuzakwenzeka ntoni kule… kule solution xa uyihitishayo*?	Good! What is going to happen to this … to this solution when you heat it?
14.	Ls:	*Amanzi azakutsha.*	The water will evaporate.
15.	T:	*Amanzi azakuthini*?	The water will what?
16.	Ls:	*Azakutsha.*	It will evaporate
17.	T:	*Ngesilungu*?	In English?
18.	Ls:	Evaporation.	Evaporation.
19.	T:	*Heke*! Water is going to evaporate, *iyavakala*?	Good! Water is going to evaporate, is it clear?
20.	T:	Now let's try and prove *ke ngoku* … ehh … if … if that prediction, *okokuba okakqe ... qajisela, ne*?	Now let's try and prove now … ehh … if … if that prediction, if … it was just guessing, okay?
21.	Ls:	Yees.	Yees.
22.	T:	*Okanye okaku ... ku ... ukuthini na kanene*? *Ukupredicta okakucingela ukuba kuright na, ne*?	Or that … that… what is it again? To predict is to imagine if it is right, okay?

Once the predictions were established, Mr. Mafunda conducted the practical demonstration to test the learners' predictions. During the demonstration Mr. Mafunda urged learners to watch the experiment carefully and elicited observations through dialogic discourse with questioning triads closely tied to learners' responses – what Wells (1999) has referred to as *contingent responsiveness*. This exploratory talk (Barnes, 1992) was conducted mainly in isiXhosa, as in the dialogic discourse exemplified in Excerpt 11.4. Once meaning had been established in isiXhosa it was again transferred to English (turns 7–9).

Excerpt 11.5 Translanguaging During Exploratory Talk While Observing an Experiment

Turn	Speaker	Speech	Gloss
1.	T:	*(learners come closer to observe the experiment) Masijonge kakuhle. Sibona ntoni ngoku phaya?*	Let's observe carefully. What are we seeing there now?
2.	Ls:	*Sibon'umphunga.*	We see vapour.
3.	T:	Heh?	Heh?
4.	Ls:	*Umphunga.*	Vapour.
5.	T:	*Sibon'umphunga. Uvelaphi la mphunga?*	We see vapour. Where does that vapour come from?
6.	Ls:	*Emanzini …*	From water ...
7.	T:	*Uvel'emanzini la mphunga uvel'emanzini.* So we call it … *intoni la mphunga lowa?* Water … water what?	That vapour comes from water it comes from water. So we call it … What is that vapour? Water … water what?
8.	Ls:	Water vapour.	Water vapour.
9.	T:	Water vapour, you see water vapour, that is ehh … water vapour, water vapour.	Water vapour, you see water vapour, that is ehh … water vapour, water vapour.

As the science knowledge was constructed over the course of the lesson, the discourse interaction patterns and translanguaging continued to be meshed as described, with new ideas explored mainly through the learners' home language, isiXhosa, in conjunction with dialogic discourse that modelled scientific thinking and processes. Once the new ideas were established, they were reiterated in English.

After the conclusion of the practical activity, Mr. Mafunda led the learners through an oral rehearsal of the written task – to write up a report of the experiment – and this was further supported by a writing frame on the chalkboard. The discourse patterns included both instructional monologue by the teacher and stretches of dialogic discourse as he elicited and modelled the scientific genre. At this stage of the lesson Mr. Mafunda still used more isiXhosa (57%) than English, but learners used more oral English (68%) as they prepared to write up the experiment in English. Mr. Mafunda also established conceptual coherence by linking the process of separating the salt and water to the generalising principle for separating mixtures: that the method for separating the substances in a mixture was based on the difference between the properties of the substances. This can be seen in Excerpt 11.6.

Excerpt 11.6 Conceptual Coherence – Linking a Particular Example to the Generalising Principle

Turn	Speaker	Speech	Gloss
1.	T:	I am talking generally, *xa usepareita imixtures yintoni na oyijongayo yeproperties*? Mhmm? *Intoni...lanto besithethe ngayo izolo*. We look at the what? …	I am talking generally: when you separate mixtures which part do you look at of properties? Mhmm? What … that thing we talked about yesterday. We looked at the what? ...
2.	Ls:	Properties …	Properties …
3.	T:	At the properties. So *iproperty esiyisebenzisileyo phaya* … we said *i-salt iyadizolva emanzini, ne*?	At the properties. So the property we used there … we said salt dissolves in water, okay?
4.	Ls:	Yees.	Yees.
5.	T:	*Heke!* And *amanzi ayathini xa uwatshisile*? *Ayaevaporaeita*. Water evaporates when heated, water evaporates when heated, *iyavakala*?	Good! And water does what when you heat it? It evaporates. Water evaporates when heated, water evaporates when heated, is it clear?

The excerpts presented above serve to illustrate how the teacher Mr. Mafunda leveraged the learners' linguistic resources in order to engage them in constructing knowledge, to provide them with access to the science content of the lesson, and to transfer that understanding to expression in English. This translanguaging practice shifted across the lesson in nuanced and responsive ways according to the stage and purpose of the lesson activity; and it was meshed with dialogic engagement and coherence of science content knowledge in ways that opened up opportunities to learn science and challenged the prevailing monolingual language ideologies in classrooms.

11.3.4 Vignette 4: Planned Translanguaging Activity: Collaborative Translation of Written Texts

This last vignette showcases a science translation activity that took place in an after-school study group in a township high school in Cape Town that formed part of a broader 9-month linguistic ethnographic study of learners' semiotic repertoires employed in meaning-making in science in this setting (Tyler, 2018). The language in education context in South Africa sketched at the start of this chapter results in very few curricular learning resources being available for science in African languages with learners routinely issued with one textbook for science in English. Efforts to ameliorate the devastating learning effects this has for African-language speakers have resulted in the publishing of a number of multilingual science and mathematics glossaries to support the school curriculum.[4] However, school budgets in most schools do not support the purchase of these glossaries, and teachers are not trained in their use. This despite South African studies that have shown the efficacy of presenting written content to multilingual children in more than one language (Heugh et al., 2017; Malherbe, 1946; Nomlomo, 2007). There were 40 copies of one such glossary in a science classroom in the school in which the study group was held, but these were not in use. One barrier to learner engagement with these glossaries is that they are often written in a register of an African language that is very foreign to the young speakers of these languages owing to the dearth of academic texts written in African languages and the cessation of all African-language content subject texts used in schooling beyond the first 3 years. The approach that unlocked the use of the glossary in this study was to include the learners in the translation process where they became producers (writers) of translation and not merely consumers (readers).

The literacy activity of translation might seem out of place in a content-learning environment such as science; however, there is a strong tradition in science education research of emphasising the literacy demands of science (Gee, 2004; Halliday & Martin, 1993; Lemke, 2004). Translation itself has been proposed as an appropriate method of deepening science understanding (Lemke, 1990). From the monolingual English-language science-learning context, Lemke proposes translation between colloquial and scientific language as a way of developing the "flexible wordings" required for understanding and using science content (p. 173). Gibbons (2006) argues for the importance of *meshed registers* in learning science, where learners use parts of colloquial registers and parts of the scientific register in talking about science. In collaborative translating, such as in the case reported on here, oral meshed registers are available for analysis. For multilingual learners who have a wide semiotic repertoire, translation becomes an even richer activity.

The study group reported on here was set up as an established translanguaging space (García & Li, 2014) in which the isiXhosa-English bilingual teens worked on understanding the science concepts of the grade 9 curriculum, as well as raising their own science questions. One of the authors, Robyn, took up the role of facilitator of the group and encouraged and modelled translanguaging in all activities. This provoked debate and discussion as the language environment of the school was strongly anglonormative (McKinney, 2017). The only named language mentioned in the school's language policy was English, and code-switching was generally frowned upon, although covertly employed by most teachers to varying degrees. The learners themselves were resistant to using isiXhosa for science and insisted in

[4] For the use of online and student-generated glossaries in higher education in South Africa see Madiba (2014) and Antia and Dyers (2019).

the earlier meetings of the study group that it could not be done. For this activity, the texts selected for translation were explanatory definitions – one of the high-status academic genres of schooling (Derewianka, 1990). Robyn printed the isiXhosa definitions of concepts in basic chemistry such as *molecule* and *element* taken from a multilingual science and mathematics glossary (Young et al., 2005) and divided the concepts between the pairs of learners. She then instructed them to make an English translation of the definitions. Yonela and Thandile[5] were given the term 'molecule' to translate together. The pairs began working on their translations. They read the isiXhosa definition, proposed and debated translations orally, wrote down their preferred version and commented on the register of isiXhosa used in the original definitions. The popular opinion on the isiXhosa register that had been used by the glossary writers was that it was too "deep" (*deep isiXhosa* refers to the formal variety spoken in rural areas) and removed from the "*isingingqi*" (local, urban variety) spoken by the teens. This resistance provoked Robyn to ask the learners to make a further translation that would approximate how the learners spoke. It was framed as "How you would explain to your friend in the taxi?" Therefore, each pair ended up with five comparable versions of the definition of their concept: the original isiXhosa version from the glossary, two English versions and two informal register versions.

Three distinct learning benefits of the translation activity can be traced through analysis of the interactional data (video and audio) and the written translations captured during the activity. First, the translation activity deepened the learners' understanding of science concepts. Yonela and Thandile worked together on the English translations of the isiXhosa definition of *molecule*. Through contestation and debate, they refined their understanding of the chemistry term. One difference between their translations pertained to a section that Yonela translated as, "One molecule of water is H2O" and Thandile translated as, 'One molecule of H2O/water has 2 hydrogens." In arguing for her translation in preference to Thandile's, Yonela explained to Thandile: "Like Thandile, *sine*-water (beat gesture) *u'ba* like one molecule *yalamanzi*" [like Thandile, we have water (beat gesture) if like one molecule of this water].

Yonela's explanation relies on a meshed oral register (Gibbons, 2006) in order for Yonela to navigate between the written English translation of the definition and the isiXhosa variety in which she is comfortable debating with her friend. Using the meshed oral register, along with a beat gesture for emphasis, Yonela is able to make a conceptual distinction between water as a *molecule* ("one molecule *yalamanzi*") and water as a *substance* ("*sine*-water"). Although this distinction could certainly benefit from being further refined and linked to other terms in chemistry, here, Yonela has progressed beyond what was offered in the class lessons in that this nuanced distinction was not dealt with in class.

[5] For a fuller discussion of Yonela and Thandile's translation process, see McKinney and Tyler (2018).

Second, the translations positioned the learners as resourceful, creative and agentic (Li, 2011) with their usually untapped semiotic resources made useful for learning science. The resources that they drew upon connected their science curriculum to their out-of-school lives. In their second translations, they were able to take up social and academic identity positions simultaneously (Ballenger, 1997). Yonela's second translation of a definition that she might give a friend in the taxi follows in brackets:

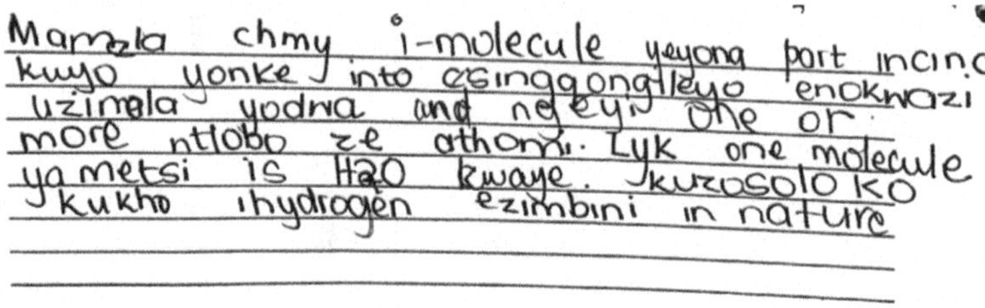
Mamela chmy i-molecule yeyona part incinci kuyo yonke into esingqongileyo enokwazi uzimela yodwa and ngeyi one or more ntlobo ze athom. Lyk one molecule ya metsi is H2O kwaye. Ikuzosolo ko kukho ihydrogen ezimbini in nature

[Listen, my friend, a molecule is the smallest part of all the things surrounding us, that is able to stand on its own and it is made up of one or more types of atoms. Like one molecule of water is H2O and there will always be two hydrogens in nature.]

Features of Yonela's social/youth repertoire are evident here in abbreviations ("chmy" for "*tjommie*', meaning friend; "*lyk*" for "like"), inclusion of lexical features of another African language, Sesotho, and English (*yametsi* and "in nature"), and everyday phrasing of scientific concepts ("*yonke into esingqongileyo*"/[all the things surrounding us] instead of "*imbumba*" [compound]). Yonela is employing a new written meshed register for doing science, which she achieves with a high degree of creativity. This achievement is notable given Yonela's disavowal, along with her peers, of her ability to write science in isiXhosa.

Third, the translation activity developed criticality in relation to language use in the learners. The group engaged in metalinguistic talk about the register used in the original isiXhosa definitions. Thandile objected to having to create an English translation of an isiXhosa definition when the isiXhosa definition had some features of English: "*Funeka sizibhale kaloku but kengoku sisiXhosa esidibene ne-English*" [we must write them but it's isiXhosa that is combined with English].

Certainly, the learners' conditioning to accept only English texts as being appropriate carriers of science meaning owing to the language conditions in their schooling up until this point played a part in their harsh criticism of the isiXhosa definitions. However, it was the learners' criticism of the isiXhosa of the definitions that provoked their second translations. Janks (2010) describes this *redesign* of a text and a task as being a pillar of critical literacy. The end product of five different written versions of a definition of each chemistry term provided an opportunity to compare and critique the different versions. It also created an opportunity for learners to discuss the use of the different parts of their semiotic repertoires in learning science, and to become aware that some have been considered to have more legitimacy than others, but that this can be challenged through task redesign.

In the translation activity learners leveraged their bilingual competence in English and isiXhosa to deepen their understanding of concepts in chemistry. In this sense the use of translation in science is shown to be a rigorous intellectual undertaking. These understandings developed in the established translanguaging space (Garcia & Li, 2014) of the study group enabled the development of *flexible wordings* (Lemke, 1990). Furthermore, producing written translations allowed learners to hone their writing skills in a high-status academic genre, namely explanatory definition. Conversations among learners as they were translating provided opportunities for criticality in relation to science language, while simultaneously developing new registers for learning science. We conclude that translation in science is a productive way of deepening conceptual understanding, provoking criticality and developing an expanded repertoire (Lin, 2015) in our learners.

11.4 Discussion: Lessons for Teacher Education

The purpose of drawing together these vignettes is to challenge monolingual ideologies and to move discussions about language and learning in bi- and multilingual classrooms away from the binaries of choosing one language over another as the language of learning and teaching, towards embracing translingual approaches that leverage all languages for learning. South Africa is a multilingual country and it makes sense to embrace this multilingualism as an asset rather than frame it as a problem in the classroom.

As stated, at present there is little in science teacher education in South Africa that prepares teachers to engage with learners' full linguistic repertoires in order to support epistemic access and engagement in learning science. What lessons for teacher education might one draw from these vignettes of practice? The first vignette demonstrates how anglonormative ideologies play out in classroom practice and the constraints that these place on learners' opportunities for meaning-making. Vignettes 2 and 3 offer examples of spontaneous translanguaging in oral classroom discourse, both in small-group work and in whole-class interactions. In these examples, translanguaging involves *teaching for transfer* (Cummins, 2008), with translanguaging employed as a scaffold to conceptual understanding and the production of oral and written science discourse in English. These are models of practice that could inform a conscious and planned translanguaging pedagogy and be adopted fairly readily, without the need for additional teaching resources. The fourth vignette offers an example of a planned translanguaging intervention, working with written bilingual science texts in ways that disrupt the taken-for-granted language/ing hierarchies in academia and in science discourse in particular.

Drawing on these vignettes, teacher education for pedagogical translanguaging in multilingual science classrooms can include the following:

- Unpacking and critiquing prevailing anglonormative and monolingual linguistic ideologies, including teachers' own ideologies, and their constraining effects on the teaching and learning of science for the majority of learners.
- Examining examples of existing classroom translanguaging practices such as those presented in this chapter and reflecting on how these can support opportunities to learn science;
- Planning and developing science lessons that take into account the particular linguistic resources of the learners and how these might be leveraged for learning science content and language.
- Developing bi- and multilingual teaching resources for science education to extend opportunities for pedagogical translanguaging into reading and producing science texts (for examples see Milligan et al., 2016).
- Opening up discussion and possibilities for assessment that engage with learners' full meaning-making resources.

Language and learning go to the heart of educational inequality in South Africa. Language ideologies rooted in coloniality and globalisation serve to entrench anglonormativity and combine with historic socio-economic disadvantage to perpetuate poor academic achievement for the majority of learners, including in science education. This is a challenge held in common with many post-colonial countries (see for example Lin, 1996; Martin, 1999; Rubagumya, 1994). Pedagogical translanguaging appears to offer a way out of the monolingual trap and a route to epistemic access and social justice in science education.

11.4.1 Transcription Conventions

⇧	Rising intonation
(…)	Audible pause

References

Ag, A., & Jørgensen, J. (2013). Ideologies, norms, and practices in youth poly-languaging. *International Journal of Bilingualism, 17*(4), 525–539.

Alexander, R. A. (2001). *Culture & pedagogy: International comparisons in primary education*. Blackwell Publishing.

Alexopoulou, E., & Driver, R. (1996). Small-group discussion in physics: Peer interaction modes in pairs and fours. *Journal of Research in Science Teaching: The Official Journal of the National Association for Research in Science Teaching, 33*(10), 1099–1114.

Antia, B. E., & Dyers, C. (2019). De-alienating the academy: Multilingual teaching as decolonial pedagogy. *Linguistics and Education, 51*, 91–100.

Ballenger, C. (1997). Social identities, moral narratives, scientific argumentation: Science talk in a bilingual classroom. *Language and Education, 11*(1), 1–14.

Bamgbose, A. (2000). *Language and exclusion: The consequences of language policies in Africa*. LIT Verlag.
Barnes, D. (1992). The role of talk in learning. In K. Norman (Ed.), *Thinking voices: The work of the National Oracy Project* (pp. 123–128). Hodder and Stoughton.
Baynham, M., & Tong King Lee. (2019). *Translation and translanguaging*. Routledge.
Bhatt, R. M., & Bolonyai, A. (2019). On the theoretical and empirical bases of translanguaging. *Working papers in Urban Language & Literacies, 254*, 1–22.
Blommaert, J. (2006). Language ideology. In E. K. Brown (Ed.), *Encyclopedia of language and linguistics* (pp. 510–522). Elsevier.
Canagarajah, S. (2018). Translingual practice as spatial repertoires: Expanding the paradigm beyond structuralist orientations. *Applied Linguistics, 39*(1), 31–54.
Chick, K. (1996). Safe-talk: Collusion in apartheid education. In H. Coleman (Ed.), *Society and the language classroom* (pp. 21–39). Cambridge University Press.
Chimbutane, F. (2011). *Rethinking bilingual education in postcolonial contexts*. Multilingual Matters.
Cummins, J. (2008). Teaching for transfer: Challenging the two solitudes assumption in bilingual education. In N. Hornberger (Ed.), *Encyclopedia of language and education* (pp. 65–75). Springer.
Department of Education. (1997). *Language in education policy, 14 July 1997*. Department of Education.
Derewianka, B. (1990). *Exploring how texts work*. Heinemann Educational Books.
Donovan, M. S., & Bransford, J. D. (2005). Introduction. In M. S. Donovan & J. D. Bransford (Eds.), *How students learn: Science in the classroom* (pp. 1–26). National Academies.
Flores, N. (2014). Let's not forget that translanguaging is a political act. *The educational linguist (blog)*. Accessed 19 July 2014.
García, O., & Li, W. (2014). *Translanguaging: Language, bilingualism and education*. Palgrave Macmillan.
Gee, J. P. (2004). Language in the science classroom: Academic social languages as the heart of school-based literacy. In E. W. Saul (Ed.), *Crossing borders in literacy and science instruction: Perspectives on theory and practice* (pp. 13–32). International Reading Association.
Gibbons, P. (2006). *Bridging discourses in the ESL classroom*. Continuum.
Grosjean, F. (1982). *Life with two languages: An introduction to bilingualism*. Harvard University Press.
Guzula, X., McKinney, C., & Tyler, R. (2016). Languaging-for-learning: Legitimising translanguaging and enabling multimodal practices in third spaces. *Southern African Linguistics and Applied Language Studies, 34*(3), 211–226.
Halliday, M. A. K., & Martin, J. R. (1993). *Writing science: Literacy and discursive power*. The Falmer Press.
Heugh, K. (2013). Multilingual education policy in South Africa constrained by theoretical and historical disconnections. *Annual Review of Applied Linguistics, 33*, 215–237.
Heugh, K., Prinsloo, C., Makgamatha, M., Diedericks, G., & Winnaar, L. (2017). Multilingualism(s) and system-wide assessment: A southern perspective. *Language and Education, 31*(3), 197–216.
Hornberger, N., & Chick, K. (2001). Co-constructing school safetime: Safetalk practices in Peruvian and South African classrooms. In M. Heller & M. Martin-Jones (Eds.), *Voices of authority: Education and linguistic difference* (pp. 31–56). Greenwood Publishing Group.
Howie, S. J. (2001). *Mathematics and science performance in grade 8 in South Africa 1998/99: TIMSS-R 1999 South Africa*. Human Sciences Research Council.
Janks, H. (2010). Language, power and pedagogies. In N. H. Hornberger & S. L. Mackay (Eds.), *Sociolinguistics and language education* (pp. 40–61). Multilingual Matters.
Jaspers, J. (2018). The transformative limits of translanguaging. *Language and Communication, 58*, 1–10.

Karlsson, A., Nygård Larsson, P., & Jakobsson, A. (2019). Multilingual students' use of translanguaging in science classrooms. *International Journal of Science Education, 41*(15), 2049–2069.

Kusters, A., Spotti, M., Swanwick, R., & Tapio, E. (2017). Beyond languages, beyond modalities: Transforming the study of semiotic repertoires. *International Journal of Multilingualism, 14*(3), 219–232.

Lemke, J. (1990). *Talking science: Language, learning and values*. Ablex Publishing Corporation.

Lemke, J. (2004). The literacies of science. In E. W. Saul (Ed.), *Crossing borders in literacy and science instruction: Perspectives on theory and practice* (pp. 33–47). International Reading Association.

Li, W. (2011). Moment analysis and translanguaging space: Discursive construction of identities by multilingual Chinese youth in Britain. *Journal of Pragmatics, 43*(5), 1222–1235.

Lin, A. M. Y. (1996). Bilingualism or linguistic segregation? Symbolic domination, resistance and code switching in Hong Kong schools. *Linguistics and Education, 8*(1), 49–84.

Lin, A. M. Y. (2015). Egalitarian bi/multilingualism and trans-semiotizing in a global world. In W. E. Wright, S. Boun, & O. Garcia (Eds.), *Handbook of bilingual and multilingual education* (pp. 19–37). Wiley-Blackwell.

Madiba, M. (2014). Promoting concept literacy through multilingual glossaries: A translanguaging approach. In C. Van Der Walt & L. Hibbert (Eds.), *Multilingual universities in South Africa: Reflecting society in higher education* (pp. 68–87). Multilingual Matters.

Makalela, L. (2016). Ubuntu translanguaging: An alternative framework for complex multilingual encounters. *Southern African Linguistics and Applied Language Studies, 34*(3), 187–196.

Makoe, P., & McKinney, C. (2014). Linguistic ideologies in multilingual South African suburban schools. *Journal of Multilingual and Multicultural Development, 35*(7), 658–673.

Malherbe, E. G. (1946). *The bilingual school: A study of bilingualism in South Africa*. Central News Agency.

Martin, P. W. (1999). Close encounters of a bilingual kind: Interactional practices in the primary classroom in Brunei. *International Journal of Educational Development, 19*(2), 127–140.

McKinney, C. (2017). *Language and power in post-colonial schooling: Ideologies in practice*. Routledge.

McKinney, C., & Tyler, R. (2018). Disinventing and reconstituting language for learning in school science. *Language and Education, 33*(2), 141–158.

Mehan, H. (1979). *Learning lessons*. Harvard University Press.

Milligan, L. O., Clegg, J., & Tikly, L. (2016). Exploring the potential for language supportive learning in English medium instruction: A Rwandan case study. *Comparative Education, 52*(3), 328–342.

Mortimer, E. F., & Scott, P. H. (2003). *Meaning making in secondary science classrooms*. Open University Press.

Msimanga, A., & Lelliott, A. (2014). Talking science in multilingual contexts in South Africa: Possibilities and challenges for engagement in learners' home languages in high school classrooms. *International Journal of Science Education, 36*(7), 1159–1183.

Msimanga, A., Denley, P., & Gumede, N. (2017). The pedagogical role of language in science teaching and learning in South Africa: A review of research 1990–2015. *African Journal of Research in Mathematics, Science and Technology Education, 21*(3), 245–255.

Mudadigwa, B., & Msimanga, A. (2019). Exploring teacher pedagogical practices that help learners make connections during the teaching of reactions in aqueous solutions at senior secondary level. *African Journal of Research in Mathematics, Science and Technology Education, 23*(3), 332–343.

Mullis, I. V. S., Martin, M. O., Foy, P., Kelly, D. L., & Fishbein, B. (2020). *TIMSS 2019 international results in mathematics and science*. Boston College, TIMSS & PIRLS International Study Center.

Nomlomo, V. S. (2007). *Science teaching and learning through the medium of English and isiXhosa: A comparative study of two primary schools in the Western Cape*. Unpublished doctoral dissertation. University of Western Cape, Cape Town.

Otheguy, R., Garcia, O., & Reid, W. (2015). Clarifying translanguaging and deconstructing named languages: A perspective from linguistics. *Applied Linguistics Review, 6*(3), 281–307.

Probyn, M. (2015). Pedagogical translanguaging: Bridging discourses in South African science classrooms. *Language and Education, 29*(3), 218–234.

Probyn, M. J. (2017). *Language and the opportunity to learn science in bilingual classrooms in the Eastern Cape, South Africa*. Unpublished doctoral dissertation, University of Cape Town, Cape Town.

Reddy, V. (2006). *Mathematics and science achievement in South African schools in TIMSS 2003*. HSRC Press.

Ricento, T. (2000). Historical and theoretical perspectives in language policy and planning. *Journal of Sociolinguistics, 4*(2), 196–213.

Rojas-Drummond, S., & Zapata, M. P. (2004). Exploratory talk, argumentation and reasoning in Mexican primary school children. *Language and Education, 18*(6), 539–557.

Rollnick, M., & Rutherford, M. (1996). The use of mother tongue and English in the learning and expression of science concepts: A classroom-based study. *International Journal of Science Education, 18*(1), 91–103.

Rubagumya, C. (1994). Introduction. In C. Rubagumya (Ed.), *Teaching and researching language in African classrooms* (pp. 1–5). Multilingual Matters.

Ruiz, R. (1984). Orientations in language planning. *National Association for Bilingual Education Journal, 8*(2), 15–34.

Thomas, W. P., & Collier, V. (1997). *School effectiveness for language minority students*. National Clearinghouse for Bilingual Education.

Tyler, R. (2018). *Semiotic repertoires in bilingual science learning: A study of learners' meaning-making practices in two sites in a Cape Town high school*. Doctoral dissertation, Faculty of Humanities, University of Cape Town.

Western Cape Education Department. (2014a). *Curriculum FET minute: DCF 0003/2014. Teaching the language of assessment across the curriculum*. https://wcedonline.westerncape.gov.za/circulars/minutes14/CMminutes/edcf3_14.html. Accessed 10 June 2016.

Western Cape Education Department. (2014b). *Language strategy 2015–2019*. https://wcedonline.westerncape.gov.za/documents/MathLanguageStrat/MathLanguage-Strategy.html. Accessed 10 June 2016.

Wells, G. (1999). *Dialogic inquiry: Towards a sociocultural practice and theory of education*. Cambridge University Press.

Williams, C. (1996). Secondary education: Teaching in the bilingual situation. In C. Williams, G. Lewis, & C. Baker (Eds.), *The language policy: Taking stock* (pp. 39–78). CAI.

Young, D., Van Der Vlugt, J., & Qanya, S. (2005). *Understanding concepts in mathematics and science: A multilingual learning and teaching resource book in English, isiXhosa, isiZulu and Afrikaans*. Maskew Miller Longman.

Professor Audrey Msimanga, 02 March 1961–29 June 2021: In memoriam

Professor Audrey Msimanga was a much loved and highly respected teacher, researcher and educational leader, and her untimely death from covid has been a devastating loss for her students, colleagues, friends and family.

Audrey was committed to social justice in education, and her research focused on the role of language in science education in low-income multilingual contexts. She forged research networks in South Africa and internationally, published widely, and was associate editor of the *Journal of Research in Science Teaching* and president-elect of the Southern African Association for Research in Mathematics, Science and Technology Education.

Audrey grew up in Zimbabwe where she graduated with a BSc and embarked on an early career in teaching science and then research in biodiversity and ornithology in Zimbabwe and Botswana. After moving to South Africa to pursue her PhD studies, she returned to teaching secondary school science. She soon moved into science teacher education at the University of the Witwatersrand, and after 2 years as head of the Faculty of Education at Sol Plaatjie University, she returned to the University of the Witwatersrand in February 2021 as head of the School of Education. Her gentle, thoughtful, wise and inclusive style of leadership endeared her to her colleagues.

Audrey leaves two sons of whom she was tremendously proud. She juggled her roles as teacher, academic, researcher, leader, and mother with distinction and grace. Hamba kahle, Audrey.

Chapter 12
Leveraging Multilingualism to Support Science Education Through Translanguaging Pedagogy

Erasmos Charamba

Abstract Research suggests that linguistically diverse students draw on their collective linguistic repertoires to meet their communicative goals in the educational space. This indicates that their language use is not strictly compartmentalized but fluid. Therefore, educators should recognize and acknowledge their students as multilingual speakers who have a multilingual repertoire and use it in social and educational contexts. The participants for this study were 87 multilingual students in 3 conveniently sampled tenth-grade Chemistry classes at one secondary school. The teachers in all three classes conduct collaborative learning activities and encourage their students to use their linguistic repertoire while working together in groups. This chapter focuses on qualitative data gathered through lesson observations and video recordings of the students as they interacted during their class and collaborative group work over 4 weeks, as well as interview responses from students and their Chemistry teachers. Drawing on Vygotsky's sociocultural theory and using sociocultural discourse analysis, the data were analyzed qualitatively to determine how students translanguage to scaffold their own and each other's understanding of science concepts. The results offer important pedagogical implications for science teachers in linguistically diverse contexts such as South Africa and advocate for translanguaging as a pedagogical tool in science teaching.

Keywords Meaning-making · Monolingualism · Multiculturalism · Multilingualism · Science education · Sociocultural theory · Translanguaging

E. Charamba (✉)
School of Education, University of the Witwatersrand, Parktown, Johannesburg, South Africa

© The Author(s), under exclusive license to Springer Nature Switzerland AG 2021

A. Jakobsson et al. (eds.), *Translanguaging in Science Education*, Sociocultural Explorations of Science Education 27,
https://doi.org/10.1007/978-3-030-82973-5_12

12.1 Introduction

Educational policy debates and research in South Africa point to an urgent ongoing need to question and cross-question how a practice (monolingual pedagogy) which allegedly results in cognitive deficiencies among students continues to be reproduced and dominates discussions about present-day educational policy and practice in the country (Charamba & Zano, 2019). This is against a backdrop of unsatisfactory academic performance by most South African students in national and international assessments. In the 2015 Trends in International Mathematics and Science Study (TIMSS) assessment, the national average score for South African ninth-grade science students was 358 points, resulting in the country attaining position 39 out of 39 countries.

TIMSS is an assessment of the mathematics and science knowledge of fourth- and eighth-grade students from selected countries around the world and South Africa participates in fifth and ninth grades (Mullis et al., 2016). In assessing students' mathematical and scientific knowledge, TIMSS 2015 used the five "international benchmarks" to scale the scores, namely, Advanced (above 625 points), High (550–625 points), Intermediate (475–550 points), Low (400–475), and Not Achieved stands at less than 400 points (Mullis et al., 2016). Results of another international assessment, Progress in International Reading Literacy Study (PIRLS), indicate that 78% of South African fourth-grade students who participated in the 2016 study could not read for comprehension (Mullis et al., 2017). PIRLS is an international study of reading comprehension achievement conducted by the International Association for the Evaluation of Educational Achievement (IEA).

In national assessments, results of the twelfth-grade pass rate for Physical Sciences between the years 2015 and 2018 stand at 58.6%; 62.0%; 65.1%; and 63.6%, respectively (Department of Education, National Senior Certificate School Subject Report, 2019). Twelfth-grade is the last grade in the South African high school education, and therefore students in this grade sit for school-leaving examinations at the end of the year (Department of Education, National Senior Certificate School Subject Report, 2019). In its analysis of these three assessments, South Africa's quality assurance body on education, Umalusi, asserted that students taught in a language other than their home language continue to experience great difficulty in comprehending concepts, interpreting questions, and drawing up responses (Stroupe et al., 2019), suggesting that more should be done with regards to the languages of instruction if students' academic performance is to improve especially in the STEM (Science, Engineering, Technology, and Mathematics) learning areas (Charamba, 2019; National Treasury Report on South African Education, 2018).

This is in line with findings of recent research which suggest that the use of languages other than students' home language for teaching and learning has several challenges and could lead to language becoming a barrier to effective learning resulting in underachievement among students (see, e.g., Buxton & Caswell, 2020; Garcia, 2019; Langman, 2014; Li, 2018; Lin, 2019; Probyn, 2019; Zhang et al., 2020). Despite many communities in the country being multilingual, students cannot bring their entire linguistic repertoire into the classroom because of school policies and practices that exclude them favoring either English or Afrikaans only. According to Vygotsky's sociocultural theory, one of the most important tools a student brings into the science classroom for learning and interactional purposes is their language (Omidire, 2019).

12.2 Theoretical Background: The Sociocultural Theory

In present-day multicultural education communities, addressing the educational needs of culturally and linguistically diverse science students tops the "to-do list" for most educators and researchers (Buxton & Caswell, 2020). The great focus being placed on pedagogical practices, such as culturally relevant (Hammond, 2014) or culturally responsive (Ishimaru, 2019) approaches that work toward just and equitable sociocultural experiences (Makoe, 2018). The sociocultural theory attempts to explain how individual mental functioning is related to cultural, institutional, and historical context; hence, the focus of the sociocultural perspective is on the roles that participation in social interactions and culturally organized activities play in influencing psychological development (Scott & Palincsar, 2013).

The original framework for the sociocultural theory rests in the work of Vygotsky (1978) in which he viewed learning as a social process and the origination of human intelligence in society or culture. The key concept in Vygotsky's theory is that cognitive development is greatly affected by social interactions an individual participates in. Vygotsky believed learning takes place through one's interaction with others (Duarte, 2019). The learned concepts are then integrated into the student's cognitive space thus facilitating the student's multifaceted development. However, Vygotsky argues that the potential for this development, especially the cognitive, is limited to what he calls a zone of proximal development (Vygotsky, 1978).

The zone of proximal development is the area of exploration for which the student is cognitively prepared for but requires scaffolding from teachers and social interactions with others to fully develop (McKinney & Tyler, 2019). The theory places great emphasis on the pivotal role played by one's participation in social interactions (Garcia, 2019) and culturally organized activities in influencing cognitive development (Li, 2018). These interactions are enabled through language use. In multicultural and multilingual science classrooms, language plays a central role in students' acquisition of scientific concepts as it is the transmitter of cultural tools, the mechanism for thinking, and the most important mental tool (Vygotsky, 1978; Zhang et al., 2020). The sociocultural theory not only looks at what the teacher brings into the science classroom but also in what the student brings as well as how the broader cultural and historical setting shapes their interaction thus acknowledging the interdependence of language, learning, and development (Krause & Prinsloo, 2016).

The sociocultural theory, therefore, is instrumental in advancing pedagogy that might rectify inequalities found in the present-day multilingual science classrooms caused by the exclusion of students' languages from the classroom (Karlsson et al., 2020). Classroom pedagogy informed by Vygotsky's sociocultural theory considers the discourse, norms, and practices associated with particular discourse and practice communities (García & Lin, 2018). From this perspective, I argue that the use of students' linguistic repertoire can be developed and made use of in the science classroom (Siry, 2014). Considering the linguistic landscape of most South African science classrooms, translanguaging as an approach to learning should be widely recommended and supported.

12.3 Translanguaging as a Pedagogical Resource in Science Education

The natural use of multiple languages together by multilingual people for communication purposes has been in practice worldwide since time immemorial (Siry, 2014). However, as a focus of research, this emerged in the twentieth-century when Cen Williams and his colleagues were investigating efficient strategies for their students to use two languages in the educational activity. They came up with the term "*trawsieithu*" (Li, 2018) to describe the process in which students would read or hear in one language (e.g., English) and then write or speak about what they would have read or heard in another (e.g., isiZulu, or vice versa). Colin Baker then translated the term into English as "translanguaging" (see Lewis et al., 2012). This, for most researchers and practicing educators, marked a paradigm shift, moving away from traditional linguistic terms such as code-switching, and code-mixing, calling into question the existence of "languages" as identifiable, distinct systems (Makoe, 2018).

The term translanguaging has been used in instructional pedagogy, in everyday social interaction, cross-modal and multimodal communication, linguistic landscape, and transgender discourse (Li, 2018; MacSwan, 2017). Translanguaging entails a process whereby a multilingual intentionally and strategically makes use of their linguistic repertoire in an integrated form for communication. In some countries, the practice has been accepted as a legitimate pedagogical approach that facilitates the scaffolding of one language by another and for meaning-making across the curriculum (Buxton & Caswell, 2020). Translanguaging, therefore, is a unitary meaning-making system in which multiple discursive practices are used to understand the bilingual world and to create a space where multilingual science students make use of their entire linguistic and semiotic repertoire in the educational space (Garcia, 2019; Licona & Kelly, 2020).

Lasagabaster and Garcia (2014) describe translanguaging as a process by which bilingual students make use of the many resources their bilingual status offers to communicate and for meaning-making purposes. It involves going between and beyond linguistic systems and structures including different modalities such as speaking, writing, and signing in or outside the classroom (Lin, 2019). It therefore breaks the artificial and ideological divides and challenges the traditional "one language at a time" mode of communication (Omidire, 2019) still dominating in South Africa, a country with 12 official languages including South African Sign language (Charamba, 2019). In defining translanguaging, another scholar, Grosjean (2019), makes use of a sports analogy on hurdles, where two athletic skills, high jump, and sprinting are involved together with other physiological processes such as breathing, sight, and so forth during the sporting event. Hurdlers use these skills as an integrated whole to excel in their sport just in the same way multilingual students use their linguistic repertoire to communicate effectively.

Translanguaging is a theory which postulates that rather than possessing a multitude of autonomous language systems, as has been traditionally thought, all users of language conveniently select and deploy particular features from a single linguistic repertoire to make meaning and to negotiate particular communicative contexts (Iversen, 2020). A three-year ethnographic study by Karlsson et al. (2020) shows that a Translanguaging Science Classroom (TSC) is an asset in appropriating a new social practice for students with limited ability to understand and express themselves in the language of instruction. Their study also suggests the use of both first and second languages has a positive effect on the continuity of science learning in multilingual classroom activities.

Focusing on a socio-cultural approach, another researcher, Duarte (2019), examined how 15-year-old multilingual students applied their various linguistic repertoires to maintain tasks in content-matter classroom activities. The study suggests that the use of multiple languages, especially the home language, occurred in cognitively challenging task-talk activities. In another research, Portoles and Marti (2017) investigated the simultaneous use of multiple languages in early learning concluded that the monolingual approach does not facilitate effective teaching and learning. In South Africa, Probyn (2019) notes that there has been a corresponding interest and research in translanguaging in South African classrooms, some of which reports on spontaneous language use in classrooms (e.g., Krause & Prinsloo, 2016; McKinney & Tyler, 2019; Probyn, 2019) while some document interventions that have adopted planned heteroglossic pedagogies that engage with students' full linguistic repertoires (e.g., Charamba, 2020a; Fortuin, 2017; Madiba, 2014; McKinney & Tyler, 2019; Msimanga et al., 2017). However, little if any, research has been carried out to explore the efficacy of translanguaging in multilingual tenth-grade South African Chemistry classrooms, hence the aim of the present study.

12.4 Research Questions

The main research questions for the study were:

(a) What is the role of language in the teaching/learning of multilingual tenth-grade Chemistry students?
(b) How can translanguaging be used in the tutelage of Chemistry and how do classroom translingual practices shape opportunities for meaning-making?

12.5 Setting and Participants

The study, being a part of a larger multilingualism project exploring the role language plays in science classrooms in Southern Africa, was conducted in Soweto, South Africa. Soweto is a township in Gauteng and borders the city's mining belt in the south. Its name is an English syllabic abbreviation for South Western Townships (Grinker & Gorelik, 2014) and was created in the 1930s when the White government started separating Blacks from Whites, creating black "townships" (Grinker, 2014). The school falls under Sect. 21 of the South African Schools Act. Schools under this section qualify for government subsidy as most of the parents fall in the low- income bracket.

This school, like all schools across the township, has a 100% black student enrolment whose home language is one of the country's nine African official languages (Sesotho, Sepedi, siSwati, Xitsonga, Setswana, Tshivenda, isiNdebele, isiXhosa, isiZulu). During the year of study (2019) the school had an enrolment of over 800 students (Grade 8–12) and three tenth-grade Chemistry classes. The classes had respective enrolments of 30; 28; and 29. All participants were multilingual with 53 of them having isiZulu as their home language and 34 had Sesotho as home language. Due to the multilingual nature of Soweto, almost everyone is fluent in more than two African languages with isiZulu, the widely spoken African language, dominating the township and the country (Grinker & Gorelik, 2014).

The official language of instruction for the school is English. The participants in the present study had received bilingual education through English and isiZulu from grade R to the third- grade, which meant that the students used both languages as a resource in all class activities. In South Africa, grade R is a class below the first-grade and enrolls children aged 5. After completing grade R, they are then enrolled in the first-grade. When the students in this study started fourth-grade, they were then introduced to monolingual pedagogy where all teaching and learning, as well as learning materials, were strictly in the English language. The three classes were conveniently sampled as they were the only Chemistry classes in the school.

All three Chemistry teachers, Mr. Anesu, Ms. Sarah, and Ms. Puleng (not their real names) hold Bachelor of Education degrees and have more than 6 years' teaching experience each. They are all multilingual, isiZulu being their home language with English and Sesotho as additional languages. Mr. Anesu had adhered faithfully to the school's policy of monolingual pedagogy for 6 years until he decided to allow his students more linguistic flexibility in their science classroom activities in the year 2018. He took this decision after "having attended workshops hosted by a local university on Decolonizing Education where I learnt of the academic benefits of translanguaging. I then decided to try it with my students as most were struggling to express themselves in English" (Mr. Anesu, Interview, 16th July 2019).

12.6 Data Collection and Analysis

This is part of a mixed-methods study that generated responses from a sample of 87 tenth-grade Chemistry students from three classes in South Africa on the role language plays in the science classroom. This part of the study focuses on qualitative data collected through non-participant classroom observation for 4 weeks. One 40-min lesson was observed in each classroom every week giving a total of 480 min of observation. The lessons were video-recorded. Video recording enabled the researcher to visualize and try to comprehend how students' authentic language use could affect their participation and understanding of concepts in the science classes (Lemke, 2012). To support the qualitative data collected through video recordings and lesson observations, interviews were conducted with teachers and students who had *translanguaged* during the observed lessons.

During the interviews, the researcher used video-elicitation techniques (Creswell, 2014) by showing the participants selected video footages where they were actively translanguaging. In the study qualitative data analysis involved a combination of both inductive and deductive analysis (Bryman, 2015), where the researcher drew deductive codes for all materials from the literature reviewed and this included: translanguaging, meaning-making, and epistemological access (Karlsson et al., 2020). Inductive codes were derived from an analysis of field notes, interview transcripts, and video footage of classroom interactions (McMillan & Schumacher, 2010) paying great attention to sections where these multilingual tenth-grade science students were actively translanguaging.

After analyzing the data, the researcher came up with themes which are also supported by prototypical quotes. All ethical considerations were observed. The participants were informed that participating in the study was voluntary. They were also informed that if they chose to participate in the study, they were free to withdraw at any time without any repercussions. To protect their identities, participants were asked to choose pseudonyms and were referred to using these names throughout the study.

12.7 Findings

In this section, I present selected examples of translanguaging practices witnessed during the lesson observations and consider how these classroom language practices shaped opportunities for meaning-making. The two headings below correspond with the themes that were found in the analysis. To illustrate this, three typical conversation sequences from the tenth-grade classes (Excerpts 12.1, 12.2, and 12.3) and interview responses from both students and teachers are chosen.

12.8 Translanguaging to Actively Scaffold Science Learning

During collaborative and whole-class interactions, translanguaging provided an avenue for the Chemistry teachers to scaffold students' language and scientific content learning simultaneously. For example, Excerpt 12.1 shows how a student's translanguaging in Mr. Anesu's classroom helps him to make a meaningful contribution to the class conversation. In this interaction, Mr. Anesu delivering a lesson on the Hydrosphere (Department of Basic Education, 2012) was explaining how some substances that dissolve in water can be harmful to living organisms. He then went on to ask the class to give examples of substances that dissolve in water. Charles raises his hand and opens the discussion.

Excerpt 12.1 Students' Conversation During One of the Lessons

Speaker	Verbatim	English translation
Charles	Chemicals *aqhamuka kuma* factory	Chemicals from factories
Philiswa	Grease is from factories kodwa *ayincibiliki*	Grease is from factories but it does not dissolve in water
Bonolo	But if you use *insipho igrease izoncibilika*	But if you use soap grease will dissolve in water
Busisiwe	But why grease *ingancibiliki ngaphandle kwensipho?*	But why doesn't grease dissolve without adding soap?
Buhle	*i-*grease, *njenge* oil, *ayisi-* polar *kanti amanzi ayi- polar*	Grease, just like oil, is non-polar whereas water is polar

In Excerpt 12.1, in giving examples of substances that dissolve in water, Charles makes use of both English and isiZulu stating that chemicals from factories dissolve in water. Philiswa joins the academic discussion by also using both languages to point out that grease (even though it's found in some factories) does not dissolve in water. Bonolo makes use of her linguistic repertoire to raise an important point saying "if you use *insipho* igrease *izoncibilika*," translated as "if you use soap grease will dissolve in the water." This raises interest in some students prompting Busisiwe to question why "grease *ingancibiliki ngaphandle kwensipho*?" (why doesn't grease dissolve in water when soap is not used?). An explanation is provided by Buhle who points out that grease, like oil, is non-polar, whereas water is polar (i-grease *njenge* oil, *ayisi-* polar *kanti amanzi ayi- polar).*

Excerpt 12.1 shows how students in Mr. Anesu's class made use of translanguaging to scaffold learning. This also presented students with a linguistic avenue to contribute during class discussions. Charles and Bonolo, for example, "before allowing translanguaging, wouldn't say a word in class. Their proficiency in English is low depriving them of an avenue to express themselves. Through translanguaging, they now participate actively in class, their performance has improved greatly" (Mr. Anesu, Interview, 24th July 2019).

In Excerpt 12.1, Bonolo manages to come up with a scientific solution to making grease dissolve in water. She correctly points out, through translanguaging that if soap is used grease will dissolve in water. This is scientifically correct since grease is non-polar while water is polar. This is the explanation given by Buhle who goes on to give an example of another non-polar substance (oil) in her "*translanguaged*" response to Busisiwe's question. To this, Mr. Anesu explains that "the class discussions are now lively and most of my students now exhibit critical thinking and analysis. Translanguaging enables them to ask thought-provoking questions, give correct scientific explanations, and has resulted in language development" (Mr. Anesu, Interview, 24th July 2019).

While teaching the same topic, Ms. Sarah asks her class to work in collaborative groups and describe, with the aid of a diagram, the water cycle. Excerpt 12.2 represents a conversation from one of the groups as they were drawing their water cycle diagram.

Excerpt 12.2 Conversation in One of the Collaborative Groups

Speaker	Verbatim	English translation
Siboniso	Let's start with *imithombo yamanzi efana nemifula namadamu*	Let's start by drawing sources of water such as rivers and dams
Mpondulo	Water evaporates *kule mithombo*	Water evaporates from these water sources
Hlengiwe	After evaporating, it condenses *kwakheke amafu*	After evaporating, it condenses forming clouds
Lebogo	After condensation let's show *imvula*	After condensation, let's show the rainfall
Salome Gugu	Show some of the rainfall collecting in water sources *enye izocwila emhlabathini* Write 'a water cycle *ayinasiqalo Noma isiphetho'*	Show some of the rainfall collecting in water sources and some sinking into the ground Write a water cycle has no beginning or end

In Excerpt 12.2, Siboniso, who was leading the group and drawing the diagram on behalf of the group, suggests they start by drawing "*imithombo yamanzi efana nemifula namadamu*" (sources of water such as rivers and dams). After drawing the sources of water, Mpondulo tells Siboniso to include arrows that show water evaporating "*kule mithombo*" (from the sources of water). In her explanation of the students' engagements in this excerpt, Ms. Sarah suggests:

> Here, Hlengiwe's contribution is two-fold. Firstly, she shows her knowledge of the processes by stating that after evaporation the next process is condensation. Secondly, she hows the group she understands what condensation is by saying 'kwakheke amafu' meaning it leads to the formation of clouds (Ms. Sarah, Interview, 31st July 2019).

In Excerpt 12.2, Lebogo makes a valuable contribution by suggesting they show rainfall after the condensation phase. Salome joins the conversation by suggesting they show some of the rainfall collecting in water sources while "*enye izocwila emhlabathini*" (while some will sink into the ground). The last contribution comes from Gugu who instructs Siboniso, through translanguaging, to write that a water cycle has no beginning or ending (*ayinasiqalo noma isiphetho).* In an interview on the 31st of July 2019, Gugu could not hide her appreciation in using both languages in the same lesson because:

> English is very difficult for me and so kwesinye isikhathi angazi ukuthi ngithini (at times I run out of words not knowing what to say) so using all languages helps. Like during that group discussion I forgot 'has no beginning' and ended up saying 'ayinasiqalo'. Using many languages also makes me understand what we are taught ekilasini (in class) and manje ngithola amamaki angcono (I am now getting better marks*)*.

In line with part of Gugu's response, Philiswa (Excerpt 12.1 line 3) highlights that "my English has improved because I learn some English words through isiZulu. For example, I learnt ukuthi ukuncibilikisa is to dissolve" (Interview, 5th August 2019).

In the other tenth-grade classroom, Ms. Puleng had asked her students to state reasons why people should pay for water they use in their houses. Presented in Excerpt 12.3 is a written answer from one of the groups.

Excerpt 12.3 Response from One Group

We have to pay for the water we use so the government can have money to build more dams, *ukuhlanzwa kwamanzi*, repairing *amapayipi aphukile*, and transport to supply water *ezindaweni ezingenamanzi ahlanzekile.*

In their response, the group states they pay for the water they use so the government can build more dams, enable water purification (*ukuhlanzwa kwamanzi), repair broken water pipes (amapayipi aphukile), and supply water to parts of the country that do not have clean, safe water (ezindaweni ezingenamanzi ahlanzekile).* In justifying why they had used translanguaging in writing their response, Jabu who was leading the group said:

> Sometimes we don't know the answer in English, so we write in isiZulu. Kodwa impendulo iyafana (but the answer is the same). The government must give us books written in isiZulu to make us understand science better. Afrikaners have books in Afrikaans. English is difficult (Interview, 6th August 2019).

The Afrikaners are a South African ethnic group who are descended from seventeenth century Dutch, German, and French settlers to South Africa and ruled South Africa from the 1940s till 1994 during the apartheid era (Grinker, 2014). Their first language is Afrikaans, and they make up approximately 5.2% of the total South African population (Statistics South Africa, 2019). During their rule, they pushed for Afrikaans to be the sole language of instruction in the country. Jabu's response was elaborated by Ms. Sarah saying:

> Afrikaans speakers are taught in their home language from grade R to university. This is the main reason Afrikaans students perform better than township students who are taught in a language different from their home language. Jabu's response is evidence of educational imbalances still prevalent in present-day South Africa. So, we are saying the schools should allow students to use all languages because we have seen this makes them understand and perform better (Ms. Sarah, Interview, 24th July 2019).

Like Excerpts 12.1 and 12.2, Excerpt 12.3 is evidence of how students made use of translanguaging to come up with sound scientific responses to questions asked. Low proficiency in the language of instruction did not hinder them from answering questions as they resorted to making use of their linguistic repertoire. The use of translanguaging, either orally or through writing, did not change the scientific correctness of students' responses. On the contrary, according to Ms. Puleng:

> Through translanguaging students can express themselves fully providing good responses. I have noticed a pleasing improvement in their performance during class activities, end of topic tests, and end of term examinations. The English language teachers in our school have also implemented translanguaging in their classes and are saying their students' written and spoken English has improved (Interview, 30th July 2019).

In the three classes observed during the study, the students' home language (isi-Zulu) was seen and taken as a resource in helping students acquire new scientific knowledge and not as a hindrance to the learning process. The home language was, thus, brought into the classrooms as a scaffold, collaborating with the language of instruction. Ms. Puleng further states that "although the school policy doesn't allow us to use other languages in the classroom, we tend to smuggle them in because of the benefits we realize from translanguaging" (Interview, 30th July 2019).

The educator goes on to suggest the need for South African schools and government "to promote pedagogy that considers translanguaging as a resource. For the two years I have allowed translanguaging in my classes I came to realize that it helps students grasp concepts better" (Ms. Puleng, Interview, 30th July 2019). In line with Ms. Puleng's assertion, research supporting the use of translanguaging has been well documented on international fora (see, e.g., Duarte, 2019; García & Lin, 2018; Karlsson et al., 2020; Li, 2018) as well as in South Africa (see, e.g., McKinney & Tyler, 2019; Msimanga et al., 2017; Omidire, 2019; Probyn, 2019).

12.9 Ways of Speaking, Writing, and Participating in a Multilingual Science Class

Basing on lessons observed during the study, there was great participation from all students during class and collaborative group discussions. The use of their linguistic repertoire opened up spaces for them to engage with one another and with the scientific matter (Garcia, 2019). One possible reason for the great engagement could be that science concepts and words often tend to be simplified when using multiple languages simultaneously (Msimanga et al., 2017), and "students find some of the terms easier to comprehend when using their linguistic repertoire. Multilingual and multimodal practices boost students' confidence, sense of belonging, and self-esteem making communication easier and more effective" (Ms. Puleng, Interview, 6th August 2019).

This was affirmed by her colleague, Ms. Sarah who also stated that most of her students "hardly talk in class if I ask them to use English language only. The use of their repertoires helps them make meaningful and informed contributions during class and group discussions" (Interview, 6th August 2019). In advocating for translanguaging pedagogy, Ms. Sarah goes on to state that "translingual practices help students grasp the scientific knowledge and I have also noticed an improvement in their command of English language" (Interview, 6th August 2019).

Translanguaging creates educational avenues where students comprehend the scientific matter discussed in the classroom, thereby being able to contribute meaningfully during academic discussions and in answering questions orally or in writing (see Excerpt 12.3). In one of his lessons on the Hydrosphere, Mr. Anesu introduced the lesson by asking his students to write down what they remembered from the previous lesson. Nkhululeko, who happened to be seated close to me, wrote:

Excerpt 12.4 Student's Response

1. Ama chemicals aqhamuka kuma factory pollute the water. This causes diseases okwenza abantu bagule.
2. Oil and grease akuchithi emanzini.
3. Amanzi ayigugu and must be conserved.

In her first response, Nkhululeko states that chemicals from factories pollute water causing diseases that make people ill (*okwenza abantu bagule*). In the second response, she states that oil and grease do not dissolve in water (*akuchithi emanzini*) and lastly water is precious (*amanzi ayigugu*) and should be conserved. Although the student used translanguaging in the same sentence, as pointed elsewhere in this chapter, this does not affect the scientific correctness of her response. As a matter of fact, "translanguaging enables my students to answer all questions and timeously. Before I allowed students to use their language repertoires, participation in class was low. Only a few students would contribute during lessons and questions were poorly answered" (Mr. Anesu, Interview, 30th July 2019).

When going through students' responses, I noticed that Nkhululeko's use of two languages in the same sentence when writing was not an isolated incident but rather the norm among most students. According to Ms. Sarah:

> This serves two purposes. Firstly, all students engage with the work written and secondly most of them get correct answers as they have a broader linguistic base to express their answers. Some even draw instead of writing. You saw what happened in that other lesson when I asked my students what they understood by Water Cycle. Some students presented their responses in a paragraph or two whereas others presented them diagrammatically. (Interview, 5th August 2019)

From this perspective, language is no longer a bounded, self-contained entity, embedded in static communicative activities (Benson, 2015). Rather, it becomes a mobile resource working together with multimodal and multisensory signs to construct meaning (Mazzaferro, 2018). Translanguaging entails going between and beyond linguistic systems and structures including different modalities such as speaking, writing, and signing in or outside the classroom (Garcia, 2019; Lin, 2019). It entails the fluid and dynamic linguistic practices that go beyond the boundaries between named societal languages, language varieties, and language and other semiotic systems (Li, 2018). In other words, it breaks the artificial and ideological divides between named languages enabling multilingual students to use their idiolect that is their full linguistic repertoire without regard for socially and politically defined language labels or boundaries (Otheguy et al., 2015).

Another researcher in multilingualism, Caruso (2018) analyzed linguistic practices in a Language and Communication policies course at the University of Algarve, in Portugal. The course professor allowed his students to use their various linguistic repertoires in asking questions, giving responses, during class and group discussions, and when writing to achieve a collective comprehension of the content. At the end of the course, the students were asked to take a structured multilingual final examination, in three languages, and results show an improvement in the academic performance of the students when spaces for translingual practices are availed. This is in conformity with García and Li's (2014) view of translanguaging as an approach to the use of one's language, bilingualism, and the instruction of bilingual students that sees language practices of bilinguals not as two autonomous language systems but as a unitary language repository.

12.10 Discussion and Implications

Concerning language for education, South African legislative policy advocates dynamic pedagogical approaches that embrace and make use of African languages as languages of instruction. According to the Department of Higher Education and Training (2015), these commitments and recommendations for multilingualism that embrace indigenous African languages in education are made in the South African Constitution (1996), the Language in Education Policy (2002), the Report on the Development of Indigenous African Languages for Use as Mediums of Instruction at University (2003), the Report of the Ministerial Committee on Transformation and Social Cohesion and the Elimination of Discrimination in Public Education Institutions (2008), the Charter for Humanities and Social Sciences (2011), and the White Paper on Post-Secondary School Education and Training.

This proclamation elevates the choice of language for educational purposes to a fundamental right with a place in the Bill of Rights. Although the country's language legislation, policy frameworks, and guidelines cited elsewhere in this chapter affirm the role of indigenous African languages, education in South Africa largely still follows a monolingual trajectory notwithstanding the multilingual nature of the classrooms (Charamba, 2020b). The present study sought to explore the role language plays in the teaching and learning of tenth-grade Chemistry students. The results of this study highlight several instances that have impacts on students' meaning-making in the multilingual science classes (see Excerpts 12.1, 12.2, and 12.3).

Translanguaging helped these multilingual students' appropriate scientific concepts and terminology. For example, in Excerpt 12.1 students managed to hold a coherent, logical discussion around substances that dissolve in water and how to make substances such as grease dissolve. Translanguaging, in all three classes, leveraged students' full linguistic repertoires in the classroom and not just the particular language that is officially used for instructional purposes in that school (English). It offered the students and their teachers a means of working toward good mastery of scientific concepts through the simultaneous use of the official language of instruction (English) and students' home language (isiZulu). As education becomes more widely available, the acceptance of translanguaging in schools opens up windows of opportunity for multilingual students to make sense of their world using all linguistic resources at their disposal (Ocampo, 2018), and this is associated with the transformation (Li, 2018), social justice, and decolonisation essential for an efficient and productive world (Palfreyman & Walt, 2017).

The sociocultural theory views human development as a socially mediated process in which students acquire their cultural values, cognition, and problem-solving strategies through collaborative dialogues amongst themselves and with more knowledgeable members of their society. It stresses the fundamental role of social interaction among community members in the development of cognition (Vygotsky, 1978), and these interactions play a central role in the process of making meaning. For maximum development of one's cognition, communication should be through a language well understood by everyone involved. Translanguaging is a metaprocess that connects linguistic practices, promotes sociolinguistic equity, allows all students to express their true identities, and leverages their true bilingualism so they may act as a whole people in their bilingual worlds (Prada & Turnbull, 2018).

Through the sociocultural lens, language plays two pivotal roles in one's cognitive development: It is the vehicle by which adults transmit information to children, and language itself becomes a very powerful tool of intellectual adaptation and advancement (Vygotsky, 1978). Translanguaging goes beyond traditional notions of bilingualism, and its strong proposition of second-language teaching and learning as its driving force is built on a heteroglossic conception of bilingualism (Zhang et al., 2020), the term referring to the ability to flexibly operate between languages available to students (García, 2019; Li, 2018). Despite a growing body of research highlighting the academic benefits of this pedagogical approach, education for multilingual science students in South Africa and some parts of the world is still typically conducted monolingually in the majority language. A frequently used argument among policymakers is that incorporating students' minority language at school risks negatively affecting the development of the majority language and content mastery in other learning areas (Ünsal, 2017).

Translanguaging in the science classroom promotes a deeper and fuller understanding for the subject matter, where scaffolding and mediated learning can take place in conjunction with more peers (Siry, 2014), thus enabling the integration of those who understand the concepts better with those who are rather struggling to understand the concepts (Omidire, 2019). In all the three classes involved in this study, translanguaging promoted collaboration among the tenth-grade Chemistry students resulting in a deeper and fuller understanding of concepts, as evidenced in the responses they were giving in class; see, for example, Excerpts 12.1, 12.2, 12.3, and 12.4. When a student's home language is not the same as the language of instruction, it could impact teaching and learning, hindering full mastery of concepts in the classroom (Childs, 2016). In this way, translanguaging provided a means of extending the use of isiZulu purposefully and systematically in the science classroom enhancing sociocultural interactions among the students and their teacher.

The students' home language was thus an important resource in the acquiring of scientific knowledge. In cases where students could not express themselves entirely in the language of instruction, they made use of their linguistic repertoire and in all cases (Excerpts 12.1, 12.2, 12.3, and 12.4) their answers were correct. While we fully accept that translanguaging pedagogy alone cannot solve the larger issues of racism, classism, colonialism, and domination of reified standard codes and unequal power relations that stigmatize students' communicative repertoires (García & Lin, 2018), it nonetheless stands a good chance of disrupting the hierarchy of languages, transforming both teachers and students' attitudes toward their diverse meaning-making resources, and enabling students' full participation in knowledge co-making (Li & Lin, 2019).

Translanguaging, as seen in the present study, transforms the classroom into a space for co-participation in the co-construction of scientific knowledge by the students and teachers thus redressing historic imbalances still existing in South African classrooms. In her study Probyn (2019) examines the process of construction of scientific knowledge in a multilingual classroom in South Africa and demonstrates how pedagogical translanguaging challenges the monoglossic and post-colonial orthodoxies. The study shows that despite significant policy changes over a quarter-century into the democratic era, the historic imbalances and inequalities of apartheid education have remained. Translanguaging, however, not only corrects historic imbalances but it also affords students the most effective and convenient pedagogical route toward knowledge acquisition.

This study suggests that translanguaging practices erase the monoglossic orientations to education that originated in the apartheid era and close the academic performance gap between majority and minority language students (see, e.g., Ms. Sarah's interview response on the 24th of July 2019). Translanguaging practices accentuate that the students and the teacher together share a joint communicative repertoire that can be seen integrated, as a whole and as a common resource (Garcia, 2019). This is used to co-construct meaning through and beyond the linguistic and semiotic resources that constitute their communicative repertoires (Childs, 2016). In the present study, translanguaging offered the teachers and students opportunities to access scientific content through the linguistic resources and communicative repertoires they bring to the science classrooms (Prada & Turnbull, 2018). The study commends the three teachers in their attempt to incorporate the students' languages into the science classroom as these are valuable learning resources.

Translanguaging requires the use of all students' meaning-making resources, for example, when students were drawing and labeling the water cycle in Excerpt 12.3. The study recommends teachers, principals, and all stakeholders in education engage and educate the greater society on the possible benefits of translanguaging in the classroom. Basing on the results of this and other studies, schools can adopt translanguaging pedagogy to promote a deeper understanding of scientific concepts and to help students improve their written and oral communication skills by tolerating the dynamic shifts from one language to the other (Karlsson et al., 2020). The above foregrounds the need for open-mindedness on the part of all stakeholders to investigate possible integration of translanguaging pedagogy into the classroom.

This study also suggests that it can no longer be business as usual in today's multilingual science classrooms as this amounts to continuously failing most of the students who have low proficiency in the language of instruction.

12.11 Conclusion

Language has been associated with a people's culture, their heritage, and the prior knowledge and experiences that students bring into the classroom when they enter the science classroom. The present study joins an international body of research that acknowledges the instrumental role language plays in the teaching and learning of science. Through translanguaging, students' full linguistic repertoire is used to facilitate effective learning in the science classroom as their various "languages" will be working in collaboration and not against each other. The present study suggests teachers move away from the current monolingual bias which in most cases results in academic underachievement by students taught in a language different from their home language.

Translanguaging in the science classroom promotes a deeper and fuller understanding of the scientific knowledge learned and helps the sustainability and development of minority languages. Translanguaging also promotes cultural cohesion, promotes creativity, instills confidence, boosts self- esteem, enhances academic performance, and promotes oneness among students, parents, and teachers (Duarte, 2019; Ishimaru, 2019; Lin, 2019; McKinney & Tyler, 2019; Omidire, 2019; Probyn, 2019).Teachers and schools that choose to ignore the complexities and dynamics of today's multilingual science classrooms are simply reinforcing past worldviews, inequality, and injustice in education (Omidire, 2019).

Embracing translanguaging in the science classroom is not a call to contest the learning and importance of the English language, the issue is that we do not have to exclude the rich linguistic repertoire students bring into the classroom. Research by educationists and sociolinguists suggests that to attain effective education and cognitive development among students, foster creativity and critical thinking, the simultaneous use of students' linguistic repertoire in the same lesson is inevitable (see, e.g., Buxton & Caswell, 2020; Duarte, 2019; Garcia, 2019; Karlsson et al., 2020; Licona & Kelly, 2020; Lin, 2019; Ünsal, 2017). In this way, we can expect deeper understanding of scientific knowledge and better academic results from our science students who are key to solving future world scientific problems.

References

Benson, P. (2015). Commenting to learn: Evidence of language and intercultural learning in comments on YouTube videos. *Language Learning & Technology, 19*(3), 88–105.

Bryman, A. (2015). *Social research methods*. Oxford University Press.

Buxton, C. A., & Caswell, L. (2020). Next generation sheltered instruction to support multilingual learners in secondary science classrooms. *Science Education, 104*(3), 555–580.

Caruso, E. (2018). Translanguaging in higher education: Using several languages for the analysis of academic content in the teaching and learning process. *Language Learning in Higher Education, 8*(1), 65–90.

Charamba, E. (2019). Learning and language: Towards a reconceptualization of their mutual interdependences in a multilingual science class. *Journal of Multilingual and Multicultural Development*. https://doi.org/10.1080/01434632.2019.1707837

Charamba, E. (2020a). Translanguaging in a multilingual class: A study of the relation between students' languages and epistemological access in science. *International Journal of Science Education, 42*(11), 1779–1798.

Charamba, E. (2020b). Pushing linguistic boundaries: Translanguaging in a bilingual science and technology classroom. *Journal of Multilingual and Multicultural Development*. https://doi.org/10.1080/01434632.2020.1783544

Charamba, E., & Zano, K. (2019). Effects of translanguaging as an intervention strategy in a South African chemistry classroom. *Bilingual Research Journal, 42*(3), 291–307.

Childs, M. (2016). Reflecting on translanguaging in multilingual classrooms: Harnessing the power of poetry and photography. *Educational Research for Social Change, 5*(1), 22–40.

Creswell, J. W. (2014). *Research design: Qualitative, quantitative, and mixed methods approaches* (4th ed.). Sage.

Department of Basic Education. (2012). *CAPS document, physical sciences*. DBE.

Department of Education, National Senior Certificate School Subject Report. (2019). Pretoria, DBE.

Department of Higher Education and Training. (2015). *Report on the use of African languages as mediums of instruction in higher education*. DHET.

Duarte, D. (2019). Translanguaging in mainstream education: A sociocultural approach. *International Journal of Bilingual Education and Bilingualism, 22*(2), 150–164.

Fortuin, E. (2017). *Developing the academic literacy of isiXhosa-speaking learners in anAfrikaans-medium school in the intermediate phase*. Unpublished MEd thesis, University of the Western Cape.

García, O. (2019). Decolonizing foreign, second, heritage and first languages: Implications for education. In D. Macedo (Ed.), *Decolonizing foreign language education* (pp. 152–168). Routledge.

García, O., & Li, W. (2014). *Translanguaging: Language, bilingualism and education*. Palgrave Pivot.

García, O., & Lin, A. (2018). English and multilingualism. In P. Seargeant, A. Hewings, & S. Pihlaja (Eds.), *Routledge handbook of English language studies* (pp. 77–92). Routledge.

Grinker, D. (2014). *Inside Soweto: Memoir of an official 1960s–1980s*. Eastern Enterprises.

Grinker, D., & Gorelik, B. (Eds.). (2014). *Inside Soweto: Memoir of an official 1960s–1980s*. Eastern Enterprises.

Grosjean, F. (2019). *A journey in languages and cultures: The life of a bicultural bilingual*. Oxford University Press.

Hammond, Z. (2014). *Culturally responsive teaching and the brain: Promoting authentic engagement and rigor among culturally and linguistically diverse students* (1st ed.). Corwin.

Ishimaru, A. M. (2019). *Just schools: Building equitable collaborations with families and communities*. Teachers College Press.

Iversen, J. Y. (2020). Pre-service teachers' translanguaging during field placement in multilingual, mainstream classrooms in Norway. *Language and Education, 34*(1), 51–65.

Karlsson, A., Nygård Larsson, P., & Jakobsson, A. (2020). The continuity of learning in a translanguaging science classroom. *Cultural Studies of Science Education, 15*(1), 1–25.
Krause, L., & Prinsloo, M. (2016). Translanguaging in a township primary school: Policy and practice. *Southern African Linguistics and Applied Language Studies, 34*(4), 347–357.
Langman, J. (2014). Translanguaging, identity and learning: Science teachers as engaged language planners. *Language Policy, 13*(1), 183–200.
Lasagabaster, D., & García, O. (2014). Translanguaging: Towards a dynamic model of bilingualism at school / Translanguaging: Hacia un modelodinámico de bilingüismoen la escuela. *Cultura y Educación, 26*(1), 557–572.
Lemke, J. (2012). Analyzing verbal data: Principles, methods, and problems. In B. Fraser, K. Tobin, & C. McRobbie (Eds.), *Second international handbook of science education* (pp. 1471–1484). Springer.
Lewis, G., Jones, B., & Baker, C. (2012). Translanguaging: Origins and development from school to street and beyond. Educational research and evaluation. *An International Journal on Theory and Practice, 18*(7), 641–654.
Li, W. (2018). Translanguaging as a practical theory of language. *Applied Linguistics, 39*(1), 9–30.
Li, W., & Lin, A. M. Y. (2019). Translanguaging classroom discourse: Pushing limits, breaking boundaries. *Classroom Discourse, 10*(3-4), 209–215.
Licona, P. R., & Kelly, G. J. (2020). Translanguaging in a middle school science classroom: Constructing scientific arguments in English and Spanish. *Cultural Studies of Science Education, 15*(2), 485–510.
Lin, A. M. Y. (2019). Theories of trans/languaging and trans-semiotizing: Implications for content-based education classrooms. *International Journal of Bilingual Education and Bilingualism, 22*(1), 5–16.
MacSwan, J. (2017). A multilingual perspective on translanguaging. *American Educational Research Journal, 54*(1), 167–201.
Madiba, M. (2014). Promoting concept literacy through multilingual glossaries: A translanguaging approach. In C. van der Walt & L. Hibbert (Eds.), *Multilingual teaching and learning in higher education in South Africa* (pp. 68–87). Multilingual Matters.
Makoe, P. (2018). Translanguaging in a monoglot context: Children mobilising and (re)positioning their multilingual repertoires as resources for learning. In G. Mazzaferro (Ed.), *Translanguaging as everyday practice* (pp. 13–30). Springer.
Mazzaferro, G. (2018). Translanguaging as everyday practice. An introduction-draft. In G. Mazzaferro (Ed.), *Translanguaging as everyday practice* (pp. 1–12). Springer International Publishing.
McKinney, C., & Tyler, R. (2019). Disinventing and reconstituting language for learning in school science. *Language and Education, 33*(2), 141–158.
McMillan, J. H., & Schumacher, S. (2010). *Research in education: Evidence-based inquiry* (7th ed.). Pearson Education.
Msimanga, A., Denley, P., & Gumede, N. (2017). The pedagogical role of language in science teaching and learning in South Africa: A review of research 1990–2015. *African Journal of Research in Mathematics, Science and Technology Education, 21*(3), 245–255.
Mullis, I. V. S., Martin, M. O., Foy, P., & Hooper, M. (2016). *TIMSS 2015 international results in mathematics*. TIMSS & PIRLS International Study Center.
Mullis, I. V. S., Martin, M. O., Foy, P., & Hooper, M. (2017). *PIRLS 2016 international results in Reading*. TIMSS & PIRLS International Study Center.
National Treasury Report on South African Education. (2018). *Department of Basic Education*. DBE.
Ocampo, A. C. (2018). Stop Forgetting Asian Americans. *Contexts, 17*(4), 76.
Omidire, M. F. (2019). Embracing multilingualism as a reality in classrooms. In M. F. Omidire (Ed.), *Multilingualism in the classroom: Teaching and learning in a challenging context*. UCT Press.

Otheguy, R., García, O., & Reid, W. (2015). Clarifying translanguaging and deconstructing named languages: A perspective from linguistics. *Applied Linguistics Review, 6*(1), 281–307.

Palfreyman, D., & Walt, C. (2017). *Academic biliteracies: Multilingual repertoires in higher education*. Multilingual matters.

Portoles, L., & Marti, O. (2017). Translanguaging as a teaching resource in early language learning of English as an additional language. *Bellaterra Journal Teaching & Learning Language& Literature, 10*(1), 61–77.

Prada, J., & Turnbull, B. (2018). The role of translanguaging in the multilingual turn: Driving philosophical and conceptual renewal in language education. *EuroAmerican Journal of Applied Linguistics and Languages, 5*(2), 8–23.

Probyn, M. (2019). Pedagogical translanguaging and the construction of science knowledge in a multilingual South African classroom: Challenging monoglossic/post-colonial orthodoxies. *Classroom Discourse, 10*(3-4), 216–236.

Scott, S. N., & Palincsar, A. S. (2013). *The historical roots of sociocultural theory*. http://www.education.com/reference/article/sociocultural-theory/. Accessed 15 Jan 2020.

Siry, C. (2014). Towards multidimensional approaches to early childhood science education. *Cultural Studies of Science Education, 9*(2), 297–304.

Statistics South Africa. (2019). *Education series volume V: Higher education and skills in South Africa, 2017*. Statistics South Africa.

Stroupe, D., Moon, J., & Michaels, S. (2019). Introduction to special issue: Epistemic tools in science education. *Science Education, 103*(4), 948–951.

Ünsal, Z. (2017). *Bilingual students' learning in science*. PhD thesis. Stockholm University.

Vygotsky, L. (1978). Mind in society. The development of higher psychological processes. In M. Cole, V. John-Steiner, S. Scribner, & E. Souberman (Eds.), *Mind in society: The development of higher psychological processes*. Harvard University Press.

Zhang, Q., Osborne, C., Shao, L., & Lin, M. (2020). A translanguaging perspective on medium of instruction in the CFL classroom. *Journal of Multilingual and Multicultural Development*. https://doi.org/10.1080/01434632.2020.1737089

The manufacturer's authorised representative in the EU is Springer Nature Customer Service Centre GmbH, Europaplatz 3, 69115 Heidelberg, Germany. If you have any concerns regarding our products, please contact ProductSafety@springernature.com

Printed and bound by CPI Group (UK) Ltd, Croydon, CR0 4YY
30/09/2026
02225706-0004